T0326408

JIAC2009 – Book of abstracts

JIAC2009
Book of abstracts

edited by:
C. Lokhorst
J.F.M. Huijsmans
R.P.M. de Louw

Wageningen Academic
P u b l i s h e r s

ISBN 978-90-8686-114-9

First published, 2009

© Wageningen Academic Publishers
The Netherlands, 2009

International Scientific Committee for selection

Prof. A. Bregt (chair)	Netherlands	EFITA
Prof. P.W.G. Groot Koerkamp (chair)	Netherlands	ECPLF
Prof. E.J. van Henten (chair)	Netherlands	ECPA
Prof. G. Beers	Netherlands	EFITA
Dr. J. Boaventura	Portugal	EFITA
Dr. G. Bonati	Italy	EFITA
Dr K. Charvat	Czech Republic	EFITA
Dr. M. Fritz	Germany	EFITA
Dr. E. Gelb	Israel	EFITA
Dr. M. Herdon	United Kingdom	EFITA
Dr. I. Houseman	United Kingdom	EFITA
Dr. S. Ninomiya	Japan	EFITA
Dr. R. Mueller	Germany	EFITA
Dr. C. Parker	United Kingdom	EFITA
Dr. R. Rognant	France	EFITA
Prof. G. Schiefer	Germany	EFITA
Dr. A. Sideridis	Greece	EFITA
Prof. Y. Thysen	Denmark	EFITA
J. Weres	Poland	EFITA
Ir. J. Wien	Netherlands	EFITA
Dr. S. Wolfert	Netherlands	EFITA
Prof. F. Zazueta	USA	EFITA
Prof. J.M. Aerts	Belgium	ECPLF
Dr. Th. Amon	Austria	ECPLF
Dr. Th. Banhazi	Australia	ECPLF
Prof. Li Baoming	China	ECPLF
Prof. D. Berckmans	Belgium	ECPLF
Dr. S. Christensen	Denmark	ECPLF
Dr. S. Cox	United Kingdom	ECPLF
Dr. T. Demmers	United Kingdom	ECPLF
Dr. M. Guarino	Italy	ECPLF
Prof. J. Hartung	Germany	ECPLF
Dr. R. Kaufmann	Switzerland	ECPLF
Prof. L. Keeling	Sweden	ECPLF
Dr. C. Lokhorst	Netherlands	ECPLF
Dr. F. Madec	France	ECPLF
Dr. E. Maltz	Israel	ECPLF
Prof. J. de Baerdemaker	Belgium	ECPA
Dr. B. Basso	Italy	ECPA
Prof. S. Blackmore	Greece	ECPA
Dr. R. Casa	Italy	ECPA
Dr. A. Fekete	Hungary	ECPA
Prof. F. Gemtos	Greece	ECPA
Dr. D. Goense	Netherlands	ECPA
Dr. G. Grenier	France	ECPA
Dr. H.-W. Griepentrog	Denmark	ECPA
Dr. H. Haapala	Finland	ECPA
Dr. J.W. Hofstee	Netherlands	ECPA

Dr. W. Hoogmoed	Netherlands	ECPA
Dr. P. Juerschik	Germany	ECPA
Dr. I. Karpinski	Germany	ECPA
Dr. A. Korsaeth	Norway	ECPA
Prof. A. McBratney	Australia	ECPA
Dr. W. Mulla	USA	ECPA
Ir. A. Nieuwenhuizen	Netherlands	ECPA
Dr. M. Norremark	Denmark	ECPA
Dr. M. Oliver	United Kingdom	ECPA
Dr. G. Rabatel	France	ECPA
Dr. J. Stafford	United Kingdom	ECPA
Dr. L.G. Torres	Spain	ECPA
Dr. G.D. Vermeulen	Netherlands	ECPA
Dr. B. Whelan	Australia	ECPA

Abstracts were used for the selection of oral and poster presentation. All corresponding authors were asked to deliver a one page abstract that will be published in this book. So, all abstracts of the oral and posters are part of this book. For the full papers of the oral presentations different proceedings for ECPLF, ECPA and EFITA are published.

Local Organizing Committee

Kees Lokhorst (chair)
Jan Huijsmans
Ramon de Louw
Yvonne van Hezik
Ingrid Luitse
Niek Botden
Mike Jacobs
Erwin van der Waal
Eldert van Henten
Daan Goense
Peter Groot Koerkamp
Arnold Bregt
Sjaak Wolfert
Jan-Erik Wien
Students Heeren XVII

Editorial

This 'JIAC2009 Book of Abstracts' contains the abstracts of the posters and papers presented at the Joint International Agricultural Conference. This conference joins the 7th EFITA conference, the 4th ECPLF conference, the 7th ECPA conference and the Field Robot Event. There is a tradition in bringing ECPA and ECPLF together. This time we have again the opportunity to bring the world of ICT specialists in agriculture (EFITA) together with the precision farming world, as we had in Montpellier in 2001. Additionally, the Field Robot Event will act as a source of inspiration and discussion in a real 'living lab' environment.

The ambitions of the organizers are (1) learning from colleagues by having scientific well produced papers, (2) bringing different groups (sub-conferences) together in 'creative' workshops: doing things together and use the power of the delegates, (3) to let industry, practice and science learn from each other and let them think about possibilities in market and innovation progress, and (4) to have an inspiring stay in the Netherlands.

To share the knowledge of the delegates there will be only one conference, and all participants receive a DVD with all papers in full and this book of abstracts. We hope you enjoy the abstracts.

The work of Precision Agriculture, Precision Livestock Farming and ICT in agriculture is entering a new phase. We see a lot of European activities which show that it is now on the EU-agenda. The harvest of running EU projects has begun, and also new interesting EU projects will be introduced during the conference. This will be the challenge for the coming years. To stimulate new activities, several cross themes and project meetings are organised during the conference.

First we want to thank all authors for their interest in this conference and for writing abstracts. Secondly, we want to thank the sponsors (gold: Animal Science Group, Plant Research International, The Ministry of Agriculture, Nature and Food Quality of the Netherlands, Challenger, Groene Kenniscooperatie, Alterra, Claas, European GNSS Supervisory Authority (GSA) silver: Vellekoop & Meesters, Hewlett-Packard, Fancom, Lely Industries, VanDrie Group, John Deere, bronze: Trimble, Mueller, eCow, Probotiq, Springer and Wiley Blackwell) for their support. Thirdly, the Animal Science Group, Plant Research International and VIAS were so kind as to bear the organisational risks. Fourthly, the editors want to thank the local organisers for their work to make this conference an ideal place for networking and exchange of stimulating ideas.

The editors

Session 01. Remote sensing

Theatre   **Session 01 no. Page**

Active sensing of the N status of wheat using optimized wavelength combination:
impact of seed rate, variety and growth stage   1   1
Jasper, J., Reusch, S. and Link, A.

Development and first tests of a mobile lab combining optical and analogical sensors
for crop monitoring in precision viticulture   2   2
Mazzetto, F., Calcante, A., Mena, A. and Vercesi, A.

Sensitivity of narrow and broad-band vegetation indexes to leaf chlorophyll
concentration in planophile crops canopies   3   3
Vincini, M. and Frazzi, E.

Comparison of methods to estimate LAI from remote sensing in maize canopies by
inversion of 1-D and 3-D models   4   4
*Casa, R., Nassetti, F., Pascucci, S., Palombo, A., D'Urso, G., Ciraolo, G., Maltese, A.,
Giordano, L. and Jones, H.G.*

Session 02. Sensing plant characteristics

Theatre   **Session 02 no. Page**

Plant leaf roughness analysis by texture classification with generalized fourier
descriptors in different dimensionality reduction context   1   5
Journaux, L., Destain, M.F., Cointault, F., Miteran, J. and Piron, A.

Evaluation of cost-effective real-time slope sensing system for wild blueberry   2   6
Zaman, Q., Schumann, A., Swain, K., Percival, D., Arshad, M. and Esau, T.

Intelligent autonomous system for the detection and treatment of fungal diseases in
arable crops   3   7
Moshou, D., Bravo, C., Vougioukas, S. and Ramon, H.

On-the-go yield and sugar sensing in grape harvester   4   8
Baguena, E.M., Barreiro, P. and Valero, C.

Session 03. Remote sensing

Theatre   **Session 03 no. Page**

Using GreenSeeker to drive variable rate application of plant growth regulators and
defoliants on cotton   1   9
Vellidis, G., Ortiz, B., Ritchie, G. and Perry, C.

Canopy temperature interpretation of thermal imagery for crop water stress
determination   2   10
Meron, M., Alchanatis, V., Cohen, Y. and Tsipris, J.

Using an active sensor to make in-season nitrogen recommendations for corn   3   11
Schmidt, J., Sripada, R., Beegle, D. and Dellinger, A.

Optical signals of oxidative stress in crops physiological state diagnostics   4   12
Kanash, E.V. and Osipov, Y.A.

Spatial patterns of wilting in sugar beet as an indicator for precision irrigation   5   13
Zhang, L., Steven, M.D., Clarke, M.L. and Jaggard, K.W.

Session 04. Remote sensing

Theatre   **Session 04 no. Page**

Development of an instrument to monitor crop growth status   1   14
Li, M., Cui, D., Li, X. and Yang, W.

Field hyperspectral imagery: a new tool for crop monitoring at plant scale   2   15
Vigneau, N., Rabatel, G. and Roumet, P.

Determination of canopy properties of winter oilseed rape using remote sensing:
techniques in field experiments   3   16
Engström, L., Söderström, M., Lindén, B., Börjesson, T. and Lorén, N.

Within-field and regional prediction of malting barley protein content using canopy
reflectance   4   17
Söderström, M., Pettersson, C.G., Börjesson, T. and Hagner, O.

Development and improvement of the air assist seeding following the map information   5   18
Chosa, T., Furuhata, M., Omine, M. and Sugiura, R.

Session 05. Biomass other than remote sensing

Theatre   **Session 05 no. Page**

Use of ultrasonic transducers for on-line biomass estimation in winter wheat   1   19
Reusch, S.

Use of a ground lidar scanner to measure vineyard leaf area and structural variability
of vines   2   20
Arnó, J., Vallès, J.M., Escolà, A., Sanz, R., Palacín, J. and Rosell, J.R.

Sensing tree canopy parameters in real time for precision fructiculture/horticulture
applications: first results   3   21
Escolà, A., Arnó, J., Sanz, R., Camp, F., Masip, J., Solanelles, F., Rosell, J.R. and Planas, S.

High-end laser rangefinder scanner in agriculture   4   22
Ehlert, D., Heisig, M. and Adamek, R.

Session 06. Plant disease

Theatre   **Session 06 no. Page**

Detection of head blight (*Fusarium* spp.) at ears of winter wheat using hyperspectral
and chlorophyll fluorescence imaging   1   23
*Bauriegel, E., Beuche, H., Dammer, K.H., Giebel, A., Herppich, W.B., Intreß, J. and
Rodemann, B.*

Early detection and discrimination of *Puccinia triticina* infestation in susceptible and
resistant wheat cultivars by chlorophyll fluorescence imaging technique   2   24
Bürling, K., Hunsche, M., Tartachnyk, I. and Noga, G.

Detection of the Tulip Breaking Virus (TBV) in tulip using spectral and vision sensors 3   25
*Polder, G., Van Der Heijden, G.W.A.M., Van Doorn, J., Van Der Schoor, R. and Baltissen,
A.H.M.C.*

Investigation into the classification of diseases of sugar beet leaves using
multispectral stereo images   4   26
Bauer, S.D., Korc, F. and Förstner, W.

Study on plant pathogenesis with help of laser-optical and poly-functional
spectrophotometers   5   27
Lisker, I. and Dminriev, A.

Session 07. Sensing plant characteristics

Theatre   **Session 07 no. Page**

MARVIN: high speed 3D imaging for seedling classification   1   28
Koenderink, N.J.J.P., Wigham, M., Golbach, F., Otten, G., Gerlich, R. and Van De Zedde, R.

Flower spatial variability in an apple orchard   2   29
Aggelopoulou, A.D., Bochtis, D., Fountas, S., Koutsostathis, A. and Gemtos, T.A.

Apple detection in natural tree canopies from multimodal images   3   30
Wachs, J.P., Alchanatis, V., Stern, H.I. and Burks, T.

Effects of seed rate and nitrogen fertilisation on cereal canopy characteristics 4   31
Kren, J., Svobodova, I., Dryslova, T., Misa, P. and Neudert, L.

Session 08. Remote sensing

Theatre   **Session 08 no. Page**

Improved modelling of maize growth by combining a biophysical model of
photosynthesis with hyperspectral remote sensing   1   32
Oppelt, N.M. and Hank, T.B.

Comparing hyperspectral vegetation indices for estimating nitrogen status of winter wheat   2   33
Li, F., Chen, X., Miao, Y., Hennig, S.D., Gnyp, M.L., Jia, L., Bareth, G. and Zhang, F.

Possibilities of cereal canopy assessment by using the NDVI   3   34
Kren, J., Lukas, V., Svobodova, I., Dryslova, T., Misa, P. and Neudert, L.

Evaluation of palm trees water status using remote thermal imaging   4   35
Cohen, Y., Alchanatis, V., Cohen, Y. and Soroker, V.

Session 09. Sensing weeds

Theatre   **Session 09 no. Page**

Classification of sugar beet and volunteer potato spectra using a neural network to select wavelengths   1   36
Nieuwenhuizen, A.T., Hofstee, J.W., Van De Zande, J.C., Meuleman, J. and Van Henten, E.J.

Automated weed detection in winter wheat by using artificial neural networks   2   37
Kluge, A. and Nordmeyer, H.

Spectral signatures of diseased sugar-beet leaves   3   38
Mahlein, A.-K., Steiner, U., Dehne, H.-W. and Oerke, E.-C.

Development of an integrated approach for weed detection in cotton, for site specific weed management   4   39
Efron, R., Alchanatis, V., Cohen, Y., Levi, A., Eizenberg, H., Yehuda, Z. and Shani, U.

Session 10. Sensing weeds

Theatre   **Session 10 no. Page**

Selectivity of weed harrowing in cereals with sensor technology in Germany   1   40
Rueda Ayala, V.P. and Gerhards, R.

Automatic derivation of weed densities from images for site-specific weed management   2   41
Weis, M. and Gerhards, R.

Evaluation of a ground-based weed mapping system in Maize   3   42
Andujar, D., Dorado, J., Ribeiro, A. and Fernandez-Quintanilla, C.

Evolution of agricultural machinery: the third way   4   43
Berducat, M., Debain, C., Lenain, R. and Cariou, C.

Session 11. Yield monitoring and ICT

Theatre   **Session 11 no. Page**

Mass flow sensor for combines with bucket conveyors   1   44
Zandonadi, R.S., Stombaugh, T.S., Queiroz, D.M. and Shearer, S.A.

Automated, low-cost yield mapping of wild blueberry fruit   2   45
Swain, K.C., Zaman, Q.U., Schumann, A.W., Percival, D.C., Dainis, N. and Esau, T.J.

Development and first tests of a farm monitoring system based on a client-server
technology   3   46
Mazzetto, F., Calcante, A. and Salomoni, F.

Agri yield management: practical solutions for profitable and sustainable agriculture
based on advanced technology   4   47
Hadders, J., Hadders, J.W.M. and Raatjes, P.

Flow behavior analysis of a rfid-tracer for traceability of grain   5   48
Steinmeier, U., Von Hörsten, D. and Lücke, W.

Session 12. Soil sensing

Theatre   **Session 12 no. Page**

Gamma ray sensor for top soil mapping: THE MOLE   1   49
Loonstra, E.H.

Combined sensor system for mapping soil properties   2   50
Mahmood, H.S., Hoogmoed, W.B. and Van Henten, E.J.

Reproducibility of different composite sampling schemes for soil phosphorus   3   51
Schirrmann, M., Domsch, H., Von Wulffen, U., Zauer, O. and Nieter, J.

The capability of non-destructive geophysical methods in precision agriculture to
capture subsoil mechanical strength   4   52
Hoefer, G. and Bachmann, J.

Comparison of different EC-mapping sensors   5   53
Lueck, E. and Ruehlmann, J.

Session 13. Soil sensing and interpretation of variability

Theatre   **Session 13 no. Page**

Interaction of tillage direction and lag distances for directional semi-variograms of
implement draught   1   54
Mclaughlin, N.B.

Mapping traffic patterns for soil compaction studies using GIS   2   55
Meijer, A.D., Heiniger, R.W. and Crozier, C.R.

Searching for the cause of variety   3   56
Jukema, J.N., Wijnholds, K.H. and Berg, Van Den, W.

Session 14. Interpretation of variability

Theatre   **Session 14 no. Page**

Delineation of site-specific management zones using geostatistics and fuzzy
clustering analysis   1   57
Castrignano, A., Fiorentino, C., Basso, B., Troccoli, A. and Pisante, M.

Mapping spatial variation in growing willow using small UAS   2   58
Rydberg, A., Hagner, O. and Aronsson, P.

iSOIL: exploring the soil as the basis for sustainable crop production and precision
farming   3   59
Van Egmond, F.M., Nüsch, A.-K., Werban, U., Sauer, U. and Dietrich, P.

Comparision of yield mapping methods and an analysis of rainfall effects on
temporal variation   4   60
Rogge, C.B.E.

The economic potential of precision farming: an interim report with regard to
nitrogen fertilization   5   61
Wagner, P.

Session 15. Geo statistics and path planning

Theatre   **Session 15 no. Page**

Optimal path planning for field operations   1   62
Hofstee, J.W., Spätjens, L.E.E.M. and IJken, H.

Combined coverage path planning for field operations   2   63
Bochtis, D.D. and Oksanen, T.

From sensor values to a map: accuracy of spatial modelling methods in agricultural
machinery works   3   64
Kaivosoja, J.

Field-scale model of the spatio-temporal vine water status in a viticulture system   4   65
Taylor, J.A., Tisseyre, B. and Lagacherie, P.

Hydropedology and pedotransfer functions (PTF)   5   66
Zacharias, S.

Session 16. Precision agriculture in regional modelling

Theatre   **Session 16 no. Page**

A multi agent simulation approach to assess agronomic income sources in the North
China Plain: case study for Quzhou county   1   67
Roth, A. and Doluschitz, R.

Sensitivity to climate change with respect to agriculture production in Hungary   2   68
Erdélyi, É.

Rotation and the temporal yield stability of landscape defined management zones: a time series analysis   3   69
Grove, J.H. and Pena-Yewtukhiw, E.M.

Bioclimatic modelling of the future ranges of crop species: an analysis of propagated uncertainty   4   70
Marinelli, M.A., Corner, R.J. and Wright, G.

Session 17. Robots and mechanisation

Theatre   **Session 17 no. Page**

A framework for motion coordination of small teams of agricultural robots   1   71
Vougioukas, S.

Model-based loading of agricultural trailers   2   72
Happich, G., Lang, T. and Harms, H.-H.

Procedures of soil farming allowing to reduce compaction   3   73
Kroulik, M., Loch, T., Kviz, Z. and Prosek, V.

Preparing a team to field robot event: educational and technological aspects   4   98
Oksanen, T., Tiusanen, J. and Kostamo, J.

Session 18. Guidance and machine performance

Theatre   **Session 18 no. Page**

Parallel guidance system for tractor-trailer system with active joint   1   74
Backman, J., Oksanen, T. and Visala, A.

A vision-guided mobile robot for precision agriculture   2   75
Ericson, S. and Åstrand, B.

An imaging system to characterise the mechanical behaviour of fertilisers in the context of centrifugal spreading   3   76
Villette, S., Piron, E., Martin, R. and Gee, C.

New high speed image acquisition system and image processing techniques for fertilizer granule trajectory determination   4   77
Hijazi, B., Cointault, F., Dubois, J., Villette, S., Vangeyte, J., Yang, F. and Paindavoine, M.

Session 19. Crop modelling

Theatre   **Session 19 no. Page**

Simulating the dynamic changes in winter wheat after grazing   1   78
Harrison, M.T., Evans, J.R. and Moore, A.D.

Modeling competition for below-ground resources and light within a winter pea (*Pisum sativum* L.) – wheat (*Triticum aestivum* L.) intercrop (Azodyn-InterCrop): towards a decision making oriented-tool   2   79
Malagoli, P.J., Naudin, C., Goulevant, G., Sester, M., Corre-Hellou, G. and Jeuffroy, M.-H.

Conductance model evaluation for predicting yield and protein content of lupin-cereal forage crops in organic farming   3   80
Azo, W.M., Davies, W.P., Lane, G.P.F. and Cannon, N.D.

Crop modelling based on the principle of maximum plant productivity   4   81
Kadaja, J. and Saue, T.

Automatic working depth control for seed drill using ISO 11783 compatible tractor   5   82
Suomi, P., Ojanne, A., Oksanen, T., Kalmari, J., Linkolehto, R. and Teye, F.

Session 20. Precision application

Theatre   **Session 20 no. Page**

Robotic control of broad-leaved dock   1   83
Van Evert, F.K., Samsom, J., Polder, G., Vijn, M., Van Dooren, H.J., Lamaker, A., Van Der Heijden, G.W.A.M., Kempenaar, C., Van Der Zalm, A. and Lotz, L.A.P.

First prototype of an automated rotary hoe for mechanical weeding of the intra-row area in row crops   2   84
Gobor, Z., Schulze Lammers, P. and Wendl, G.

Auto boom control to avoid spraying pre-defined areas   3   85
Mickelåker, J. and Svensson, S.A.

Image processing algorithms for a selective vineyard robotic sprayer   4   86
Berenstein, R., Edan, Y., Ben-Shachar, O., Shapiro, A. and Bechar, A.

WURking: a small sized autonomous robot for the farm of the future   5   87
Van Henten, E.J., Van Asselt, C.J., Bakker, T., Blaauw, S.K., Govers, M.H.A.M., Hofstee, J.W., Jansen, R., Nieuwenhuizen, A.T., Speetjens, S.L., Stigter, J.D., Van Straten, G. and Van Willigenburg, L.G.

Session 21. Crop and regional modelling

Theatre   **Session 21 no. Page**

A supply/demand, single-organ crop growth model   1   88
Seginer, I. and Gent, M.

Prediction of within field cotton yield losses caused by the southern root-knot nematode with cropping system model-CROPGRO-cotton   2   89
Ortiz, B., Hoogenboom, G., Vellidis, G., Boote, K. and Perry, C.

Use of geographic information systems in crop protection warning service   3   90
Zeuner, T. and Kleinhenz, B.

Plant-specific and canopy density spraying to control fungal diseases in bed-grown crops   4   91
Van De Zande, J.C., Achten, V.T.J.M., Schepers, H.T.A.M., Van Der Lans, A., Michielsen, J.M.G.P., Stallinga, H. and Van Velde, P.

Session 22. ICT in precision agriculture

Theatre   **Session 22 no. Page**

Data collection and two-way communication to support decision making by pest scouts   1   92
Hetzroni, A., Meron, M., Frier, I. and Magrisso, Y.

Prototype system of monitoring farm operation with wearable device and field server   2   93
Fukatsu, T., Sugahara, K., Nanseki, T. and Ninomiya, S.

Information requirements and data sources for automated irrigation control and supervision in tree crops   3   94
Casadesus, J.

Hands free registration of tillage activities in agriculture   4   95
Janssen, H., Van Der Wal, T., Beek, A., Van Rossum, J. and Uyterlinde, M.

A multi-level modelling approach for food supply chains using the Unified Modeling Language (UML)   5   96
Lehmann, R., Fritz, M. and Schiefer, G.

Session 23. Robots

Theatre   **Session 23 no. Page**

BoniRob: an autonomous field robot platform for individual plant phenotyping   1   97
Ruckelshausen, A., Biber, P., Dorna, M., Gremmes, H., Klose, R., Linz, A., Rahe, F., Resch, R., Thiel, M., Trautz, D. and Weiss, U.

Safe and reliable: further development of a field robot   2   99
Griepentrog, H.W., Andersen, N.A., Andersen, J.C., Blanke, M., Heinemann, O., Nielsen, J., Pedersen, S.M., Ravn, O., Madsen, T. and Wulfsohn, D.

Simple tunable control for automatic guidance of four-wheel steered vehicles   3   100
Bakker, T., Van Asselt, C.J., Bontsema, J., Müller, J. and Van Straten, G.

Development of a small agricultural field inspection vehicle   4   101
Gottschalk, R., Burgos-Artizzu, X.P. and Ribeiro, A.

Session 24. Precision application

Theatre   **Session 24 no. Page**

Assessing the potential of automatic section control   1   102
Stombaugh, T.S., Zandonadi, R.S., Dowdy, T. and Shearer, S.A.

Performance of auto-boom control for agricultural sprayers   2   103
Molin, J.P., Reynaldo, E.F. and Povh, F.P.

The development of a computer vision based and real-time plant tracking system for dot spraying   3   104
Nørremark, M., Olsen, H.J. and Lund, I.

SensiSpray: site-specific precise dosing of pesticides by on-line sensing   4   105
Van De Zande, J.C., Achten, V.T.J.M., Kempenaar, C., Michielsen, J.M.G.P., Van Der Schans, D., Stallinga, H. and Van Velde, P.

Session 25. Advanced ICT methodologies

Theatre   **Session 25 no. Page**

Comparing companies and strategies: a genetic algorithms approach   1   106
Hennen, W.H.G.J.

Prioritizing agriculture research projects emphasizing on analytical hierarchical process (AHP)   2   107
Mortazavi, M., Rajabbeigi, M. and Mokhtari, H.

Using formal concept analysis to calculate similarities among corn diseases   3   108
Souza, K.X.S., Massruhá, S.M.F.S. and Fernandes, V.M.

Probability distribution on the causal bacteria of clinical mastitis in dairy cows using naive Bayesian networks   4   109
Steeneveld, W., Van Der Gaag, L.C., Barkema, H.W. and Hogeveen, H.

Session 26. Advanced ICT methodologies

Theatre   **Session 26 no. Page**

Modeling an automated system to identify and classify stingless bees using the wing morphometry: a pattern recognition approach   1   110
Bueno, J., Francoy, T., Imperatriz-Fonseca, V. and Saraiva, A.

Image analysis and 3D geometry modeling in investigating agri-food product properties   2   111
Weres, J., Nowakowski, K., Rogacki, P. and Jarysz, M.

A technical opportunity index based on a fuzzy footprint of the machine for site-specific management: application to viticulture   3   112
Paoli, J.N., Tisseyre, B., Strauss, O. and Mcbratney, A.B.

Expedient spatial differentiation of technologies of precise agriculture according to productivity factors   4   113
Uskov, A.O. and Zaharian, J.G.

Session 27. Economic modeling

Theatre   **Session 27 no.** **Page**

Methodology to estimate economic levels of profitability of precision agriculture :
simulation for crop systems in Haute-Normandie   1   114
Bourgain, O. and Llorens, J.-M.

Evolution from 2002 to 2008 of the economic interest of precision agriculture in
fertilisation management   2   115
Perrin, O. and Llorens, J.-M.

Modeling precision dairy farming technology investment decisions   3   116
Bewley, J.M., Boehlje, M.D., Gray, A.W., Hogeveen, H., Eicher, S.D. and Schutz, M.M.

Economic modeling of livestock diseases: a decision support tool   4   117
Bennett, R.M., Mcclement, I. and Mcfarlane, I.D.

Integrated assessment of economic and environmental risk of agricultural production
by FAPS-DB system   5   118
Nanseki, T., Sato, M., Marte, W.E. and Xu, Y.

Session 28. Information exchange

Theatre   **Session 28 no.** **Page**

Harmonisation of data exchange in the IACS subsidies request procedure   1   119
Schmitz, M., Martini, D., Frisch, J. and Kunisch, M.

agroXML: a standard for data exchange in agriculture   2   120
Kunisch, M., Martini, D., Schmitz, M. and Frisch, J.

Enabling integration of distributed data for agricultural software applications using
agroXML   3   121
Martini, D., Schmitz, M., Frisch, J. and Kunisch, M.

Populating an observatory of e-government services for rural areas: analysing the
Greek services   4   122
Manouselis, N., Tzikopoulos, A., Costopoulou, C.O. and Sideridis, A.

Session 29. Information systems development

Theatre   **Session 29 no.** **Page**

Information modelling in dairy production   1   123
Lindstrøm, J. and Sørensen, C.G.

Computational architecture of OTAG prototype   2   124
Visoli, M. and Ternes, S.

Robust performance: principles and potential application in livestock production systems   3   125
Van Der Veen, A.A., Ten Napel, J., Oosting, S.J., Bontsema, J., Van Der Zijpp, A.J. and Groot Koerkamp, P.W.G.

Applying enterprise application architectures in integrated modelling   4   126
Knapen, M.J.R.

Session 30. Information systems development

Theatre   **Session 30 no. Page**

The integration of open standards for enterprise service bus in a solution for agribusiness and enviromental applications development   1   127
Murakami, E., Santana, F.S., Stange, R.L. and Saraiva, A.M.

Evolution of a SOA-based architecture for agroenvironmental purposes integrating GIS services to an ESB environment   2   128
Santana, F.S., Stange, R.L., Gushiken, I.Y., Murakami, E. and Saraiva, A.M.

Business process modeling of the pesticide life cycle: a service-oriented approach   3   129
Wolfert, J., Matocha, D., Verloop, C.M. and Beulens, A.J.M.

Plantform: smart, agile and integrated enterprise information systems for ornamental horticulture   4   130
Verloop, C.M., Verdouw, C.N. and Van Der Hoeven, R.

Designing an information and communication technology system to train private agricultural insurance brokers in Iran   5   131
Omidi Najafabadi, M., Farajollah Hosseini, J., Mirdamadi, M. and Moghadasi, R.

Session 31. Information systems development

Theatre   **Session 31 no. Page**

Architecture of information systems for automated arable farming   1   132
Thysen, I.

Implementing systems usability evaluation in the design process of active farm management information system   2   133
Norros, L., Pesonen, L., Suomi, P. and Sørensen, C.

Migration of decision support systems to the mobile environment   3   134
Karetsos, S. and Sideridis, A.

Optimization of maize harvest by mobile ad-hoc network communication   4   135
Jensen, A.L., Hammarberg, J., Dobers, R., Kristensen, L.M. and Thysen, I.

The vineyard digital dashboard   5   136
Neto, M.C., Lopes, C., Maia, J. and Fernandes, L.M.

Session 32. E-trust

Theatre   **Session 32 no. Page**

Importance of trust building elements in B2B agri-food chains   1   137
Meixner, O., Ameseder, C., Haas, R., Canavari, M., Fritz, M. and Hofstede, G.J.

A process perspective on trust building in B2B: evidence from the agri-food sector   2   138
Hofstede, G.J., Oosterkamp, E. and Fritz, M.

Potential application of ICT in agricultural and food sector: performing trust and
new trade relationships in e-business environments   3   139
Fernandez, C., De Felipe, I. and Briz, J.

Interchangeability of traditional trust elements by electronic trust elements:
implementation and evaluation in the case of an e-business platform   4   140
Haas, R., Meixner, O., Ameseder, C., Canavari, M., Cantore, N. and Fritz, M.

Session 33. Environmental management

Theatre   **Session 33 no. Page**

AQUARIUS: a simulation program for evaluating water quality closure rules for
shellfish industry   1   141
Conte, F.S. and Ahmadi, A.

A multi-criteria decision support model for assessing conservation practice adoption   2   142
Vellidis, G., Lowrance, R., Crow, S., Bosch, D. and Murphy, P.

Multivariable decision-making method in landscape planning: watermills in
Pápateszér, Hungary   3   143
Barabás, J., Fülöp, G.Y. and Jombach, S.

Evaluation of SWAT for a small watershed in Eastern Canada   4   144
Ahmad, H., Havard, P., Jamieson, R., Madani, A. and Zaman, Q.

Land use conversion and the hydrological functions effects during the dry season in
Atibainha watershed, Brazil   5   145
Rosa Pereira, V., Teixeira Filho, J. and Dal Fabbro, I.M.

Session 34. Decision support systems

Theatre   **Session 34 no. Page**

Decision support for grazing management: evaluation of suitability and estimation of
potential on alpine pastures for sheep and goats   1   146
Blaschka, A., Guggenberger, T. and Ringdorfer, F.

A decision support system to predict plot infestation with soil-borne pathogens   2   147
Goldstein, E., Hetzroni, A., Cohen, Y., Tsror, L. and Zig, U.

Increasing profit of a co-digestation plant utilizing real time process data   3   148
Van Riel, J.W., Andre, G. and Timmerman, M.

Improvement of work processes in *Gypsophila* flowers   4   149
Bechar, A., Lanir, T., Ruhrberg, Y. and Edan, Y.

Session 35. Decision support systems

Theatre   **Session 35 no. Page**

The development of a sensor and crop growth model based decision tool for site
specific nitrogen application   1   150
Goense, D. and De Boer, J.

Fuzzy logic inference system (FIS) based on soil and crop growth information for
optimum N rate on corn   2   151
Tremblay, N., Bouroubi, M.Y., Panneton, B., Vigneault, P. and Bélec, C.

SMA, an agrometeorological system for crop monitoring   3   152
Faria, R.T., Tsukahara, R.Y., Silva, F.F., Gomes, C.D., Caramori, P.H. and Silva, D.A.B.

How long should your corn-cob pipe be: modeling yield*quality interactions in
Sweetcorn   4   153
Taylor, J.A., Hedges, S. and Whelan, B.M.

Session 36. Technology and ICT adoption

Theatre   **Session 36 no. Page**

To determine the challenges in application of ICT for agricultural extension in Iran   1   154
Farajallah Hosseini, S. and Niknamee, M.

CAP subsidy claims: trying to encourage French farmers to abandon paper forms
and to use the Internet   2   155
Waksman, G., Holl, C. and Coffion, R.

Business adoption of broadband internet in the South West of England   3   156
Warren, M.F.

Italian rural ICT (computer and internet):deployment and accessibility   4   157
Namdarian, I.

Session 37. Technology and ICT adoption

Theatre   **Session 37 no. Page**

Stimulating interest in and adoption of precision agriculture methods on small farm
operations   1   158
Yohn, C.W., Basden, T.J., Rayburn, E.B., Pena - Yewtukhiw, E.M. and Fullen, J.T.

M-learning in agriculture: potential and barriers   2   159
Hansen, J.P. and Hejndorf, P.

E-learning for agriculture and food sector in Poland   3   160
Wozniakowski, M., Orlowski, A. and Wozniakowski, T.

Thinking styles of agriculturalists   4   161
Singh, S. and Lubbe, S.

Session 38. Precision mineral management

Theatre   **Session 38 no. Page**

Optimization model for variable rate application in extensive crops in Chile: the
effects of fertilizer distribution within the field   1   162
Ortega, R.A., Muñoz, R.E. and Acosta, L.E.

Sensitivity analysis of site-specific fertilizer application   2   163
Gebbers, R., Herbst, R. and Wenkel, K.-O.

Precision manure management across site-specific management zones in the Western
Great Plains of the USA   3   164
Moshia, M., Khosla, R., Westfall, D., Davis, J. and Reich, R.

Sensitivity analysis of soil nutrient mapping   4   165
Gebbers, R., Herbst, R. and Wenkel, K.-O.

Session 39. Precision agriculture management

Theatre   **Session 39 no. Page**

Irrigation management zones for precision viticulture according to intra-field variability 1   166
Vallés-Bigordà, D., Martínez-Casasnovas, J.A. and Ramos, M.C.

Impact of fertigation by sprinkler irrigation on variability of crop performance   2   167
Nouri, H., Razavi Najafabadi, J., Amin, M. and Aimrun, W.

Spatial variability of cover crop performance in row crops   3   168
Munoz, J.D., Kravchenko, A., Snapp, S. and Gehl, R.

Precision agriculture in an olive trees plantation in Southern Greece   4   169
*Fountas, S., Bouloulis, C., Paraskevopoulos, A., Nanos, G., Gemtos, T., Wulfsohn, D. and
Galanis, M.*

Influence of site specific herbicide application on the economic threshold of the
chemical weed control   5   170
Gutjahr, C., Weis, M., Sökefeld, M. and Gerhards, R.

Session 40. Living labs

Theatre   **Session 40 no. Page**

Rural living labs: used based innovations for rural areas   1   171
Zurita, L.

KodA: from knowledge to practice for Dutch arable farming   2   172
Paree, P.G.A., Van Gurp, H. and Wolfert, J.

Farmers as co-developers of innovative precision farming systems   3   173
Eastwood, C.R., Chapman, D.F. and Paine, M.S.

Living lab "information management in agri-food supply chain networks"   4   174
Verloop, C.M., Wolfert, J. and Beulens, A.J.M.

Session 41. Traceability

Theatre   **Session 41 no. Page**

Traceability of food of animal origin: major findings of the interdisciplinary project
IT FoodTrace   1   175
Doluschitz, R. and Engler, B.

A traceability system for the food service industry   2   176
Rogge, C.B.E. and Becker, T.C.

Improving information management in organic pork production chains   3   177
Hoffmann, C. and Doluschitz, R.

Logistic study on the recall of non-conform perishable produce through the supply
chain by means of discrete event simulation model   4   178
Busato, P. and Berruto, R.

Session 42. Information sharing in supply chains

Theatre   **Session 42 no. Page**

How are public authorities and food enterprises interwoven: state of the art of
information transmission in the food sector   1   179
Otto, J., Frost, M. and Doluschitz, R.

Information technologies and transparency in agri-food supply chains   2   180
Frentrup, M. and Theuvsen, L.

Interaction models in the fresh fruit and vegetable supply chain using new
technologies for sustainability and quality preservation   3   181
Reiche, R., Fritz, M. and Schiefer, G.

Data warehouse for soybeans and corn market on Brazil 4 182
Elias Correa, F., Pizzigatti Corrêa, P.L., Aparecido Alves, L.R. and Saraiva, A.M.

Session 43. Spatial modelling

Theatre   **Session 43 no. Page**

Spatial optimisation as a tool to maximise profit on mixed-enterprise farms within
environmental constraints   1   183
Cristia, V., Houin, R. and Betteridge, K.

Spatial variability of spikelet sterility in temperate rice in Chile   2   184
Ortega, R.A., Del Solar, D.E. and Acevedo, E.

Investigations on the management of small scale variability of soil nutrients   3   185
Herbst, R., Schneider, M. and Wagner, P.

Selective harvesting zones from remote sensing images and yield data and relation to
grape quality parameters for precision viticulture   4   186
Agelet-Fernández, J., Martínez-Casasnovas, J.A. and Arnó, J.

Heterogeneity analysis on multiple scales: a new insight in precision agriculture   5   187
Karydas, C.G., Zalidis, G.C., Tsatsarelis, K.A., Misopolinos, N.L. and Silleos, N.G.

Session 44. Spatial modelling

Theatre   **Session 44 no. Page**

Geographic management and monitoring of livestock disease events   1   188
Janssen, H., Staritsky, I. and Vanmeulebrouk, B.

Creating space for biodiversity by planning swath patterns and field margins using
accurate geometry   2   189
De Bruin, S., Heijting, S., Klompe, A., Lerink, P., Vonk, M.B. and Van Der Wal, T.

Agricultural-GIS-Sphere: an innovative expert system for national renewable energy
and food planning   3   190
Guggenberger, T., Bartelme, N. and Leithold, A.

Auto-steer navigation profitability and its influence on management practices: a
whole farm analysis   4   191
Shockley, J.M., Dillon, C. and Stombaugh, T.

Improving decision support in plant prodution with GIS   5   192
Endler, M. and Roehrig, M.

Session 45. Supply chain modelling

Theatre **Session 45 no. Page**

ICT and the blooming bloom trade 1 193
Fleming, E., Mueller, R.A.E. and Thiemann, F.

The design of a marketing information system to enhance the competitiveness of
agricultural commodities: some evidence from the grain sector 2 194
Meyer, C., Fritz, M. and Schiefer, G.

Use of discrete event simulation model to study the logistic of a supply-chain for
fresh produce district 3 195
Berruto, R., Busato, P. and Brandimarte, P.

Supply chain optimization of rapeseed as biomass applied on the Danish conditions 4 196
Sambra, A., Sørensen, C.G. and Kristensen, E.F.

Session 46. Improving information transfer to aid (beet)growers

Theatre **Session 46 no. Page**

Information transfer in the European beet growing countries 1 197
Maassen, J. and other members IIRB Working Group

UK beet industry experiences in the provision of internet based communications,
decision support tools and contract administration facilities 2 198
Bee, P.

Audiovisual tools as integrated part of information transfer strategies 3 199
Kämmerling, B.

The four horsemen of innovation: learning styles, entrepreneurship, attitudes and
knowledge network 4 200
Miedema, J. and Faber, N.

Session 47. E-learning and e-content

Theatre **Session 47 no. Page**

Towards teacher competence on metadata and online resources: the case of
agricultural learning resources 1 201
Manouselis, N., Sotiriou, S., Tzikopoulos, A., Costopoulou, C. and Sideridis, A.

Research on open source e-learning tools and agricultural applications 2 202
Lengyel, P. and Herdon, M.

Virtual form of education in a lifelong learning: chance for countryside 3 203
Jarolimek, J., Vanek, J., Havlicek, Z., Silerova, E. and Simek, P.

Designing e-learning courses in WebCT environment: a case study   4   204
Põldaru, R. and Roots, J.

Road mapping of curriculum developments and training experiences in agricultural
informatics education   5   205
Herdon, M.

Session 48. Portals and media

Theatre   **Session 48 no. Page**

Agricultural web TV in France   1   206
Waksman, G., Burriel, C., Holl, C. and Masselin-Silvin, S.

Interoperable metadata for a federation of learning repositories on organic
agriculture and agroecology   2   207
Manouselis, N., Palavitsinis, N. and Kastrantas, K.

"WebInfo": the new website for the national centre of the Danish agricultural
advisory service   3   208
Hørning, A., Hansen, J.P., Hansen, N.F., Kjær, K.B. and Bundgaard, E.

Agrarian WWW portal   4   209
Simek, P., Jarolimek, J., Vanek, J., Silerova, E. and Havlicek, Z.

Session 49. Feeding dairy

Theatre   **Session 49 no. Page**

Precision concentrate rationing to the dairy cow using on-line daily milk
composition sensor, milk yield and body weight   1   210
Maltz, E., Antler, A., Halachmi, I. and Schmilovitch, Z.

Evaluation of a dynamic linear model to estimate the daily individual milk yield
response on concentrate intake and milking interval length of dairy cows   2   211
André, G., Bleumer, E.J.B. and Van Duinkerken, G.

Implementation of an application for daily individual concentrate feeding in
commercial software for use on dairy farms   3   212
Bleumer, E.J.B., André, G. and Van Duinkerken, G.

An approach to precisely calculate variable dosing of highly nutritious and energetic
animal feed   4   213
Ortiz-Laurel, H. and Rossel, D.

Session 50. Data quality and poultry applications

Theatre   **Session 50 no. Page**

A study on the cause and effect of lameness in broiler chickens   1   214
Cangar, O., Everaert, N., De Ketelaere, B., Bahr, C., Decuypere, E. and Berckmans, D.

Integrated ecological hotspot identification of organic egg production 2 215
Dekker, S.E.M., De Boer, I.J.M., Aarnink, A.J.A. and Groot Koerkamp, P.W.G.

Automated monitoring of milk meters 3 216
De Mol, R.M. and André, G.

Masuring and modelling heat production by hen eggs 4 217
Hakimhashemi, M., Eren Özcan, S., Exadaktylos, V. and Berckmans, D.

Session 51. Livestock environment

Theatre **Session 51 no. Page**

Dust reduction in poultry houses by spraying rapeseed oil 1 218
Aarnink, A.J.A., Van Harn, J., Winkel, A., De Buisonje, F.E., Van Hattum, T.G. and Ogink, N.W.M.

Simulating the effect of forced pit ventilation on ammonia emission from a naturally
ventilated cow house with CFD 2 219
Sapounas, A.A., Campen, J.B., Smits, M.C.J. and Dooren, H.J.C.

Development of new methods and strategies for monitoring operational performance
of emission mitigation technology at livestock operations 3 220
Ogink, N.W.M., Melse, R.W. and Mosquera, J.

Development and evaluation of two ISOagriNET compliant systems for measuring
environment and consumption data in animal housing systems 4 221
Kuhlmann, A., Herd, D., Gallmann, E., Rößler, B. and Jungbluth, T.

Dust emission factors and physical properties in order to develop dispersion models 5 222
Nannen, C. and Büscher, W.

Session 52. Wireless sensing

Theatre **Session 52 no. Page**

A wireless network for measuring rumen pH in Dairy cows 1 223
Goense, D., Houwers, W., Müller, H.C., Unsenos, D. and Wehren, W.

Measuring rumen pH and temperature by an indwelling and wireless data
transmitting unit and application under different feeding conditions 2 224
Gasteiner, J., Fallast, M., Rosenkranz, S., Häusler, J., Schneider, K. and Guggenberger, T.

Recording of tracking behaviour of dairy cows with wireless technologies 3 225
Ipema, A.H., Bleumer, E.J.B., Hogewerf, P.H., Lokhorst, C., De Mol, R.M., Janssen, H. and Van Der Wal, T.

Estimating impact on clover-grass yield caused by traffic intensities 4 226
Jørgensen, R.N., Sørensen, C.G., Green, O. and Kristensen, K.

Impelementation of herd management system with wireless sensor networks 5 227
Wu, T., Goo, S., Goh, H., Kwong, K., Michie, C. and Andonovic, I.

Session 53. Fencing

Theatre	Session 53 no.	Page

Could virtual fences work without giving cows electric shocks? — 1 — 228
Umstatter, C., Tailleur, C., Ross, D. and Haskell, M.J.

A method for managing cow traffic in a pastoral automatic milking system — 2 — 229
Jago, J., Bright, K., Jensen, R. and Dela Rue, B.

Stakeless fencing for mountain pastures — 3 — 230
Monod, M.O., Faure, P., Moiroux, L. and Rameau, P.

Pastures from space: evaluating satellite-based pasture measurement for Australian dairy farmers — 4 — 231
Eastwood, C.R., Mata, G., Handcock, R. and Kenny, S.

Session 54. Dairy fertility and calving management

Theatre	Session 54 no.	Page

Cow body shape and automation of condition scoring — 1 — 232
Halachmi, I., Polak, P., Roberts, D.J., Boyce, R.E. and Klopcic, M.

Identifying changes in dairy cow behaviour to predict calving — 2 — 233
Macrae, A.I., Miedema, H.M., Dywer, C. and Cockram, M.S.

Combination of activity and lying/standing data for detection of oestrus in cows — 3 — 234
Jónsson, R.I., Blanke, M., Poulsen, N.K., Højsgaard, S. and Munksgaard, L.

Mathematical optimization to improve cows' artificial insemination services — 4 — 235
Halachmi, I., Shneider, B., Gilad, D. and Eben Chaime, M.

A new generation of fertility monitoring in cattle herds — 5 — 236
Balzer, H.-U., Kultus, K. and Köhler, S.

Session 55. Animal identification

Theatre	Session 55 no.	Page

First results of a large field trial regarding electronic tagging of sheep in Germany — 1 — 237
Bauer, U., Kilian, M., Harms, J. and Wendl, G.

Using a wide electronic pop hole based on RFID-technology with high-frequency transponders to monitor the ranging behaviour of laying hens in alternative housing systems — 2 — 238
Thurner, S., Pauli, S., Wendl, G. and Preisinger, R.

Using injectable transponders for sheep identification — 3 — 239
Hogewerf, P.H., Ipema, A.H., Binnendijk, G.P., Lambooij, E. and Schuiling, H.J.

Application of RFID technology in cow and herd management in dairy herds in Canada   4   240
Murray, B.B., Rumbles, I. and Rodenburg, J.

Electronic ear tags for tracing fattening pigs according to housing and production system   5   241
Burose, F., Jungbluth, T. and Zähner, M.

Session 56. Pigs

Theatre   **Session 56 no. Page**

Real-time measurement of pig activity in practical conditions   1   242
Leroy, T., Borgonovo, F., Costa, A., Aerts, J.-M., Guarino, M. and Berckmans, D.

Active feeding control and environmental enrichment with call-feeding-stations   2   243
Manteuffel, G.

Automatic detection of pig vocalization as a management aid in precision livestock
farming   3   244
Schön, P.C., Düpjan, S. and Manteuffel, G.

Session 57. Future Farming

Theatre   **Session 57 no. Page**

FutureFarm: the European farm of tomorrow   1   245
Blackmore, B.S.

Management strategies and practices for precision agriculture operations   2   246
Fountas, S., Pedersen, S., Blackmore, S., Chatzinikos, A., Sorensen, C., Pesonen, L., Basso, B. and Nash, E.

Can compliance to crop production standards be automatically assessed?   3   247
Nash, E.J., Fountas, S. and Vatsanidou, A.

Typology of precision farming technologies suitable for EU-Farms   4   248
Schwarz, J., Werner, A.B. and Dreger, F.

Session 58. Mastitis detection

Theatre   **Session 58 no. Page**

Decision tree induction as an automated detection tool for clinical mastitis using data
from six Dutch dairy herds milking with an automatic milking system   1   249
Kamphuis, C., Mollenhorst, H., Feelders, A. and Hogeveen, H.

Mastitis detection: visual observation compared to inline, quarter and milking SCC   2   250
Mollenhorst, H., Van Der Tol, P.P.J. and Hogeveen, H.

Inline SCC monitoring improves clinical mastitis detection in an automatic milking system   3   251
Hogeveen, H., Kamphuis, C., Sherlock, R., Jago, J. and Mein, G.

Use of a cow-specific probability of having clinical mastitis to determine the predictive value positive of automatic milking systems   4   252
Steeneveld, W., Barkema, H.W. and Hogeveen, H.

Session 59. Future Farming

Theatre   **Session 59 no. Page**

Analysis of external drivers for farm management and their influences on farm management information systems   1   253
Charvat, K. and Gnip, P.

Potential savings and economic benefits in arable farming from better information management   2   254
Pedersen, S.M., Fountas, S., Sørensen, C.G., Pesonen, L., Basso, B., Ørum, J.E. and Blackmore, S.

Crop models provide the "desired extra information" to reduce farmer's risk in decision making: the case of nitrogen application rates   3   255
Basso, B., Fountas, S., Sartori, L., Pedersen, S.M., Sorensen, C., Pesonen, L., Werner, A. and Blackmore, S.

Technologies for a standardised information infrastructure to assist compliance to crop production standards   4   256
Nash, E.J., Nikkilä, R., Pesonen, L., Sørenson, C.G. and Fountas, S.

FutureFarm and the future of precision agriculture in Europe   5   257
Lowenberg-Deboer, J. and Griffin, T.

Session 60. European relevance for precision agriculture

Theatre   **Session 60 no. Page**

Future GNSS: farmers navigate towards trusted farming   1   258
Lokers, R.M. and Van Der Wal, T.

New GPS based methods accredited by the EC for area measurement   2   259
Grzebellus, M.

URM as tool for Shared Environmental Information System (SEIS)   3   260
Charvat, K., Kafka, S. and Cepicky, J.

System analysis of management information systems for the future   4   261
Sørensen, C.G., Fountas, S., Pesonen, L., Basso, B., Pedersen, S.M. and Nash, E.

The implementation of the European Common Agricultural Policy and spatial data infrastructures   5   262
Van Der Wal, T., Kay, S. and Devos, W.

Session 61. Locomotion

Theatre   **Session 61 no. Page**

Recording and analysis of locomotion in dairy cows with 3D accelerometers   1   263
De Mol, R.M., Lammers, R.J.H., Pompe, J.C.A.M., Hogewerf, P.H. and Ipema, A.H.

An intelligent wireless accelerometer system for measuring gait features and lying
time in dairy cows   2   264
Pastell, M., Hakojärvi, M., Hänninen, L. and Tiusanen, J.

Recording of dairy cow behaviour with wireless accelerometers   3   265
De Mol, R.M., Bleumer, E.J.B., Hogewerf, P.H. and Ipema, A.H.

Approach to model based motion scoring for lameness detection in dairy cattle 4   266
Pluk, A., Bahr, C., Maertens, W., Vangeyte, J., Sonck, B. and Berckmans, D.

Session 62. Constraint commonalities of adoption of technological innovation

Theatre   **Session 62 no. Page**

ICT adoption trends in agriculture   1   267
Gelb, E.

Session 63. ICT and the New Food Economy

Theatre   **Session 63 no. Page**

The impact of ICT on the food economy   1   268
Bunte, F., Groeneveld, R., Hofstede, G.J., Top, J. and Wolfert, S.

Session 64. Poster session

Poster   **Session 64 no. Page**

Precision Agriculture

Spatial variability of soil properties and the occurrence of soil-borne pests in sugar beet 1   269
Scholz, C., Patzold, S. and Welp, G.

Ground-based integration sensor and instrumentation system for measuring crop
conditions   2   270
Lan, Y., Zhang, H., Lacey, R., Huang, Y. and Hoffmann, W.

Differentiation of several leaf rust (*Puccinia rocondita*) severity classes in winter
wheat (*Triticum aestivum*) with *in situ* hyperspectral data   3   271
Mewes, T., Franke, J. and Menz, G.

Optimization of radiation mode for plants with different geometrical arrangement   4   272
Rakut'ko, S.

Analyze of temporal variations in NDVI within-field spatial variability as a possible
basis for precision farming   5   273
Nováková, M., Halas, J. and Scholtz, P.

Small scale spatial variability of soil physical properties of a ferralsol with three
different land uses   6   274
Paz-Ferreiro, J., Pereira De Almeida, V. and Alves, M.C.

Spy-See: advanced vision system for phenotyping in greenhouses   7   275
Polder, G., Van Der Heijden, G.W.A.M., Glasbey, C.A. and Dieleman, J.A.

Modelling crop root growth and biomass accumulation   8   276
Hautala, M. and Hakojärvi, M.

Variable-rate lime application for Louisiana sugarcane production systems   9   277
Johnson, R. and Viator, H.

Use of aerial imaging and electrical conductivity for spatial variability mapping of
soil conditions   10   278
Lukas, V., Neudert, L. and Kren, J.

Evaluating water status in irrigated grapevines and olives using thermal imaging   11   279
*Alchanatis, V., Cohen, Y., Sprinstin, M., Naor, A., Meron, M., Cohen, S., Ben-Gal, A., Agam,
N., Yermiyahu, U. and Dag, A.*

The effect of dew on reflectance obtained from ground active optical sensors   12   280
Povh, F.P., Gimenez, L.M. and Molin, J.P.

Yield-SAFE for water management: calibration and validation for maize in
Mediterranean and Atlantic regions   13   281
Mayus, M., Palma, J., Topcu, S., Herzog, F. and Van Keulen, H.

Using vegetation indices to determine peanut maturity   14   282
Vellidis, G., Ortiz, B., Beasley, J. and Rucker, K.

Mapping available soil micronutrients in an Atlantic agricultural landscape   15   283
Paz-Ferreiro, J. and Vieira, S.R.

Comparison between the tillage force and the cone penetrometer data   16   284
Csiba, M., Milics, G., Barthalos, P. and Nemenyi, M.

The software for the mobile information-measuring complexes   17   285
Petrushin, A.F.

Mobile measuring-calculating agro-meteorological complex   18   286
Efimov, A.E., Danilova, T.N., Kozyreva, L.V., Kochegarov, S.F. and Uskov, I.B.

Development of a yield mapping system for precision agriculture   19   287
Zhang, M., Li, M., Liu, G. and Wang, M.

Method of determination of thermal properties of soil   20   288
Ivanova, K.F.

Real-time analysis of soil parameters with NIR spectroscopy   21   289
Li, M. and Zheng, L.

Development of an online measurement system to soil EC   22   290
Zhao, Y., Li, M., Zhang, J. and Zhang, M.

Using X-ray method for prediction of field seeds germination in precision agriculture   23   291
Arkhipov, M.V., Gusakova, L.P., Velikanov, L.P., Vilichko, A.K., Alferova, D.V. and
Zheludkov, A.G.

A new self-excited oscillator method for measuring soil water content and elecrtical
conductivity in technologies of precision agriculture   24   292
Ananyev, I.P.

The analysis and application of navigation system for autonomous vehicle in agriculture 25   293
Zhang, M., Liu, G. and Zhou, J.J.

Positioning and control technologies of automatic navigation system of agricultural
machinery   26   294
Gang, L. and Xiangjian, M.

Spatial prediction of wheat yield and grain production using terrain attributes by
Artificial Neural Network (ANN)   27   295
Ayoubi, S., Nouruzi, M., Jalalian, A., Deghani, A.A. and Khademi, H.

Diagnosing crop growth and nitrogen nitritional status around panicle initiation stage
of rice with digital camera image analysis   28   296
Lee, K.J., Choi, D.H. and Lee, B.W.

Environment monitoring system based on the wireless sensor network   29   297
Li, L. and Haixia, L.

A new sensor investigation for in-field soil nitrate nutrient rapid detection   30   298
Zhang, M., Ang, S., Wang, M.H. and Nguyen, C.

Effect of organic manure and nitrogen level on sugar beet (Beta [vulgaris) yield and
root nitrate content   31   299
Abd El Lateef, E.M.

Radio-controlled complex of agricultural crops aeromonitoring for information
support of precision agriculture technology   32   300
Slinchuk, S.G., Petrushin, A.F. and Aivazov, G.S.

In situ and at real time characterization of the heterogeneity of soil and the dynamical
behavior vehicle   33   301
Chanet, M., Marionneau, A. and Seger, M.

Participatory-decision support system for irrigation management based on earth
observation methodologies and open-source webGIS tool: a case study from Italy   34   302
Osann Jochum, M.A., Belmonte Calera, A., Nino, P., Lupia, F. and Vanino, S.

Assessment of weed cover: a study of the consistency and accuracy of human
perception   35   303
Andujar, D., Ribeiro, A., Fernandez-Quintanilla, C. and Dorado, J.

An inexpensive process for 350-1100 nm wavelength aerial photography for precision agriculture 36 304
Pudełko, R. and Igras, J.

Low altitude aerial photography used in remote sensing for monitoring the expansion of diseases 37 305
Nieróbca, A., Pudełko, R. and Kozyra, J.

Use of RADAR measured precipitation data in disease forecast models 38 306
Jung, J. and Kleinhenz, B.

Geophilus electricus: a new soil mapping system 39 307
Ruehlmann, J., Lueck, E. and Spangenberg, U.

Using optical remote sensing for estimating canopy water content 40 308
Clevers, J.G.P.W.

Assessment of soil properties on cropping farms with *Geophilus electricus* 41 309
Spangenberg, U., Lueck, E. and Ruehlmann, J.

Kopernikus inside of EarthlookCz project 42 310
Charvat, K., Horak, P. and Vlk, M.

Interconnection of altitude of stationary GPS observation points and soil moisture with formation of winter wheat grain yield 43 311
Lapins, D., Dinaburga, G., Plume, A., Berzins, A. and Kopmanis, J.

Modelling spatio-temporal weed population dynamics for the development of strategies in site specific weed control 44 312
Nordmeyer, H., Richter, O. and Sandt, N.

Spatial and temporal variability of N, P and K in a rice field 45 313
Morales, L.A. and Paz-Ferreiro, J.

Modeling case study of expected maize yield quantity 46 314
Boksai, D.B. and Erdélyi, É.E.

Precision Livestock Farming

Decision making for cattle using movement-based pattern recognition 47 315
Godsk, T. and Kjærgaard, M.B.

Prototype system to recognize agricultural operations automatically based on RFID 48 316
Sugahara, K., Nanseki, T. and Fukatsu, T.

Controlling of silage crop compaction for validation of silage quality and aerobic stability in case of silage bagging technology 49 317
Maack, C. and Büscher, W.

Prospects for the application of genomic markers in precision livestock farming 50 318
Smits, M.A., Calus, M.P. and Veerkamp, R.F.

Measuring of vacuum near teatend inside the teatcup with piezoresistive pressure transmitters and usage of output signal for steering milking machines 51 319
Ströbel, U., Rose, S. and Brunsch, R.

Diurnal feeding patterns of individual dairy cows fed with an automatic precision
feeding system   52   320
Abbink, N., André, G., Bleumer, E.J.B. and Pompe, J.C.A.M.

Estimation of feed intake of dairy cows by means of feeding behaviour caracteristics   53   321
Azizi, O., Kaufmann, O. and Hasselmann, L.

The effect of different geometrical wind tunnels in aerodynamic characteristics and
ammonia mass transfer process in aqueous solution   54   322
Saha, C.K., Zhang, G., Ye, Z., Rong, L. and Strøm, J.S.

Monitoring pig's activity using manual visual labelling   55   323
Borgonovo, F., Leroy, T., Costa, A., Aerts, J.-M., Berckmans, D. and Guarino, M.

Online quantification of the excitement of a stallion exposed to an oestrous mare   56   324
Avezzù, V., Jansen, F., Guarino, M., Pecile, A.M., Quanten, S. and Berckmans, D.

Water saving through smarter irrigation in Australian dairy farming: use of
intelligent irrigation controller and wireless sensor network   57   325
*Dassanayake, D., Malano, H., Dassanayake, K.B., Langford, J., Douglas, P. and Dunn,
G.M.*

Mathematical modelling of infectious diseases: transmission of foot-and-mouth
disease in Pantanal Matogrossense region, Brazil   58   326
Ternes, S., Aguilar, R., Maidana, N. and Yang, H.

Feeding soy hulls to robotically milked high-yielding dairy cows increased milk
production   59   327
Halachmi, I., Shoshani, E., Solomon, R., Maltz, E. and Miron, J.

Analysis of capacity reserves in automatic milking systems   60   328
Harms, J. and Wendl, G.

Deterministic grass model   61   329
Huzsvai, L.

Urine sensors for sheep and cattle   62   330
Betteridge, K., Carter, M.L. and Costall, D.A.

Effects of different ratios of nonfiber carbohydrate to rumen degradable protein on
the performance, blood and rumen samples of Holstein cows   63   331
Rafiee, H.

Recording of water intake of suckler cows to detect forthcoming calving   64   332
Mačuhová, J., Jais, C. and Oppermann, P.

Near body temperature measurements in broilers with a wireless sensor   65   333
Bleumer, E.J.B., Hogewerf, P.H. and Ipema, A.H.

EFITA

Postharvest food participatory technology development among rural farmers in Nigeria   66   334
Abiodun, A.A. and Ogundele, B.A.

Adoption of automation and information technologies in relation to milking and animal monitoring on New Zealand dairy farms 67 335
Dela Rue, B. and Jago, J.

Chain-wide communication exchange in the German pork industry: an empirical analysis 68 336
Plumeyer, C.H. and Theuvsen, L.

Utilization and comparison of three ground based sensors for nitrogen managment in irrigated corn 69 337
Shaver, T.M., Westfall, D.G. and Khosla, R.

Implementation of recording of good agricultural practices in an agro-entrepreneurial information system 70 338
Caceres, V.

Discrimination of a soil-borne disease complex due to simultaneous infection of Heterodera schachtii and Rhizoctonia solani in sugar beet 71 339
Hillnhütter, C., Sikora, R.A. and Oerke, E.-C.

Mobile internet in e-government: the case study of Hungary 72 340
Szilagyi, R. and Szilagyi, E.

Spatial variability in irrigated cotton in relation to soil electroconductivity 73 341
Bauer, P., Stone, K. and Busscher, W.

Environmental benefits of ecologically friendly paddy fields: a choice experiment approach 74 342
Aizaki, H.

Usage of information technology over the combine harvester users: a case study for Antalya region 75 343
Yilmaz, D., Canakci, M. and Akinci, I.

The economic impact of contract farming: a case study in Maharashtra state of India 76 344
Kulkarni, S., Grethe, H. and Van Huylenbroeck, G.

Determination of supporting and repressive elements of the integrated animal health system 77 345
Fick, J. and Doluschitz, R.

Cost and benefit aspects of it-based quality assurance and traceability systems: results of a Delphi-survey 78 346
Roth, M. and Doluschitz, R.

Scientific based expert models, a method to downsizing information to many farmers 79 347
Mayer, W.

Prototype of an academic ISO11783 compatible task controller 80 348
Ojanne, A., Kaivosoja, J., Suomi, P., Nikkilä, R., Kalmari, J. and Oksanen, T.

Assessing whole-farm risk positions by means of a risk barometer: an internet application for arable and poultry farmers 81 349
Van Asseldonk, M., Baltussen, W., Hennen, W. and Van Horne, P.

Variable rate seeding for French wheat production: profitability and production risk management potential   82   350
Dillon, C.R., Gandonou, J. and Shockley, J.

The potential benefit of site-specific phosphorus and potassium management for small fields   83   351
Pena-Yewtukhiw, E.M., Rayburn, E.B., Yohn, C.W. and Fullen, J.T.

ICT mass customization in flowers & food: feasibility of service-oriented BPM platforms   84   352
Verdouw, C.N., Verloop, C.M. and Ten Voorde, H.

Agro living lab: a new platform for user-centered design in agriculture   85   353
Haapala, H.E.S.

SARI®, computer software for sectioning and assessment remote images for precision agriculture uses   86   354
Garcia-Torres, L., Gómez-Candón, D., Caballero-Novella, J.J., Peña-Barragán, J.M., Jurado-Exposito, M., Castillejo-Gonzalez, I., García-Ferrer, A. and López-Granados, F.

Hybrid wireless networks for advanced communication services in agriculture   87   355
Boffety, D., Chanet, J.P., Hou, K.M., André, G., Amamra, A., De Sousa, G. and Jacquot, A.

Multi scale analysis of the factors influencing wheat quality   88   356
Har Gil, D., Svoray, T. and Bonfil, D.J.

CAPRICORN: a windows program for formulating and evaluating rations for goats   89   357
Ahmadi, A. and Robinson, P.H.

Computer model use in decision making system for planning and management of agrotechnologies   90   358
Yakushev, V.V.

Methodological approaches to estimating of an optimal time instant of agrotechnical practices   91   359
Bure, V.M. and Yakushev, V.P.

Investigation on relations between organic matter, Zn and Cu content in some soils of Guilan   92   360
Norouzi, M., Naraghi, M., Alidoust, E., Honarmand, M. and Forghani, A.

Optimal cropping patterns in Saudi Arabia using mathematical sector model   93   361
Al-Abdulkader, A.M., Al-Amoud, A.I., Awad, F.S., Al-Tokhais, A.S., Basahi, J.M., Al-Moshailih, A.M., Al-Dakheel, Y.Y., Alazba, A.A. and Al-Hamed, S.A.

Agronix, neural network for fertilisation management   94   362
Megna, A., Spada, V., Campisi, R. and Manzini, V.

Geo-fertilizer advice modeled using BPM and SOA   95   363
Verloop, C.M., Wolfert, J. and Beulens, A.J.M.

A new algorithm to use oblique view images as a low cost remote sensing system   96   364
Hartmann, K., Lilienthal, H. and Schnug, E.

Fostering quality of argumentative computer-supported collaborative learning within academic education in the agri-food sciences 97 365
Noroozi, O., Biemans, H., Mulder, M. and Chizari, M.

Analysis the role of Information and Communication Technologies (ICTs) in improvement of food security Iran's rural households 98 366
Lashgarara, F., Mirdamadi, S.M. and Farajjolah Hosseini, S.J.

E-government services for famers and experience in the North-East Hungary region 99 367
Szenas, S.Z. and Herdon, M.

Water delivery optimization program, of Jiroft Dam irrigation networks, by using genetic algorithm 100 368
Rahnama, M.B. and Jahanshae, P.

Designing, implementation and institutionalizing the national agricultural innovation system 101 369
Mortazavi, M. and Ranaie, H.

The analysis of modeling methods applied to forest growth research 102 370
Komasilovs, V. and Arhipova, I.

The role of e-signature implementation in regional development 103 371
Zacepins, A. and Arhipova, I.

Get and hold clients on your web project 104 372
Razgale, V.

A data warehouse for decision suport in the Brazilian beef industry 105 373
Massruhá, S., Barioni, L., Lacerda, T., Lima, H., Da Silva, O. and Narciso, M.

Studying the importance of web portals in an agriculture educational Institution through the evaluation of the web portal of ITVHE 106 374
Mokhtari Aski, H., Rajabbeigi, M. and Mortazavi, M.

Assessing factors effecting on application of internet and website in educational and research of graduate students in college of agriculture 107 375
Mirdamadi, M. and Alimoradian, P.

Precise agriculture and mezo- and microclimate production agroecology system 108 376
Nasonov, D.V. and Uskov, I.B.

Web platform for simulation of agricultural mechanization technologies 109 377
Muraru, V.M., Pirna, I., Cardei, P., Muraru Ionel, C. and Sfaru, R.

A fine-tuned phosphorus strategy for sustainable production 110 378
Söderström, M., Ulén, B. and Stenberg, M.

A system to increase safety and security of agricultural products by image information 111 379
Tanaka, K. and Hirafuji, M.

An analysis of ICT adoption trends on Irish farms 112 380
Murphy, D.J.

Options for precision horticulture in Gala orchards based on site-specific
relationships between environmental factors and harvest production parameters 113 381
Manfrini, L., Corelli Grappadelli, L. and Taylor, J.A.

A fuzzy inference segmentation algorithm for the delineation of management zones
and classes 114 382
Pedroso, M., Guilleume, S., Charnomordic, B., Taylor, J. and Tisseyre, B.

An integrative approach of using satellite based information for precision farming:
Talking.Fields 115 383
Bach, H., Migdall, S. and Ohl, N.

Advanced web technologies for environment-related mashups 116 384
Maria Koukouli, M.K. and Alexander Sideridis, A.S.

Towards a data model for the Integration of LADM and LPIS 117 385
Inan, H.I., Milenov, P., Van Oosterom, P., Sagris, V., Zevenbergen, J. and Yomralioglu, T.

The use of video segmentation technique to build a lightweight tool for remote
monitoring of agricultural process through Internet 118 386
Hirakawa, A.R., Saraiva, A.M. and Amancio, S.M.

Wine traceability and wireless sensor networks 119 387
Gogliano Sobrinho, O. and Cugnasca, C.E.

Electronic claiming of SAPS: the Hungarian case 120 388
Csótó, M. and Székely, L.

Efficient knowledge transfer for advisers and farmers 121 389
Quendler, E. and Boxberger, J.

I3S: integrated solution support system for water stress mitigation 122 390
Kassahun, A., Krause, A. and Roosenschoon, O.

Technological advances in developing Web applications for the agri-food sector 123 391
Weres, J., Kozłowski, R.J., Kluza, T. and Mueller, W.

An Internet-based decision-support system for winter oilseed rape protection 124 392
Kozłowski, R.J.

Neural image analysis for agricultural products 125 393
Nowakowski, K.

Neural identification of selected kinds of insects based on computer technology for
the image analysis 126 394
Boniecki, P., Piekarska-Boniecka, H. and Nowakowski, K.

Utilization of the computer image analysis and artificial neural network models for
measuring lamb's intramuscular fat level content 127 395
Przybylak, A. and Boniecki, P.

In situ sensors for agriculture 128 396
Charvat, K., Gnip, P., Jezek, J. and Musil, M.

PCR detection for mapping in-field variation of soil-borne pathogens 129 397
Jonsson, A., Almquist, C. and Wallenhammar, A.-C.

Expert system for crop selection in Punjab region 130 398
Jassar, S. and Sawhney, B.

Application of multispectral and hyperspectral vegetation indices for prediction of
yield and grain quality of spring barley in Hungary 131 399
*Milics, G., Burai, P., Lénárt, C.S., Tamás, J., Papp, Z., Deákvári, J., Kovács, L., Fenyvesi, L.
and Neményi, M.*

Analysis of the intra-field variability in vineyards for the definition of management
zones by means multi-variant analysis 132 400
Díez-Galera, Y., Martínez-Casasnovas, J.A., Rosell, J.R. and Arnó, J.

The use of OpenMI in model based integrated assessments 133 401
Knapen, M.J.R.

ICT adoption constraints in horticulture: comparison of the ISHS and ILVO
questionnaire results to the EFITA baseline data sets 134 402
Taragola, N., Van Lierde, D. and Gelb, E.

Microsprayer accuracy for application of glyphosate on weed potato plants between
sugar 135 403
Nieuwenhuizen, A.T., Hofstee, J.W., Van De Zande, J.C. and Van Henten, E.J.

Active sensing of the N status of wheat using optimized wavelength combination: impact of seed rate, variety and growth stage

Jasper, J., Reusch, S. and Link, A., Yara International ASA, Research Centre Hanninghof, Hanninghof 35, 48249 Duelmen, Germany

Variable rate nitrogen fertilizer application needs efficient determination of the nitrogen nutrition status of crops with high spatial and temporal resolution. A suitable approach to get this information fast and at low cost is proximal sensing of the light that is reflected from the crop canopy, either using the sunlight (passive sensors) or artificial light sources (active sensors). Although there is an increasing interest in active crop sensing, the knowledge on the appropriate choice of wavebands to be collected and vegetation indices to be calculated in order to get a reliable estimate of the crop N status has mainly been derived from experiments where passive field-spectrometer systems have been used to collect multi-spectral reflectance data. Whether the findings in those experiments are applicable to active remote sensing is at least questionable. This is because artificial light sources produce a divergent beam that penetrates crop canopies less deep compared to the parallel beam of the sun. The impact of canopy structure on light reflectance might therefore be different for active and passive crop sensors and may affect the suitability of spectral indices for N uptake estimation and their susceptibility to agronomic factors like crop variety and seeding rate. Previous research has indicated that an optimized spectral index based on one waveband in the NIR and the other close to the infrared shoulder of canopy reflectance might be superior to standard NDVI indices for both passive and active sensors. Since only little is known about the effect of agricultural practices on active measurements using different spectral indices, the accuracy of N uptake estimations by active remote sensing was investigated in two winter wheat field trials with varying N-rates, contrasting varieties and different seed densities in 2005 and 2007. As for passive canopy sensors, the sensitivity of active N uptake measurements depends on the choice of appropriate wavelength combinations. The optimized spectral index turned out to be superior to indices that combine wavelengths taken from the near infrared and the visible spectral range. It was found to be largely unaffected by variety properties and seed rates and showed less saturation at high N uptake levels later in the season, compared to the NDVI. The results furthermore indicate that the relationship between active reflectance measurement and N uptake depends on the physiological growth stage of the crop, regardless of the spectral indices used. Thus, algorithms to derive nitrogen fertilizer recommendations from the measurements need to be developed for N top dressings at respective specific growth stages separately.

Development and first tests of a mobile lab combining optical and analogical sensors for crop monitoring in precision viticulture

Mazzetto, F.[1], Calcante, A.[1], Mena, A.[1] and Vercesi, A.[2], [1]University of Milan, Institute of Agricultural Engineering, via Celoria 2, 20133 Milan, Italy, [2]University of Milan, Institute of Plant Pathology, via Celoria 2, 20133 Milan, Italy

Actually remote-sensing (RS) is the most diffused technique used to realize cultural monitoring in Precision Viticulture systems. This paper considers the possibility to integrate RS information with those obtained by ground sensing technologies employed directly in vineyard. To this aim a mobile lab is developed; it implements: a) a commercial optical sensor – the GreenSeeker RT100 – able to calculate NDVI and Red/NIR indices in real time and normally used on cover crops, and b) ultrasonic sensors for row thickness estimation. The aim of this work is to evaluate canopy health and vigour status in the same time. During 2007-2008 campaign, tests were carried out in a commercial vineyard through specific on board-tractor instrumentation. They aimed to evaluate the monitoring system performance regarding disease appearance and development, and vegetative vigour variations due to different phenological status. Tests were carried out from May to August and involved two groups of Pinot Grigio rows: - Group A: 3 rows never treated with agrochemicals during the productive season; - Group B: 4 rows treated with usual defence strategy at regular time. Tests consisted in mobile surveys carried out through a mobile lab equipped with: a) two coupled GreenSeeker (at 1.16 m in height from ground level), b) three couples of ultrasonic sensors, c) a DGPS receiver to georefer data collected while travelling in vineyard. By mounting ultrasonic sensors at different heights (0.9, 1.3, 1.7 m) the site specific canopy volume can be derived from the measured horizontal distance to the foliage. Contextually parcels have been identified on the same rows; manual morphological and physiological observations have been conducted to characterize vegetation phytosanitary status. Optical data have been processed through the IDW method in order to obtain NDVI maps. In this case, ultrasonic data have been used to identify each monitored row and to leave out turning manoeuvre data points. Then NDVI and ultrasonic data referred to the parcels have been compared. Measurements repeatability has been verified through the realisation of consecutive passages of the mobile lab along the same rows both according the same direction and the reverse one, during the same test. Both NDVI values and ultrasonic data show a high repeatability (R^2=0.81 and R^2=0.77, respectively). Then NDVI maps of all rows and different parcels have been produced. They clearly show difference in vegetation conditions evolution in the two examined groups and low vegetative vigour in correspondence of areas infected by Plasmopara viticola, confirmed by manual relieves. Maps of infestation intensity have been produced according to the results of pathological manual surveys. The comparison between infestation and NDVI maps confirmed the real vine phytosanitary status even if at a qualitative level. Concerning ultrasonic sensors, canopy semi-thickness has been calculated as the difference between the sensor distance from vegetative wall and half inter-row-space. The comparison between NDVI values and canopy semi-thickness calculated on parcels highlights lack of vegetation along the rows corresponding to areas presenting low NDVI values (indicating critical vegetation conditions) but absence of infestation. In this way the identification of areas really interested by disease is possible.

Sensitivity of narrow and broad-band vegetation indexes to leaf chlorophyll concentration in planophile crops canopies

Vincini, M. and Frazzi, E., Università Cattolica del Sacro Cuore, via Emilia Parmense 84, 29100 Piacenza, Italy

Both narrow-band and broad-band vegetation indices (VI) can be effectively used, due to their sensitivity to leaf area index (LAI) and, secondarily, to leaf chlorophyll concentration, to evaluate the overall photosynthetic capacity of a canopy as expressed by canopy chlorophyll density (CCD). CCD is a measure of photosynthetic potential at the canopy level sensitive to soil N availability. However, only narrow-band VI, requiring high-spectral resolution reflectance data, have proven to be specifically sensitivity to leaf chlorophyll concentration at the canopy scale, an effective indicator of nutritional stress. We have proposed the chlorophyll vegetation index (CVI), a broad-band vegetation index specifically sensitive to leaf chlorophyll concentration at the canopy scale. The present work addresses the comparison of the sensitivity of the broad-band CVI and of different specifically proposed narrow-band VI], to leaf clorophill concentration at the canopy scale in planophile (i.e., with low average leaf angle) crops. The comparison is conducted by the analysis of a large synthetic data-set obtained by the SAILH+PROSPECT coupled leaf and canopy reflectance model used in the direct mode. Synthetic data are used for a sensitivity analysis of the different VI based on a sensitivity function obtained according to the method proposed by Ji and Peters. The sensitivity function is calculated as the ratio of the first derivative of the regression function – using leaf chlorophyll concentration as the independent variable (x) and the VI values as the dependent variable – and the standard error $\sigma_{\hat{y}}$ of the predicted value ($\hat{y}$).

Comparison of methods to estimate LAI from remote sensing in maize canopies by inversion of 1-D and 3-D models

Casa, R.[1], Nassetti, F.[1], Pascucci, S.[2], Palombo, A.[2], D'urso, G.[3], Ciraolo, G.[4], Maltese, A.[4], Giordano, L.[5] and Jones, H.G.[6], [1]Università della Tuscia, Produzione Vegetale, Via San Camillo de Lellis, 01100 Viterbo, Italy, [2]C.N.R., Laboratorio Aereo Ricerche Ambientali, Via del Fosso del Cavaliere, 100, 00133 Roma, Italy, [3]Università di Napoli, Ingegneria Agraria e Agronomia del Territorio, Via Università 100, 80055 Portici, Napoli, Italy, [4]Università di Palermo, Ingegneria Idraulica ed Applicazioni Ambientali, Viale delle Scienze, 90128 Palermo, Italy, [5]ENEA, Biotec-Des, Via Anguillarese, 301, 00123 S.Maria di Galeria, Roma, Italy, [6]University of Dundee, Division of Plant Science, Invergowrie, Dundee DD2 5DA, United Kingdom

Leaf area index (LAI) and leaf angle distribution (LAD) are needed for monitoring the status of crops and diagnosing possible stress conditions, as well as for the estimation of yield through the use of crop growth models. Estimation of these parameters through remote sensing is a cost-effective way for obtaining rapidly information on these properties. The inversion of canopy reflectance models is widely employed for the retrieval of such information. The accuracy of the estimates depends greatly on the realism of models canopy representation and the a-priori knowledge of vegetation characteristics. The objective of the present work was to compare the use of different type of models. This assessment was based on the use of specifically gathered field data. Measurement campaigns were carried out in maize at different growth stages in 2007 and 2008. Multiple-look-angle remote sensing data were acquired by using a field sensor positioning system based on a extending bipod arm with levelling head and clinometer, having the possibility of holding 2-3 collimated sensors. Spectral radiance data were acquired using an Analytical Spectral Devices (ASD) spectroradiometer in the range 350-2500 nm and converted to absolute reflectance using a calibrated Spectralon panel. Simultaneously, collimated red-NIR images were acquired employing the Dycam ADC camera. Biophysical characteristics of maize plants were collected in correspondence of the remote sensing data acquisitions. These included direct measurement of leaf angles, using an electronic clinometer (in 2007), single leaves lengths, widths and area, and LAI. In 2008 leaf angles were measured by processing digital photographs of plants taken against a white background by using the 'silhouette' method. Leaf disks were collected for N and chlorophyll analysis. Separately soil spectral reflectance data were also acquired. Images were classified using a supervised classification procedure in order to obtain the shaded and sunlit fractions of the image components (leaves and soil), necessary for the inversion of 3-D models of plant canopies. A classical 1-D model (PROSAIL) was inverted using the multiangular spectral reflectance data. Classified images fraction data were employed in the inversion of a 3-D ray-tracing model of a plant canopy, by using a look-up-table procedure. The results show that the accuracy of the estimates depends greatly on an adequate sampling of the angular variability of spectral reflectance or image fraction components. However the greatest uncertainty in the model inversion procedures are introduced by the assumptions concerning the values of ancillary model parameters requiring previous knowledge on crop canopy properties.

Plant leaf roughness analysis by texture classification with generalized fourier descriptors in different dimensionality reduction context

Journaux, L.[1], Destain, M.F.[2], Cointault, F.[1], Miteran, J.[3] and Piron, A.[2], [1]Enesad, Engineering Sciences, 26 Bd Dr Petitjean BP 87999, 21079 Dijon Cedex, France, [2]Faculté Universitaire des Sciences Agronomiques de Gembloux, Unité de Mécanique et Construction, 2 passage des déportés, 5030 Gembloux, Belgium, [3]Université de Bourgogne, Le2i, Avenue Alain Savary BP 47870, 21078 Dijon Cedex, France

In precision spraying research, one objective is to minimize the volume of phytosanitary product in order to be more environmentally respectful with more effective plant treatments. Thus the main goal is to be sure the spray reaches the target, to reduce losses that occur at the application. If mechanisms of losses by drift are now well known, losses due to runoffs on leaf are still poorly understood. These last one are related to adhesion mechanisms of liquid on a surface for which specific models have been developed and showed that the leaf roughness is the predominant factor. However, in this context, little robust works have been carried out on the characterization of the plant leaf roughness. Taking the computer vision viewpoint, we propose to characterize the leaf roughness of different plant leaf images (1242 texture images) through the performance of a combination of spatio-frequential texture feature extraction and 6 classification methods like Support Vector Machine. For feature extraction we consider the invariants called Generalized Fourier Descriptors (GFD), which provide robust but high dimensional texture features. Unfortunately, in classification context, high-dimensional data are often redundant, highly correlated and suffer from the problem of the 'Hughes Phenomenon', which causes inaccurate classification. To improve the classification performance of the original features we combine the classification steps with a selection of 13 linear and nonlinear Dimensionality Reduction (DR) techniques (from the classical Principal Component Analysis (PCA) to recent methods as Laplacian Eigenmaps (LE) or Kernel Discriminant Analysis (KDA)) which transform high-dimensional data into a meaningful representation of reduced dimensionality. The results show that in any case, the SVM classifier outperforms all other classification methods using the original feature space. However, we experimentally demonstrated that some DR methods improve final classification performances for natural texture features, and we proposed a rank classification of these methods. The best DR methods are the KDA (0.4% error rate) and LE (1.25% error rate) even if the standard PCA (2.35% error rate) still offers a good tradeoff between computation time and performances. We conclude on the well performance of leaf roughness characterization through the combination of GFD for feature extraction, SVM for classification method and KDA for dimensionality reduction.

Evaluation of cost-effective real-time slope sensing system for wild blueberry

Zaman, Q.[1], Schumann, A.[2], Swain, K.[1], Percival, D.[1], Arshad, M.[1] and Esau, T.[1], [1]Nova Scotia Agricultural College, P.O. Box 550, B2N 5E3, Truro,Nova Scotia, Canada, [2]CREC, University of Florida, 700 Experiment Station Road, FL 33850, Lake Alfred, Florida, USA

Wild blueberry fields are developed from native stands on deforested farmland by removing competing vegetation. The crop is unique as it is native to northeastern North America and has never been cultivated. The majority of fields are situated in acidic soils that are low in mineral nutrients, and have significant bare spots, weed patches and gentle to severe topography. Soil surface properties are a critical component used by producers in developing wild blueberry management plans. Topographic properties are not yet routinely used to guide within wild blueberry field management and preventing erosion. Spatial mapping of soil variables, topographic features and fruit yield will help to manage the field inputs in a site-specific fashion. An automated cost-effective slope sensing system comprising of tilt sensor, Trimble AgGPS332 and laptop was developed and mounted on an ATV for real-time slope measurement and mapping in wild blueberry fields. A tilt sensor was developed from 3-axis accelerometers (Parallax, Inc., Rocklin, Calf. USA) and would be capable of sensing the tilt of a vehicle in any orientation on a slope. The configuration used two accelerometers, one vertically and one horizontally, placed in a custom made thin metal box. The custom software for reading DGPS and tilt sensor output, calculating slope, and storing the data in a central database was written in Delphi 5.0 for a 32-bit Windows operating system. The database table could be easily viewed in the 'browse/edit' page of the program. The tilt sensor could be calibrated using the option 'calibrate' on the main screen of the software before taking the measurements in the field. Slope data together with matching coordinates could then be viewed and easily exported by copying to the Windows clipboard, and pasting to a spreadsheet or GIS program for further processing or mapping. Four commercial wild blueberry fields were surveyed in Central Nova Scotia to evaluate the performance of the system. The automatically-sensed slopes (SS) were compared with manually measured slopes (MS) at 20 randomly selected points in each field to examine the accuracy of the system. Selected soil properties, leaf nutrients and fruit samples were also taken at 60 control points in the zones of selected fields where slopes were different. The system measured slope reliably in selected wild blueberry fields. The relationship between MS and SS was highly significant (R^2 ranged from 0.95 to 0.99) in all selected fields. The SS map showed substantial variation in slope in the fields. MS slopes were overlaid on SS maps and matched with the SS throughout the fields. Therefore, the use of low-cost and reliable accelerometers with a Trimble AgGPS is a reasonable alternative for a cost-effective slope sensing system compared to the expensive RTK-DGPS. The accelerometers allow real-time slope measurement which could be used to regulate agrochemical applications and safety features on farm vehicles. The SS real-time maps can be used to adjust vehicle speed at particular slopes to avoid accidents in wild blueberry fields having highly variable slopes. The soil properties, leaf nutrients, and fruit yield data will be correlated to the slopes in different zones of the selected fields. Slope data could help to develop prescription maps for site-specific application of agrochemicals. These results will be presented in paper.

Intelligent autonomous system for the detection and treatment of fungal diseases in arable crops
Moshou, D.[1], Bravo, C.[2], Vougioukas, S.[1] and Ramon, H.[2], [1]Aristotle University of Thessaloniki, Agricultural Engineering Lab, Faculty of Agriculture, P.O. 275, 54124, Thessaloniki, Greece, [2]Katholieke Universiteit Leuven, Division MeBioS, Kasteelpark Arenberg 30, 3001 Heverlee, Belgium

Crop protection represents one of the highest costs for farm budgets. Farmers are under increasing social and economic pressure to dramatically reduce pesticide use. Although most weed and disease infestations occur in patches, the most widely used practice is to spray pesticides uniformly. In this paper, the development phases of a ground-based real-time remote sensing system are described. The proposed system can be carried by tractors or robotic platforms. This prototype system makes possible the detection of plant diseases automatically in arable crops at an early stage of disease development, even before the diseases are visibly detectable, during field operations. The methodology uses differences in reflectance and fluorescence properties between healthy and diseased plants. The diseases under investigation were Yellow Rust (Puccinia Striiformis) and Septoria Graminicola. Hyperspectral reflectance, fluorescence imaging, and multispectral reflectance imaging techniques were developed for simultaneous acquisition in the same canopy. New fluorescence acquisition techniques were developed, experimental platforms were constructed, and the advantage of using sensor fusion was proven. An intelligent multisensor fusion decision system based on neural networks was developed aiming at predicting the presence of diseases or plant stresses, in order to treat the diseases in a spatially variable way. A robust multi-sensor platform integrating optical sensing, GPS and a data processing unit was constructed and calibrated. The functionallity of automatic disease sensing and detection devices is crucial in order to conceive a site-specific spraying strategy against fungal foliar diseases. Furthermore, field tests were carried out to optimise the functioning of the multi-sensor disease-detection device. An overview is provided on how disease presence data are processed in order to enable an automatic site-specific spraying strategy in winter wheat. Furthermore, mapping of diseases based on automated optical sensing and intelligent prediction gave a spatially variable recommendation for spraying.

On-the-go yield and sugar sensing in grape harvester

Baguena, E.M., Barreiro, P. and Valero, C., Laboratorio de Propiedades Físicas y Tecnologías Avanzadas en Agroalimentación, Rural Engineering, Avda. Complutense, s/n, 28040 Madrid, Spain

This paper summarised the results of a joint R+D project between university and industry. The study was developed at the Alt Penèdes region, in Barcelona, during the 2005, 2006 and 2007 (on 3, 22, 69 fields restively). The quality sensors set-up in year 2007, mounted on a New Holland SB55 grape harvester, were: two load cells, one refractometer, an ambient temperature prove and a GPS antenna, while in 2006 only the load cells and the GPS performed properly. The method used for this study is as follows: 1. Data recording from GPS and Logger (the latter is use for according and digitalising the sensor signal); 2. Wireless download of data to a PC; 3. Automatic data integration in a simple file; 4. Lane automatic identification due trajectory angles, machine forward speed determination, worked time calculation, adjustment lane masic flow, kg/m, and total amount harvested, kg/unload grapes, characteristic determination about quantity of soluble solids and temperature during harvest; 5. Data broadcasting through GPRS to the winery; 6. Comparison of transmitted data with the invoice of the winery containers. After the season was finished, a data post processing was performed in order to a assess the causes of isolated incidences that were registered in 10 fields. Also a recalibration of the sensors for future seasons was performed. At current stage R^2 of 0.988 is found between winery and in field yield data. Beside georeference data were gathered and compare to the remote photos in 'Instituto Cartográfico de Cataluña'. Site-specific yield maps and speed maps have being computer while broad soluble solid information is available due to slight dysfunctions of the grape juice pumping system towards to the refractometer.

Using GreenSeeker to drive variable rate application of plant growth regulators and defoliants on cotton

Vellidis, G.[1], Ortiz, B.[2], Ritchie, G.[3] and Perry, C.[1], [1]University of Georgia, Biological & Agricultural Engineering, 2329 Rainwater Rd., Tifton GA 31793-0748, USA, [2]Auburn University, Agronomy and Soils, 202 Funchess Hall, Auburn AL 36849-5412, USA, [3]University of Georgia, Crop & Soil Sciences, NESPAL Building, Tifton GA 31793-0748, USA

Plant growth regulators (PGRs) are applied to cotton to maintain a balance between vegetative and reproductive growth. There are many techniques used to determine the timing and application rate of PGRs but they are traditionally applied at a constant rate. Chemical defoliants are applied prior to harvest to promote leaf drop and boll opening. Like PGRs, defoliants are traditionally applied at a constant rate. Our earlier work has documented the uneven distribution of plant biomass in cotton fields. This uneven distribution is a result of variability in soil parameters such as nutrients, moisture, pH, texture and variability in microclimate and disease and pest pressures. Common sense as well as recent research suggests that variable rate application (VRA) of PGRs and defoliants compensates for the uneven distribution of plant biomass and is a good management practice. This is true at several levels. For example, applying more PGR or defoliant to a section of the field with high biomass and less to a section with low biomass will result in more uniform plant growth or defoliation. In contrast, constant rate applications frequently result in over-application or under-application and subsequently uneven growth or defoliation. This in turn sometimes requires further applications. PGRs and defoliants are a major expense for cotton growers and inefficient use can drive up production costs. Uniform growth and defoliation also results in higher picking efficiency, higher fiber quality, and ultimately in higher yields. VRA also has environmental benefits as chemicals are applied where needed at the rates needed. In this study we evaluate the use of variable rate application of PGRs and defoliant on cotton grown in Georgia, USA. We use the GreenSeeker sensing system to measure NDVI in three producer cotton fields weekly during the growing season. A GreenSeeker RT200 on-the-go variable rate application and mapping system consisting of six GreenSeeker sensors was installed on a JD 6700 high-clearance sprayer. The NDVI data are used to create NDVI maps of the fields. The Management Zone Analyst software is used to group the data into management zones. The producer then ground-truths the maps and recommends application rates for PGR. At the end of the growing season, the same will be done with defoliant. To quantify the impact of VRA, we established an experimental block within each of the three fields. Each block contains 3 replicates of 3 treatments. Each replicate consists of an 18-row strip running the length of the field. The three treatments are: control, variable rate application (VRA) of PGRs only, and VRA of defoliants only. Under the control treatment, both PGR and defoliant is applied at a constant rate. In the other two treatments, either the PGR or defoliant is applied variably while the other input is applied at a constant rate. Within the strips, cotton growth parameters such as plant height, leaf area index, and nodes above last flower are quantified biweekly. These data, along with yield data which will be collected at the end of the growing season, will be used to assess the effectiveness of VRA in cotton. This paper will describe the experiment and present the results.

Canopy temperature interpretation of thermal imagery for crop water stress determination

Meron, M.[1], Alchanatis, V.[2], Cohen, Y.[2] and Tsipris, J.[1], [1]MIGAl Galilee Technology Center, Crop Ecology Laboratory, P.O. Box 831, Kiryat Shmona 11016, Israel, [2]Institute of Agricultural Engineering, Sensing, Information and Mechanization Engineering, ARO - The Volcani Center P.O. Box 6, Bet Dagan 50250, Israel

Crop water stress determination methods from canopy temperatures, derived from the surface energy balance equations, treat the canopy temperature as a single value. simplifying the real life canopy temperature distribution into the 'big-leaf' paradigm, under assumption that the canopy behaves as a single homogeneous virtual leaf, covering the surface. In experimental situations involving stress evaluation based on energy balance models, leaves were usually selected for measurements as most resembling the 'big leaf' such as healthy, sunlit, fully expanded, mature but nor senescent. Introduction of very high resolution thermal imagery, 0.01 to 0.3 m pixel size, acquired from low altitude platforms, enables finely detailed measurement of the whole canopy, raising the question of how to interpret the data, and how to select the relevant temperatures. One approach is to select the sunlit leaves (SL methods), confirming to the conventional methods. However thermal imagery alone lacks part of the information, and needs additional canopy marking or synchronized visible imagery, making the process complicated and expensive. The other approach is to use full frame pixel statistics (STAT method), without pattern recognition, by selecting the mean temperature of the cold fraction from the pixel histogram, assuming that the 'big leaf' may consist of a mixture of shaded and sunlit leaves. That greatly simplifies processing, important consideration in field application. In a series of experiments conducted in cotton and vine grapes, both approaches were tested in parallel. Ground referenced thermal and visible images were overlapped and sunlit leaves were selected for crop temperature evaluation. The pixel histograms of the same images were analyzed for the mean temperatures of the lowest 25%, 33% and 50% of the pixels, after discarding soil related, higher than air temperature + 7 °C pixels. Several crop water stress indices (CWSI) were compared to leaf and stem water potentials (LWP) and stomatal conductance (gL). Results were reported separately before. In this work the partial results of the cotton and grapes experiments were compared and analyzed. Good agreement was found between both SL and STAT methods, demonstrating the suitability of both methods in canopy temperature evaluation for crop water stress determination.

Using an active sensor to make in-season nitrogen recommendations for corn

Schmidt, J.[1], Sripada, R.[2], Beegle, D.[3] and Dellinger, A.[4], [1]USDA - Agricultural Research Service, Pasture Systems & Watershed Management Research Unit, Building 3702 Curtin Road, University Park, Pennsylvania 16802, USA, [2]Canaan Valley Institute, Pasture Systems & Watershed Management Research Unit, Building 3702 Curtin Road, University Park, Pennsylvania 16802, USA, [3]The Pennsylvania State University, Department of Crop and Soil Sciences, 116 ASI Building, University Park, Pennsylvania 16802, USA, [4]USDA - Natural Resources Conservation Service, 1383 Arcadia Road, Room 200, Lancaster, PA 17601, USA

An active crop canopy reflectance sensor could increase N-use efficiency in corn (*Zea mays* L.), if temporal and spatial variability in soil N availability and plant demand are adequately accounted for with an in-season application. Our objective was to evaluate the success of using an active sensor for making N recommendations to corn. Seven increments of in-season N fertilizer (0 to 280 kg/ha) were applied to corn at each of 23 sites during four years. These sites were selected to represent a wide range of previous crop, preplant manure or fertilizer applications, or soil water characteristics (landscape position) conditions typical in the mid-Atlantic region of the US. Canopy reflectance in the 590 nm and 880 nm wavelengths, soil samples, and chlorophyll meter measurements were collected at the 7-leaf growth stage. Relative Green Normalized Difference Vegetation Index (RGNDVI) was also determined, as GNDVI(0N) / GNDVI (280 kg N applied at planting). Grain yield was determined at harvest. Economic Optimum N Rate (EONR) was determined using a quadratic-plateau yield response function. Results from the first two years indicated that EONR was strongly related (r^2=0.76) to RGNDVI in a linear-floor type relationship. For these two years of data, EONR was as strongly related (or better) to RGNDVI than results from a late season stalk nitrate test (r^2=0.42), an analytical laboratory's preseason N recommendation (r^2=0.39), and results from an in-season chlorophyll meter reading (r^2=0.73). An in-season soil nitrate test (PSNT) was the only 'traditional' approach to making N recommendations that performed (r^2=0.78) as well as the active sensor; however, this latter approach is not accommodating to within-field spatial variability. Results from the 15 additional sites during the next two years were evaluated in the context of the initial relationship developed using the active sensor. The active sensor was as good as or better than 'traditional' approaches to making N recommendations for corn, and accommodates temporal and spatial variability in an easy-to-use method.

Optical signals of oxidative stress in crops physiological state diagnostics
Kanash, E.V. and Osipov, Y.A., Agrophysical Research Institute, 14 Grazhdansky prosp., 195220 Saint-Petersburg, Russian Federation

Accumulation of oxygen active forms and resulting oxidative stress are nonspecific response of plants to deterioration of an environmental conditions (water or nutrition deficiency, high UV-B irradiance, increased temperatures of vegetation, etc.). The aim of this study is selection of criteria for oxidative stress diagnostics, estimation of the physiological state of plants and detection of mineral nutrition deficiency at early stages of its occurrence. Spectra of reflection and colour characteristic of leaves were registered with HR2000 fiber-optical spectrometer (Ocean Optics, USA). Canopy colour was determined using radio-controlled aerial camera. To investigate relationship between net productivity of plants and various reflection indices, linear and multiple regression was performed. We found that a highly reliable indicator of the appearance of oxidative stress and, hence, of mineral nutrition deficiency is change of parameters describing the activity of 'down regulation' photosystem II processes. Optical criteria (indices) applied in mineral nutrition deficiency diagnostics allow estimating a potential capacity of plants to assimilate solar energy, namely: photosynthetic system capacity (ChlRI- chlorophyll reflection index); and, efficiency of light energy transformation in photochemical processes of photosynthesis (SIPI – the sum of carotenoids to the sum of chlorophyll ratio, PRI – photosynthetic radiation-use-efficiency, ARI and FRI – anthocyanins and flavonols contents, R_{800} – light scattering criterion dependent upon surface characteristics and structure of the leaf). The inference about the depression of plants and deterioration of their physiological state under the impact of stressor strongly limiting their growth can be made by registering reduction of chlorophyll reflection index measured for each crop and variety in optimal conditions. When the impact of stressor is poorly expressed and at early stages of stress development when chlorophyll concentration does not vary or varies slightly, the plant depression is detected by the increase in SIPI,PRI,ARI, and R_{800} indicating reduction in photosynthetic radiation-use-efficiency and growth inhibition.

Spatial patterns of wilting in sugar beet as an indicator for precision irrigation

Zhang, L.[1], Steven, M.D.[1], Clarke, M.L.[1] and Jaggard, K.W.[2], [1]University of Nottingham, School of Geography, University Park, Nottingham NG7 2RD, United Kingdom, [2]Brooms Barn Research Station, Higham, Bury St Edmunds, Suffolk IP28 6NP, United Kingdom

The application of precision irrigation requires sensors to map within-field variations of water requirement. Conventional remote sensing techniques provide only a shallow estimate of water status, which is of little value. We therefore tested the ability of a sensitive crop – sugar beet – to act as an intermediate sensor providing an integrated measure of water status throughout its rooting depth. Archive aerial photographs and satellite imagery of Eastern England show crop patterns resulting from ice age soil movements associated with frost and thaw processes. We found that the patterns are spatially consistent, reappearing in the same location in successive years and that, when sugar beet is present, are strongly associated with spatial variations in late-summer wilting. Field sampling of soil cores to 1m depth established that within-field wilting zones are significantly associated with coarser, less water-retentive soils and that stress class, determined by classification of the digitised images, is weakly correlated with total available water ($r^2 \approx 0.4$). These results suggest that wilting in sugar beet can be used as an intermediate sensor for quantifying potential soil water availability within the root zone. Within-field stress maps generated in one year could be applied as a strategic tool allowing precision irrigation to be applied to high-value crops in following years, helping to make more sustainable use of water resources.

Development of an instrument to monitor crop growth status

Li, M., Cui, D., Li, X. and Yang, W., China Agricultural University, Key Laboratory of Modern Precision Agriculture System Integration Research, CAU 125, Qinghua Donglu 17, Haidian District, Beijing 100083, China

Nitrogen content of the crop is an important index used to evaluate the growth status and predict the yield and the quality of the crop. So it is necessary to real time detect and diagnose the nitrogen status of the crop, and to develop a portable crop growth monitor. Our object is to design an optical sensor to measure spectral reflectance of the plant canopy for determining N status in the plants. The developed instrument was composed of the sensor and the controller. The sensor was designed with four channels representing different spectral bands. The channel 3 and channel 4 measured the sunlight, while the channel 1 and channel 2 measured the reflected light from the plant canopy. The angle of solar incidence can make the output of photodiode changing. In order to avoid this influence, the milky diffuse glass was used as the optical window of the channel 3 and channel 4. The optical mirror was chose as the window of the channel 1 and channel 2. When using this detector to measure reflectance ratios, the sensor was lift on the top of the canopy and the output of the sensor was transmitted to the controller. Then the controller dealt with the output and saved the data. Each channel had the same structure. It was made up of two parts. Each channel consisted of an optical window, a washer, a filter, and a sensor. The channel 1 and the channel 3 had the same filters and sensors. The wavelength of the filter was 610 nm and the electric eye was made up of Si. And the wavelength of the filter in the channel 2 and channel 4 was 1,220 nm. The photodiode in those was made up of InGaAs. The wavebands were determined based on the research report. The sunlight went though the filter and reached the photodiode. Firstly, the calibration tests of the device were carried out. The weather was fine. The standard board was put on the ground out of the shadow, which the reflected light form was measured by the monitor Each measurement was replicated five times and the average was taken as the final result. From the test results we found that there was a good correlation between the solar illumination and the output voltages of four channels measured by the monitor. To evaluate performance of the monitor, we did field text in a wheat field. The results measured by portable monitor show well. It was obvious that the reflected light from the plant wais certainly measured by the monitor developed. The portable crop growth monitor based on optical principle was developed and tested in the field. The Spectral reflectance was calculated using the monitor's outputs and showed a close correlation with the result measured by Crop Scan. This study was supported by 863 program (2006AA10Z202).

Field hyperspectral imagery: a new tool for crop monitoring at plant scale

Vigneau, N.[1], Rabatel, G.[1] and Roumet, P.[2], [1]Cemagref, UMR ITAP, 361 rue Jean François Breton, F-34196 Montpellier, France, [2]INRA, UMR Diversité & Adaptation des Plantes Cultivées (DIA-PC), Domaine de Melgueil, 34130 Mauguio, France

Spectral reflectance in NIR and visible domain is often used to obtain information about crops in precision agriculture, because high spectral resolution in this domain allows to access detailed physical and physiological plant properties. Spectral reflectance acquisition in agronomy is mainly based on two different approaches: airborne or satellite imagery and spectroscopy. Both have advantages and disadvantages. Airborne or satellite imagery allows to obtain data about a large area but the spatial resolution is not high enough to study phenomena at plant scale. On the contrary, spectroscopy allows to work at plant scale but collecting spectra is a tedious and time-consuming work. Some approaches in precision agriculture would need the advantages of both. It is the case for phenotype characterisation which requires information about physiological status for a great number of plots on a large area and on the whole growth cycle. Field hyperspectral imagery can answer to this need. However, to keep benefits of high spatial resolution, local lighting perturbations and consequences of outdoor conditions must be taken into account. In the study presented here, we used a pushbroom CCD camera (HySpex VNIR 1,600 - 160, Norsk Elektro Optikk) fitted on a motorised rail fixed on a tractor to take images of wheat plots. The camera have a spectral range of 600 nm (from 0.4 µm to 1 µm) with a spectral resolution of 3.7 nm. Data are digitalised on 12 bits. Each image represents about 0.30 m across track by 1.50 m along track. The spatial resolution across track is 0.2 mm at 1 m above the canopy. The spatial resolution along track, depending on the motion speed, can be adjusted between 0.2 and 5 mm. Images have been radiometrically corrected with data given by camera constructor to obtain signal in radiance. Correction in reflectance was also necessary to compare images taken at different dates and with different illumination conditions. For that purpose a ceramic plate (less expensive and less fragile than spectralon) was settled in the camera field of view to be included in the wheat plot image. This ceramic has been calibrated with spectralon in laboratory to obtain its absolute reflectance, so that it allows afterwards to recover the level and spectral composition of the outdoor lighting during the image shooting.Once the images have been corrected with respect to illumination conditions, they provide us with spectral data that can be assimilated to leaf reflectance spectra, (through a scalar factor depending on the leaf orientation). Moreover, the high spatial resolution guarantees most of the pixels to be pure pixels. Thus, these data can be directly compared with spectrometric measurements. As a first application, the hyperspectral images have been exploited to build leaf nitrogen content cartography using a chemometrical model resulting from spectrometric data. These results will be presented and discussed. Further developments will concern the refinement of illumination correction, taking into account the effect of the ceramic BRDF, as well as the effect of local secondary reflections through an appropriate model.

Determination of canopy properties of winter oilseed rape using remote sensing: techniques in field experiments

Engström, L.[1], Söderström, M.[1], Lindén, B.[1], Börjesson, T.[2] and Lorén, N.[3], [1]Swedish University of Agricultural Sciences, Dep. of soil and environment, Div. of precision agriculture, P.O. Box 234, SE-532 23 Skara, Sweden, [2]Lantmännen, Business development, Grain, SE-531 87 Lidköping, Sweden, [3]SIK – The Swedish Insitute for Food and Biotechnology, Structure and Material DesignBox, Box 5401, SE-402 29 Göteborg, Sweden

Measuring canopy reflectance in winter oilseed rape has proved to be a fast and non-destructive tool for monitoring canopy properties in precision agriculture. The use of vegetation indices of visible and near-infrared (NIR) light allow for prediction of quantative growth properties directly at the field level. In field experiments determining plant density, crop biomass and N content are important for understanding crop development and crop management but also work-intensive and often erroneous. In two-year field experiments of oilseed rape from 2007 to 2009, in southwestern Sweden, investigations were carried out in order to evaluate the use of the handheld Yara N-Sensor for determination of biomass and N-content instead of N-analysis of crop samples in the laboratory. The use of digital photography to determine plant density was also investigated with the purpose to replace time consuming counts of plants by hand in these field experiments. The experiment was performed in 40 sampling areas of 12x15 m distributed on four different fields with varying plant density. N-sensor measurements and digital photos were obtained in late autumn, early spring simultaneously with counts of plants and measurements of DM and N-content in crop samples from 2 m^2 within each sampling area. The N-content of the crop in late autumn in 2007 (17/10) was 18 kg N ha^{-1}, on average, and ranged from 3 to 81 kg N ha^{-1} (std. dev.=15). The r^2 values for the relationship (fitted power function) between sensor values and biomass in kg DM ha^{-1} and N-content (kg N ha^{-1}) were 0.93 and 0.92, respectively. The area fraction of ground photographs covered by crop estimated by the image analysis algorithm was well correlated (r^2=0.95) with the N content of the manually cut crop samples. Using image analysis, fewer plants were detected than with manual counting. This was accentuated when the plant density increased. Therefore a logarithmic curve best described the relationship between manual counting and image analysis. In these trials the cost for N-sensor measurements was 38% of the cost for cutting the crop and analysing the N-content in a laboratory. The results indicate that the method could be a faster and less costly way to determine the DM- and N-content of winter oilseed rape in field trials.

Within-field and regional prediction of malting barley protein content using canopy reflectance

Söderström, M.[1], Pettersson, C.G.[2], Börjesson, T.[2] and Hagner, O.[3], [1]Swedish University of Agricultural Sciences, Dept. of Soil and Environment, P.O. Box 234, SE-532 23 Skara, Sweden, [2]Lantmännen, Business development, Grain, Östra hamnen, SE-531 87 Lidköping, Sweden, [3]Swedish University of Agricultural Sciences, Dept. of Forest Resource Management, Remote Sensing Laboratory, SE-901 83 Umeå, Sweden

The production of high quality malting barley for the brewery industry is important for farmers. It is also a challenging crop for the producer since quality criteria is rather strict. The protein content should range between 9.5 and 11.5%. This criterion is not always reached in Sweden mainly due to varying weather conditions. Also the within-field variability of protein content may be considerable. If the pattern of variation would be known before the harvest, it would make it easier for the farmer to deliver grain with the right quality. A combination of remote sensing data from handheld or tractor-borne equipment, meteorological data and soil data has shown to be useful for explaining both quantity and quality of malting barley. This has resulted in different prediction models, one based on the TCARI vegetation index on data at crop development stage DC 37 and another estimation model at DC 69. The aim of this project is twofold: 1) to test and further develop these field models for prognosis of malting barley protein content based on measurements of the Yara N-Sensor system; 2) to investigate the possibility to use satellite data to upscale the local predictions to cover malting barley fields in a larger region. The project was implemented in two test areas: one in central Sweden and the other in the southwestern part. In the former area, frequently too high protein levels have been recorded, whereas the opposite is common in the latter district. Manually collected control and calibration data (just before harvest) as well as reflectance data from the Yara N-Sensor (at DC 37 and 69) were obtained from a number of malting barley fields in 2007 and 2008. In addition, 10- and 20-metre resolution SPOT satellite scenes registered as close in time as possible were acquired. Model predictions were made for the calibration fields by both Yara N-Sensor data and satellite imagery. Predictions were also made for validation fields. Initial results indicate that the available prediction from the ground sensor is useful for locating areas within fields with low or high protein content, but calibration samples were needed in order to correctly adjust the predicted concentrations. Satellite data was also useful for protein predictions in malting barley even if the wavelengths shown to be the best on the ground measurements are not available. In an operational test during the second year, the protein model assisted a farmer with 75 ha of malting barley distributed on ten fields with large variation in grain quality with planning and management of the harvest.

Development and improvement of the air assist seeding following the map information
Chosa, T.[1], Furuhata, M.[1], Omine, M.[2] and Sugiura, R.[2], [1]National Agricultural Reserch Center, Advanced Loaland Rice Farming Reserch Team, Inada 1-2-1, Joetsu, Niigata, 943-0193, Japan, [2]National Agricultural Research Center for Kyushu Okinawa Region, 6651-2 Yokoichi-machi, Miyakonojo, Miyazaki, 885-0091, Japan

Precision farming involves site-specific crop management based on information technology to manage temporal and spatial variability, given that agricultural products are affected by variability. Studies have also monitored field and growth variability. Variable rate application and decision-making systems have also been discussed. These cropping systems are evaluated by monitoring the yield variability, cost improvement, or value-added quality. The cropping cycle of monitoring, analyzing, taking action, and evaluating is useful for introducing new cropping systems. Any new system has to be evaluated both statistically and in terms of the variability. We are studying a new rice seeding technology called air-assist seeding, designed to reduce production costs and to save labor. The new machine has been developed and initial agronomic research has been conducted. Field trials are now being conducted. The new cropping system needs to be evaluated in terms of both its potential and spatial variability. This study further developed air-assist seeding for precision farming. This study examined the spatial variability and site-specific management of the field trials. The growth variability of a rice paddy was monitored using an unmanned helicopter equipped with a digital camera over two seasons. The first year, a standard commercial camera was used. In the second year, a plant cover ratio camera with an attached near infrared filter was also used in order to discuss details of the growth variability. In the first year, some areas with poor growth were observed. Three reasons for this were formulated: 1) unexpected movement of the seeding machine; 2) the effects of field inclination; and 3) differences in the physical characteristics of the soil, i.e., the soil surface before seeding was too soft in the areas of poor growth. Before the second season, the seeding machine was improved, the effects of field inclination were considered, and the seeding rate was changed in areas of poor growth in order to stabilize the initial growth. The growth variability was reduced in the second year, although there were still some areas of poor growth. Further improvement is required and data on yield variability must be considered

Use of ultrasonic transducers for on-line biomass estimation in winter wheat
Reusch, S., Yara International ASA, Research Centre Hanninghof, Hanninghof 35, 48249 Dülmen, Germany

The in-season aboveground biomass is a key parameter to describe the crop status and the spatial variability of a field. Variable-rate applications of mineral fertilizers, growth regulators or fungicides can be based on this parameter. In the recent years various mainly optical sensors have been successfully developed and calibrated for crop biomass, but these devices are often costly, typically read a composite signal of biomass and chlorophyll and sometimes only allow measurements with a small footprint. Standard industrial ultrasonic transducers are comparably cheap, can measure a reasonable footprint and the signal solely depends on the canopy structure. Because of this, a study has been carried out to explore the capabilities and limitations of ultrasonic transducers for biomass estimation in winter wheat. An ultrasonic transducer was mounted on a vehicle, pointing straight downward to the crop. While the vehicle was moving, the device continuously sent short ultrasonic pulses. After sending each pulse, the sensor sequentially registered the echoes reflected by the individual leaf layers and the ground. Resulting echograms were averaged and stored for later analysis. Measurements were taken in winter wheat trials at multiple times during the 2007 and 2008 growing season. The trial plots comprised different nitrogen levels, cultivars and seed densities. Immediately after the measurements, a sample area was harvested from each plot and the dry matter was determined. From the echograms characteristic parameters were extracted and related to the crop biomass. Results show a very good relationship between sensor readings and dry matter (R^2 between 0.85 and 0.93, depending on cultivar). The relationship was not affected by crop density, but by cultivar. Cultivar effects correlated with the cultivar's typical crop height. Repeated measurements within the growing season showed that the relationship slightly shifts with growth stage. From the results it is suggested that both the growth stage and the cultivar should be taken into account if the dry matter has to be predicted from the sensor readings. The above findings suggest that with standard industrial ultrasonic transducers it is possible to remotely determine the above-ground biomass quickly, easily and at low cost.

Use of a ground lidar scanner to measure vineyard leaf area and structural variability of vines
Arnó, J., Vallès, J.M., Escolà, A., Sanz, R., Palacín, J. and Rosell, J.R., Universitat de Lleida, Agro-forestry Engineering, Rovira Roure, 191, 25198 Lleida, Spain

The measure of vine vigour by on-the-go proximal sensors can be an indirect method to determine grape yield and grape quality. There are several indexes related with foliage characteristics of vines. The Leaf Area Index is probably the most used in viticulture. Nowadays, the Leaf Area Index (LAI) is estimated in a satisfactory way in orchards by means of ground based canopy volume measurements using LIDAR scanners, and assuming that there is a significant correlation between the volume and the foliage surface of trees. In this work, geometric and structural parameters of vines have been computed from scanned data. The most important are the height of vines (H), the cross-sectional area (A), the canopy volume (V) and, as an index related with foliage density, the Tree Area Index (TAI). Light Detection and Ranging (LIDAR) sensors operate based on the measurement of time a laser pulse takes to travel from the sensor to the target, in this case, the vegetation surface of vines and back to the sensor. For each interception, the sensor determines the radial distance (r) between the intercepted point and LIDAR position and the angular coordinate (θ) of this intercepted point according to an adequate reference system. Because the laser beam is pulsed by the LIDAR with different angular values in each vertical scan, field data are organised as a matrix of polar coordinates (r, θ) of intercepted points. Therefore, each scan defines a matrix of data and the displacement of LIDAR along the row defines several scans with their corresponding matrixes of polar coordinates. To be correct, the LIDAR used is a low-cost general-purpose LMS-200 model (Sick, Düsseldorf, Germany), with an accuracy of ±15 mm in a range up to 8 m, scanning angle of 180° and angular resolution selectable at 1° (used in this work), 0.5° or 0.25°. To assess the feasibility of LIDAR in precision viticulture, several field tests were made in a block (Merlot) using a tractor-mounted LIDAR system that traversed the crop in the parallel direction to the row of vines. Delimited a within-row of 4 m length, two repeated measurements in both sides were made and this methodology was repeated four times coinciding with different stages of vines. The sensor had a standard RS232 serial port for data transfer and the MATLAB software was used for data acquisition and process support. In order to contrast the relationships between the indexes built from sensor data and real values of Leaf Area Index (LAI), the scanned vines were defoliated after LIDAR measurements to obtain LAI values corresponding to sections of 4 m, 2 m and 1 m length, respectively. Linear regression analysis showed good correlation (R^2=0.7282) between canopy volume (V) and experimental values of LAI for 1 m long sections. Nevertheless, the best predictor of LAI was the Tree Area Index (R^2=0.9188) formulated as the ratio between crop detected area and ground area and assuming, when the parameter is calculated, that laser beam transmission probability within vines could be approximated by means of the Poisson probability model or Beer's law. Since there is a significant variability of foliage distribution along vine rows, LIDAR sensor detects canopy structure differences and can predict foliage density if adequate crop parameters are built from sensor data. Therefore, LIDAR sensor is revealed as a feasible technology for monitoring within-field leaf area variability.

Sensing tree canopy parameters in real time for precision fructiculture/horticulture applications: first results

Escolà, A.[1], Arnó, J.[1], Sanz, R.[1], Camp, F.[2], Masip, J.[1], Solanelles, F.[2], Rosell, J.R.[1] and Planas, S.[1],
[1]University of Lleida, Agroforestry engineering, Av. Rovira Roure 191, 25198 Lleida, Spain, [2]Generalitat de Catalunya, DAR - Centre de Mecanització Agrària, Av. Rovira Roure 191, 25198 Lleida, Spain

Since ground based lidar sensors started to be used in canopy characterization of fruit orchards, many studies have been done to correlate its measurements with trusted methodologies. Tree height, width, volume, foliar area, foliage density and other parameters have been estimated from lidar data. Most of these techniques need multiple vertical scans to obtain information about individual trees or the whole orchard with statistical significance. However, this information is neither obtained nor used in real time (RT). Some work has recently been done on processing lidar data in RT to implement variable rate technologies (VRT) in fruit growing taking into account intra-row canopy variability. Foliage density, foliar area and canopy volume are estimated from different approaches to be used in dose adjustment for plant protection products (PPP) foliar applications. In this paper, a methodology is described to precisely assess the foliar area of a single vertical lidar scan and a statistical model is developed to estimate it from lidar data. The lidar used is a SICK LMS-200 sensor configured to perform 181 measurements in a vertical line with 1 degree angular resolution. The trial was done in a *Pyrus communis* L. cv. 'Conference' orchard. First step to assess the foliar area of a single scan, is to exactly determine the leaves that intercept the laser beam. Since the wavelength of the laser is 905nm, a nightshot camcorder was used to see the spots on leaves in total dark conditions. Second step is to collect the spotted leaves. This is achieved by using a rotary visible laser pointer used for levelling in the construction industry (LEICA RUGBY 55 construction laser). This device creates a visible red line that has to be placed in a manner to perfectly overlap the vertical laser spot line seen by means of the camcorder. After this procedure, leaves contained in the vertical scan are clearly visible and can be easily collected. The trial started at 10:00 pm and finished at 4:00 am in order to work in the dark. 10 vertical scans were performed and leaves were collected and stored according to 2 different heights. The result of the sampling was 10 lower and 9 upper samples. The statistical analysis consisted in applying a linear multivariate regression model with a stepwise selection method to avoid collinearity problems. Two models were built, one for the total amount of 19 samples (partial scans) and another for the 10 full vertical scans. In the former, foliar area has a statistically significant correlation with canopy height and canopy half-cross-section with an R^2 of 0.775. According to the latter, foliar area has a statistically significant correlation with the same variables and a new one, the average canopy half-width, with an R^2 of 0.969. The results of this work encourage the research and implementation of VRT in precision fructiculture/horticulture, especially in precision spray of PPP. It is possible to estimate foliar area from lidar data in real time with acceptable accuracy and thus to calculate foliage density, understood as the ratio between foliar surface and canopy volume. Both variables are important parameters in dose adjustment of PPP but many other applications have to be considered when canopy characterization is needed.

High-end laser rangefinder scanner in agriculture

Ehlert, D., Heisig, M. and Adamek, R., Leibniz-Institute for Agricultural Engineering Potsdam-Bornim (ATB), Engineering for Crop Production, Max-Eyth-Allee 100, 14469, Germany; dehlert@atb-potsdam.de

Knowledge of site-specific crop parameters such as plant height, coverage, and biomass density is indispensable for optimising crop management and harvesting processes. Vehicle based sensors for measuring crop parameters are essential prerequisite to gather this information. In the last years laser rangefinder sensors have became more and more in the focus of agricultural engineering practice and research. In the sector of market available agricultural machinery laser rangefinder are installed already onto the market available combine harvesters for detection of crop edges to optimise the cutting width and to perform auto guidance. In the field of research, market available laser rangefinders were investigated in horticulture and in agriculture. In agriculture Thösing et al. made a first test to measure crop mass in an oat stand. Kirk et al. estimated in a comparative study the canopy structure from laser range measurements and computer vision. For vehicle based measuring of crop biomass, market available laser rangefinder were analysed and tested in pilot to measure grown crop biomass in oilseed rape, winter rye, winter wheat and grassland. Lenaerts et al. 2008 predicted crop stand density using LIDAR-Sensors. The laser rangefinder models investigated did to meet full the demands resulting from specific agricultural conditions. A new high-end laser rangefinders scanner (ibeo-ALASCA XT) was developed specifically for drivers' assistance and autonomous guiding in road vehicles. This sensor system was tested in 2008 regarding the potential for agricultural purposes. The tests were focussed on the measuring properties under defined specific conditions and in field crop stands. The sensor achieved very good results in surveying crop height and crop biomass for far measuring distances. Furthermore, tram lines, crop stand edges and swaths were detected clearly to support the autonomous guidance of sprayers, spreaders and harvesting machines. Taking into account the expected mass production and from this resulting price reduction a high agricultural application potential (inclusive obstacle detection) can be concluded.

Detection og head blight (*Fusarium* spp.) at ears of winter wheat using hyperspectral and chlorophyll fluorescence imaging

Bauriegel, E.[1], Beuche, H.[1], Dammer, K.H.[1], Giebel, A.[1], Herppich, W.B.[1], Intreß, J.[1] and Rodemann, B.[2],
[1]Leibnitz-Institute for Agricultural Engineering, Max-Eyth-Allee 100, 14469 Potsdam, Germany, [2]Julius Kühn Instiute, Messeweg11/12, 38104 Braunschweig, Germany

New regulations for critical values of Fusarium toxins Deoxynivalenol, Zearalenon and Fumonisine in cereals and cereal products have been applied since June 2005 in the European Union. For selective mycotoxin analyses of the cereal product information about disease infestation within the field of origin could be helpful. Therefore in the future an early detection of disease symptoms at the ears together with the corresponding geographic position becomes more important. In a research project methods of digital image analyses were tested for detection of the head blight disease. For hyperspectral image acquisition a laboratory measuring system was developed. Software components for hardware control, pre-processing and saving of image information as well as for calibration of the hyperspectral image scanner were coded using LabView©. Time series of healthy and Fusarium artificial infected ears within the wavelength range from 400 nm to 1000 nm were recorded. The hyperspectral image analyses software ENVI was applied to define disease specific signatures. Using the classification method Spectral Angle Mapper (SAM) healthy and diseased ear tissue had been distinguished. Simultaneous to hyperspectral images chlorophyll fluorescence image acquisition of healthy and Fusarium infected ears and a visual rating was performed. Using the FluorCam 690 MF device the photosynthetic efficiency of healthy and diseased ear tissue was measured. The influence of the disease infestation on the photosynthetic efficiency was determined. A shifting of photosynthetic efficiency of healthy compared to diseased ears was obtained between sixth and eleventh day after artificial disease inoculation. The severity of disease symptoms were correlated with the photosynthetic efficiency.

Early detection and discrimination of *Puccinia triticina* infestation in susceptible and resistant wheat cultivars by chlorophyll fluorescence imaging technique

Bürling, K., Hunsche, M., Tartachnyk, I. and Noga, G., University of Bonn, Institute of Crop Science and Resource Conservation, Horticultural Science, Auf dem Hügel 6, 53121 Bonn, Germany

In modern agriculture, there is need for a rapid and objective screening method for pathogen resistance of new cultivars. Using the PAM-fluorescence-imaging technique as a non-destructive method for such purposes we hypothesised, that not only detection but also discrimination of wheat (*Triticum aestivum*) cultivars differing in their level of resistance against the economically important leaf rust (*Puccinia triticina*) pathogen can be accomplished. Experiments were done using the cultivars Dekan and Retro as representatives for a susceptible and a resistant genotype, respectively. Plants were grown under controlled conditions at a 14/10 h (day/night) photoperiod, temperature of $20/15 \pm 2$ °C, a relative humidity of 75/80 $\pm$ 10% and a light intensity of 200 μmol m^{-2} s^{-1} in a climate chamber. At the 3-leaf stage, individual leaves were inoculated with Puccinia triticina by applying microdroplets (two 6μl droplets per leaf) of the spore suspension. During the inoculation period of 24h, plants were kept under a plastic cover and saturated relative humidity. After applying different concentrations (800-100,000 spores ml^{-1}; 1:5 dilution), it was revealed that 100,000 spores ml^{-1} was the best concentration for further screening studies. Fluorescence measurements (PAM-Imaging, Heinz-Walz GmbH) were carried out every day at the same time until the first small red-brown pustules appeared in the centre of chlorotic spots. Data processing included choosing the most sensitive parameter of the collected images from fluorescence kinetic measurements as well as the appropriate handling of image information. The focus was primarily on the quantum yield of non-regulated energy dissipation in PSII (Y(NO)) which seemed to be better suited to provide information on the spatial and temporal changes in the physiology of different cultivars earlier than other fluorescence parameters. Two and 3 days after inoculation, the susceptible cultivar showed more pronounced changes of 15.5% and 14.8%, whereas the resistant cultivar showed 4.3% and 4.6%. Six days after inoculation when chlorotic spots became apparent in both cultivars, Dekan showed a stronger response of 14.1% than Retro (7%). The PAM-fluorescence-imaging technique therefore shows potential for early discrimination of *Puccinia triticina* infection between susceptible and resistant wheat cultivars.

Detection of the Tulip Breaking Virus (TBV) in tulip using spectral and vision sensors

Polder, G.[1], Van Der Heijden, G.W.A.M.[1], Van Doorn, J.[2], Van Der Schoor, R.[3] and Baltissen, A.H.M.C.[2], [1]Wageningen UR, Biometris, P.O. Box 100, 6700 AC, Wageningen, Netherlands, [2]Wageningen UR, Applied Plant Research, P.O. Box 85, 2160 AB, Lisse, Netherlands, [3]Wageningen UR, Plant Research International, P.O. Box 16, 6700 AA, Wageningen, Netherlands

Tulip and other bulbous ornamental crops are plagued by viral diseases, causing a reduction of the quantity and especially the quality of the product. Visual selection of suspected plants in the field is used to keep the disease under control. In order to reduce the high labor costs our objective was to develop a robot system which automatically detects and removes diseased plants. As a first step a feasibility study was carried out to test the detection performance of several optical sensors on virus symptoms in tulips under laboratory conditions. Three tulip varieties were cultured in open containers in the field. Early in the growing season each individual plant was assessed for the presence of TBV symptoms. Afterwards, corresponding leafs were measured using four different vision sensors: 1) A RGB Color camera 2) A spectrophotometer (400-2,400 nm) 3) A spectral camera (400-900 nm) 4) Chlorophyl fluorescence camera (MIPS system) An Elisa (Enzyme ImmunoAssay) analysis using TBV-specific antisera and a validated protocol was carried out on the same leaves. These measurements were used as reference analysis. A large number of parameters were extracted from the raw sensor data. Linear discriminant analysis using leave one out cross validation was used to predict whether a plant is healthy or diseased. It was found that the error of visual assessment of symptoms differed only 9% in the best case (cultivar Yokohama) in comparison to the ELISA values. The errors for the cultivar Yokohama for the different sensors were: 1) RGB Color camera: 11%; 2) Spectrophotometer (400-2,400 nm): 14%; 3) Spectral camera (400-900 nm): 9%; 4) Chlorophyl fluorescence camera (MIPS system): 31%. For the other tulip cultivars the results were more or less similar. The analysis of the spectrophotometer data shows that the most important features (wavelengths) are in the visual range (below 1000 nm). This implies that for practical implementation no expensive infra-red sensors are needed, which improves the economical feasibility of the system. The fact that the performance of these techniques performs similar as the crop expert is very promising and gives reason for follow up with as final goal an autonomous robot for detection and removal of diseased tulip plants.

Investigation into the classification of diseases of sugar beet leaves using multispectral stereo images

Bauer, S.D., Korc, F. and Förstner, W., Institute of Geodesy and Geoinformation, Department of Photogrammetry, Nussallee 15, 53115 Bonn, Germany; sabine.bauer@uni-bonn.de

This paper reports on methods for the automatic detection and classification of leaf diseases based on high resolution multispectral stereo images. Leaf diseases are economically important as they could cause a yield loss. Early and reliable detection of leaf diseases therefore is of utmost practical relevance - especially in the context of precision agriculture for localized treatment with fungicides. Our interest is the analysis of sugar beet due to their economical impact. Leaves of sugar beet may be infected by several diseases, such as rust (*Uromyces betae*), powdery mildew (*Erysiphe betae*) and other leaf spot diseases (*Cercospora beticola* and *Ramularia beticola*). In order to allow in vivo analysis we exploit the 3D-structure of the leaves. This enables to compensate for illumination and occlusion effects. Therefore we investigate multispectral stereo images. In order to obtain best classification results we apply conditional random fields. In contrast to pixel based classifiers we are able to model the local context and contrary to object centered classifiers we simultaneously segment and classify the image. In a first investigation we analyse multispectral stereo images of single leaves taken in a lab under well controlled illumination conditions. This diminishes the effect of the 3D-structure. For each leaf we take four RGB images ('FujiFilm FinePix S5600', 2,592 x 1,944, 1px equates 0.0967 mm) and one infrared image ('Tetracam ADC', 1,280 x 1,024, 1px are 0.2123 mm) from an altitude of 30 cm and different positions. The infrared camera has the channels RED, GREEN and NIR (700 - 950 nm). The photographed sugar beet leaves are healthy or either infected with the the leaf spot pathogen Cercospora beticola or with the rust fungus Uromyces betae. We calibrate and orientate the cameras and in this way fuse the five images. The derived leaf surface model obtains seven features per leaf surface element (blue, 2x green, 2x red, infrared, depth). In this examination the depth information helps us to separate the leaf from the background. We compare three classification methods: (1) pixelwise maximum a posteriori classification (MAP), (2) objectwise MAP (3) global MAP and global maximum posterior marginal classification using the spatial context within a conditional random field model. Training uses the expectation maximization algorithm to represent the mixture density of the features and MAP approach based on pseudo-likelihood approxi- mate scheme to train the parameters of the conditional random field model. We evaluate the methods based on a few hundred annotated leaves. We especially investigate the precision of the classification for the *Cercospora* leaf spot and the rust in various development stages. We also investigate the relevance of the chosen features and the effect of image resolution. First results indicate an admirable error rate of below 1 percent for *C. beticola* and a significant improvement when using the spatial context with computation times down to below 2 seconds on 0.5 MB image sections.

Study on plant pathogenesis with help of laser-optical and poly-functional spectrophotometers
Lisker, I.[1] and Dminriev, A.[2], [1]Agrophysical Research Institut of Russifn Agricultural Academy, 14 Grazhdansky pr., 195220 St-Peterburg, Russian Federation, [2]All-Russian Research Institut of Plant Protecnijn of Russifn Agricultural Academy, 3 Podbelsky r., 196608, Russian Federation

Structure of new poly-functional laser-optical and spectral photometers are described, which were designed in Agrophysical Institute and differ from the known instruments in that they can measure simultaneously all the effects of direct interaction of monochromatic irrandiance with the object investigated: incident irradiance, mirror and diffuse reflectance, transient and absorbed irradiance, and indicatrises of diffuse reflectance at various inclination angles in vertical and azimuth (360°). Both methods permit to study change of chlorophyll concentration in tissues of intact plants – sensitive indicator of their pathogenesis. By these techniques diseases of barley and wheat can be discovered some days before the visual inspection appears. They also permit to evaluate status and germination ability of tomato dry seeds.

MARVIN: high speed 3D imaging for seedling classification

Koenderink, N.J.J.P., Wigham, M., Golbach, F., Otten, G., Gerlich, R. and Van De Zedde, R., Wageningen UR, GreenVision, Bornsesteeg 50, 6708 PD Wageningen, Netherlands; Nicole.Koenderink@wur.nl

In horticulture, seedlings are assessed on their expected productivity of fruits or vegetables to significantly increase the total yield of the whole crop. At present, quality grading is a manual, costly process. Experts have been trained to take many quality factors into account and use all information to make a balanced decision on the quality class a plant belongs to. Automation of this seedling inspection process is complicated, due to the complex sorting rules and the inherent natural variation between individual plants. A consortium of seed companies and plant growers have asked us to design a computer vision system that determines both simple quality criteria such as leaf area and stem length, and more complex criteria such as scrawniness, rough plants, and defects on the leaves and stem. Moreover, the consortium requires a system that is fast enough and that can accurately assess approximately 20,000 seedlings per hour. Based on these requirements we have developed MARVIN, a 3D knowledge-based computer vision application. In this paper, we focus on the high speed 3D image acquisition module that records plants and delivers 3D point clouds as output. The objective of our research is to develop a fast 3D image acquisition module that can record 20,000 seedlings per hour in sufficient detail to obtain an accurate and objective quality assessment of the recorded plants. To the best of our knowledge, no 3D image acquisition techniques exist that are precise and fast enough for this task. An additional requirement is that the image acquisition module should be robust enough to work in an industrial setting. The 3D plant model is created based on the information from a range of 20 to 30 RGB cameras. We have developed an image acquisition technique based on volumetric intersection, the shape-from-silhouette algorithm and 3D visual hull computation. The use of 30 cameras leads to a non-trivial calibration procedure. A calibration procedure has been developed. This will be incorporated into the MARVIN-machine in such a way that an operator can easily calibrate the system when required. We have shown that the proposed 3D acquisition technique is accurate and fast enough to record plants for the seedling inspection task. The image acquisition solution together with the knowledge-based segmentation and classification task will result in the summer of 2010 in a robotised seedling handling and inspection machine that will be robust enough to function in an industrial setting.

Flower spatial variability in an apple orchard

Aggelopoulou, A.D.[1], Bochtis, D.[2], Fountas, S.[1], Koutsostathis, A.[1] and Gemtos, T.A.[1], [1]University of Thessaly, Crop Production and Rural Environment, Fytokou Str., 38446 N. Ionia, Greece, [2]University of Thessaloniki, Agricultural Engineering, Aristotle University of Thessaloniki, 54124 Thessaloniki, Greece

Precision pomology is the application of site-specific management into orchards. Orchards are regarded as high value crops and the application of site-specific management has high potential. The objectives of this study were (1) to study the flower variability in an apple orchard and (2) to examine the correlation between the number of flowers and the yield of the current year. The research was carried out in a commercial 5 ha apple orchard, located in Agia area, Central Greece. The orchard included two apple cultivars, Red Chief which was the main cultivar and Golden Delicious which was the pollinator. The between-row spacing of the trees was 3.5 m and the intra-row 2 m. Trees were trained as free palmette. In April of 2007 when the trees were in full blossom 300 photos of whole trees were taken in a grid of 20 m x 7m. The location of those trees was recorded using a hand-held computer with GPS in order to create the flower map. In September of 2007 yield mapping was carried out measuring yield per ten trees and recording the position in the centre of the ten trees. A prediction method was developed, based on image processing. The implemented algorithm converts the intensity images into binary images using a threshold based method. The texture of each photo is then correlated to the yield of the corresponding orchard sector. The evaluation of the method, as well as the optimization of the algorithm's parameters, was carried out using (repetitively) two randomly generated sub-sets of the total photos. The first set was used for the building of the correlation model while the second one was used for the model's validation. The results showed significant variability of the number of flowers and the yield of the trees (CV~50%). From the validation of the proposed method, the mean error on the yield prediction was found to be in the order of 10%. These results indicate that potential yield might be predicted early in the season from flowering maps, which is very important for the farmer and fruit industry. Other potential uses of the flower map include planning site-specific chemical thinning of flowers and site specific fertilization of the orchard based on flower variability. Furthermore, the low computational requirements of the proposed prediction algorithm make feasible the real time (on-the-go) planning of the previous applications.

Apple detection in natural tree canopies from multimodal images

Wachs, J.P.[1], Alchanatis, V.[2], Stern, H.I.[1] and Burks, T.[3], [1]Ben-Gurion University of the Negev, Department of Industrial Engineering and Management, P.O. Box 653, Be'er-Sheva 84105, Israel, [2]ARO - Volcani Center, Institute of Agricultural Engineering, P.O. Box 6, Bet Dagan 50250, Israel, [3]University of Florida, Agricultural and Biological Engineering, 225 Frazier-Rogers Hall, P.O. Box 110570, Gainesville, FL, 110570, USA; victor@volcani.agri.gov.il

In orchards, yield mapping and harvesting are labor intensive tasks and constitute the largest expense in orchards management in general, and specifically in apple orchards. Hence there is a need to develop autonomous fruit detection and robotic picking systems. The first step in such systems is the detection of green apples in the tree canopy using machine vision algorithms. In this work we developed a system that recognizes green apples within a tree canopy (sometimes occluded by leaves and branches) using thermal infra-red and color images. Thermal infra-red provides clues regarding the physical structure and location of the apples based on their temperature (leaves accumulate less heat and radiate faster than apples), while color images provide evidence of circular shape. Images were acquired using two cameras, one color and one thermal, attached on to the other so that they have approximately the same field of view. Initially, the optimal registration parameters between the images of the two cameras were obtained, using maximization of mutual information. Haar features were then applied separately to color and to thermal infra-red images through a process called Boosting, to detect apples from the background. A contribution reported in this work, is the voting scheme added to the output of the RGB Haar detector which reduces false alarms without affecting the recognition rate. The system performance was evaluated by comparing the number of apples that were manually counted in the digital images, to the number of apples that were automatically detected by the developed system. The resulting classifiers alone can partially recognize the on-trees apples however when combined together, the recognition accuracy is increased. The combination approach shows that the recognition accuracy was increased (74%) compared to the conventional approach of detection using either the color (66%) or the IR (52%) modalities alone. One interesting feature of the methodology is that the three main processes: registration, Haar feature detection in RGB and IR are independent and hence can be easily parallelized by assigning each process to a different CPU.

Effects of seed rate and nitrogen fertilisation on cereal canopy characteristics

Kren, J.[1,2], Svobodova, I.[2], Dryslova, T.[1], Misa, P.[2] and Neudert, L.[1], [1]Mendel University of Agriculture and Forestry in Brno, Zemedelska 1, 613 00 Brno, Czech Republic, [2]Agrotest Fyto, Ltd., Havlickova 2787/121, 767 01 Kromeriz, Czech Republic

The state of the cereal stand and its structure reflect variability in soil conditions as well as cropping treatments. Shoots and stems with spikes are the most often assessed units of cereal stand structure. Their size and number per area unit is changing during the plants growth within the stand. Understanding the rules of cereal canopy development can contribute to the improvement of crop management practices. During three years (2005-2007) small-plot field experiments with winter wheat and spring barley were carried out in two locations in the Czech Republic differing in soil and climatic conditions. Variants with contrast stand structure (different seeding rates and N rates) were investigated at growth stages BBCH 22-25, 31, 37, 55, 61, 87 and 91. In sampling plots, squares of the size 0.25 m^2 (0.5 x 0.5 m) were marked out to obtain plants for analyses of stand structure and nutritional status at assessed growth stages. Analyses involved determination of: (1) numbers and weight of individual plants and tillers, (2) fresh and dry matter weight of the above-ground biomass, (3) nitrogen uptake by the above ground biomass. Results were processed by using of basic statistical characteristics, analysis of variance and correlation analysis. The changing plant density and availability of nitrogen result in both crops in changes in numbers and size of tillers. Two types of relationships can be observed in a cereal stand - relationships among plants which are more or less of a random nature and intra-plant relationships which are dependent on the hierarchic structure of plants. Plant responses to certain conditions are expressed in changes of intra-plant relationships, which are reflected in variability of their tillers. Due to a higher density of stand caused by higher seed rate or higher dose of nitrogen, the competition within the stand increased and influenced the variability of size of plants and tillers. A higher inter-plant competition was expressed by lower values of the coefficient of variation (CV) of the plant weight and the number of tillers per plant. On the other hand, intra-plant competition increased values of the CV of the shoot weight. These effects were expressed at most in variants with higher seed rate in the BBCH 31 stage. During stem elongation the differentiation of tillers into two groups, i.e. vegetative and potentially productive, take place. This was accelerated by the lack of nitrogen or by a higher plant density. The separation of shoots into two groups enabled to determine the proportion of the potentially productive biomass in the total above-ground biomass. A low variability of this proportion (CV ranged from 7.06 to 15.79% and from 7.37 to 17.46% for winter wheat and spring barley, respectively) gives a possibility to use it as the indicator of the effectiveness of inputs into the crop cultivation and of a crop productive potential. These observations may contribute to the development of more efficient methods of canopy control and to an estimation of 'productive part' of above-ground biomass by using canopy spectral characteristics. This study was supported by the Czech Science Foundation of the Czech Republic, project No. 521/05/2299 and by the Research plan No. MSM6215648905, which is financed by the Ministry of Education, Youth and Sports of the Czech Republic.

Improved modelling of maize growth by combining a biophysical model of photosynthesis with hyperspectral remote sensing

Oppelt, N.M. and Hank, T.B., Ludwig-Maximilians-Universität München, Geography, Luisenstr 37, 80333 Munich, Germany

With the increasing public awareness of the deteriorating nutrition situation and the increasing energy demand of a rapidly growing world population, questions of agricultural productivity have gained importance in many different scientific publications. Especially the management efficiency of the available arable land is currently investigated. Computer aided modelling techniques here provide a promising method for the assessment of the potential of agricultural production for different natural environments. In order to map crop primary production, complex model approaches are required that are taking into account the actual physical and chemical processes, which are steering the exchange of mass and energy between the crop and the atmosphere. However, these model approaches are restricted to the accuracy of the input data, which naturally is unable to include unexpected events such as the incidence of pests and diseases, crop failure due to windbreaks, the effects of hailstorms or mechanically inflicted crop damages that are due to the applied machinery. The study presented here investigates the potential of improvement for a physically based multiscale model approach, when the static input data is enhanced by dynamic remote sensing information. The landsurface model PROMET (Processes of Radiation, Mass and Energy Transfer) was generally applied, while the remote sensing input data was derived from hyperspectral data of the CHRIS (Compact High Resolution Imaging Spectrometer) sensor, which is operated by ESA (European Space Agency). The PROMET model, whose multi layer vegetation routine basically applies the Farquhar et al. photosynthesis approach, was set up to a field scale model run (10 x 10m) for a test acre tilled with maize (*Zea mays*) of the cultivar 'Magister', mapping the crop development of the season 2005. During the model run, information on the absorptive capacity of the leaves for two canopy layers (top, sunlit layer and bottom, shaded layer) was updated from remote sensing measurements, where angular CHRIS images were available. Control data could be acquired through intensive field sampling campaigns that monitored the development of the stand throughout the vegetation period of the year 2005 also accompanying the satellite overflights. While the model without additional dynamic input data was able to reasonably reproduce the average net primary production of the crop, the spatial heterogeneity in the field was severely underestimated. The combination of remote sensing information with the vegetation model led to a significant improvement of the spatial heterogeneity of the crop development in the model, which again entailed an overall improvement of the model results in comparison to measured reference data. Although the integration of the remote sensing product still offers appealing possibilities of development and improvement, the results are promising and will be further pursued in future scientific endeavours.

Comparing hyperspectral vegetation indices for estimating nitrogen status of winter wheat

Li, F.[1,2], Chen, X.[1], Miao, Y.[1], Hennig, S.D.[3], Gnyp, M.L.[3], Jia, L.[4], Bareth, G.[3] and Zhang, F.[1], [1]China Agricultural University, Department of Plant Nutrition, No. 2, Yuanmingyuan West Road, 100193, Beijing, China, [2]Inner Mongolia Agricultural University, College of Ecology & Environmental Science, 306 Zhaowuda Road, 010019, Hohhot, China, [3]University of Cologne, Department of Georgraphy, Albertus-Magnus-Platz, 50923 Köln, Germany, [4]Hebei Academy of Agricultural and Forestry Sciences, Institute of Agriculture Resource & Environment, No. 598, Heping West Road, 050051, Shijiazhuang, China

Hyperspectral remote sensing is a promising tool for timely monitoring of crop nitrogen (N) status and in-season crop N management. Many hyperspectral vegetation indices have been published, but research is lacking to compare these indices for estimating winter wheat (*Triticum aestivum* L.) N status at different growth stages using a common dataset. The objective of this study is to identify promising hyperspectral vegetation indices for estimating winter wheat N concentration and compare them with a series of published vegetation indices using field data from farmers' fields. Three field experiments involving three winter wheat varieties and 3-6 N rates were conducted with cooperative farmers from 2005-2007 in Shandong Province, China. Three classes of indices were tested involving all possible two band combinations from 350 nm to 1075 nm: simple ratio index (SRI), difference index (DI), and normalized difference index (NDI). The identified promising indices were then compared with a wide range of published hyperspectral vegetation indices related to plant N or chlorophyll estimation. Results indicated that the best performing indices for estimating wheat N concentration mainly used bands from 350-450 nm, near red edge and they generally had higher R^2 than published indices. The estimation power in Feekes growth stage 8-10 was higher than that in Feekes growth stage 4-7. The performance of different hyperspectral vegetation indices was significantly influenced by growth stages and years, and two or more indices may be needed for estimating plant N concentration in the fields across different stages and environments.

Possibilities of cereal canopy assessment by using the NDVI

Kren, J.[1,2], Lukas, V.[1], Svobodova, I.[2], Dryslova, T.[1], Misa, P.[2] and Neudert, L.[1], [1]Mendel University of Agriculture and Forestry in Brno, Zemedelska 1, 613 00 Brno, Czech Republic, [2]Agrotest Fyto, Ltd., Havlickova 2787/121, 767 01 Kromeriz, Czech Republic

A disadvantage of traditional approaches to canopy management based on growth analysis (counting numbers of plants and tillers per unit area) is high labour consumption of area estimation of stand heterogeneity. Therefore possibilities of improving such an assessment for variable application of cropping treatments were verified. During three years (2005-2007) small-plot field experiments with winter wheat and spring barley were carried out in two locations in the Czech Republic differing in soil and climatic conditions. Variants with contrast stand structure (different seeding rates and N rates) were investigated at growth stages BBCH 22-25, 31, 37, 55, 61, and 87. The listed assessments were performed: (1) the NDVI using imaging of 0.25 m^2 (0,5 x 0,5 m) of the stand from the 5 m height (1 pixel = 2 mm) with a camera MS3100 (DuncanTech), (2) on the imaged stand areas numbers of plants and numbers of tillers per plants, weight of individual tillers, the fresh and dry weight of the aboveground biomass, nutrient and chlorophyll contents in plant samples, (3) yield components in BBCH 91. Results were processed by using of elementary statistical characteristics, analysis of variance and correlation analysis. Statistically significant effects of year, locality, experimental variant and growth stage on values of NDVI were found out. Increased NDVI values indicated: (1) greater amount of biomass and its dry matter per stand unit area, above all within the period of stem elongation, (2) higher average weight of plants, a higher numbers of tillers per plant, and a higher numbers of plants per stand unit area within the period of tillering, (3) higher average weight of tillers and higher numbers of tillers per stand unit area at the beginning of the stem elongation, (4) more intensive green colouration of the canopy, which indicated higher nitrogen uptake. Increased heterogeneity of the stand (assessed by means of coefficient of variation for the weight of plants and tillers) reduced values of NDVI. This can be explained by a greater occurrence of bare soil surface in uneven stands during early stages of growth till the full canopy closure (usually till BBCH 31). In this period, the values of NDVI are more affected by quality of stand establishment (i.e. of uniform distribution of individual plants). When using NDVI, similar values may indicate either a greater amount of aboveground biomass with a deficit of nitrogen or its lower amount in a good nutritional status. It is also rather difficult to decide if the aboveground biomass consists of a greater number of less tillering plants or, on the contrary, of a smaller number of plants with more tillers. This means that the effect of structural and physiological characteristics of the canopy on values of NDVI requires a further investigation. Nevertheless, the present level of knowledge enables a use of NDVI, especially for the evaluation of heterogeneity of cereal stand within the fields. A great advantage is a possibility of a quick and areal evaluation of the canopy and a more flexible crop management. This study was supported by the Czech Science Foundation of the Czech Republic, project No. 521/05/2299 and by the Research plan No. MSM6215648905, which is financed by the Ministry of Education, Youth and Sports of the Czech Republic.

Evaluation of palm trees water status using remote thermal imaging

Cohen, Y.[1], Alchanatis, V.[1], Cohen, Y.[2] and Soroker, V.[3], [1]Agricultural research organization (ARO), Volcani Center, Institute of agricultural engineering, P.O. Box 6, 50250, Bet-Dagan, Israel, [2]Agricultural research organization (ARO), Volcani Center, Plant Sciences, P.O. Box 6, 50250, Bet-Dagan, Israel, [3]Agricultural research organization (ARO), Volcani Center, Entomology and the units of Nematology and Chemistry, P.O. Box 6, 50250, Bet-Dagan, Israel

As quality water becomes scarcer, it is becoming increasingly important on a global scale to manage crop irrigations wisely. The challenge is to increase water-use efficiency while maintaining high quality crops and minimizing the amount of water applied. Date palm (*Phoenix dactylifera* L.) is a major tree crop in and regions of the Middle East and North Africa, having an important impact on the economy of many countries in these regions. While date palm is thermophilic and a drought tolerant plant, water deficiency significantly diminishes growth rate and reduce date production and quality. Monitoring date palms water status is the first step toward wise-irrigation management. Canopy temperature has long been recognized as an indicator of crop water status. High-resolution thermal imaging systems have been used to evaluate water status of several crops and orchards, e.g. cotton, corn, vineyards and olives. The objective of the present work was to investigate the potential to evaluate and map water status of individual date palms for future use in site specific irrigation management. Thermal images of date palm trees under two irrigation levels (commercial and 80% deficit) were acquired from an elevated stage reaching about 5m above the trees canopy. Leaf temperature of single trees canopy was extracted from the thermal images. The temperature of the commercial irrigation level was lower than that of the water stressed palms by 2-3 degrees. Based on these results airborne thermal images were acquired over a number of palm orchards along the Jordan valley. In few of the orchards, two irrigation levels were applied (commercial and 80% deficit). The thermal images were processed to map the canopy temperature of the palm trees. First, palm trees canopy was extracted mainly from soil. A watershed image processing algorithm was employed to find the low temperature sinks that represent the canopy. Then, binary image was transformed to polygons of canopy outlines. A random set of trees (polygons) was selected from each plot, allowing for a minimal distance between the selected trees (calculated by the semivariogram of the original thermal image). Statistical analysis of the temperature of the randomly selected sets revealed that the canopy temperature in the reduced irrigated plots was significantly higher than that of the commercially irrigated one. The results of this work indicate the potential use of airborne thermal images to assess the water status of irrigated palm trees and as a potential tool for irrigation management.

Classification of sugar beet and volunteer potato spectra using a neural network to select wavelengths
Nieuwenhuizen, A.T.[1], Hofstee, J.W.[1], Van De Zande, J.C.[2], Meuleman, J.[2] and Van Henten, E.J.[1], [1]Wageningen University, Farm Technology Group, P.O. Box 17, 6700 EW Wageningen, Netherlands, [2]Plant Research International, Field Technology Innovations, P.O. Box 16, 6700 AA Wageningen, Netherlands

Volunteer potato plants are an important weed in sugar beet crops in the Netherlands. As a consequence, much attention is paid to the control of these weeds. However, to efficiently detect these weeds, automated methods need to be developed to reduce the labor requirements and secure future removal and control practices. Therefore, imaging techniques were proposed for detection of volunteer potatoes in between sugar beet plants. Multispectral recordings of five sugar beet and five volunteer potato plants were taken in 2006 and 2007 at three different growth stages. The recordings registered vegetation reflection at 167 wavelength variables between 450 and 1,665 nanometer and a minimum of 100 spectra were recorded for each plant that was measured. One spectrum measurement is the reflection of 1 mm^2 vegetation. The resulting spectra were identified as sugar beet and volunteer potato spectra using a variable selection method followed by a classification. A variable selection method was chosen as for future detection systems specific discriminating wavelengths need to be identified. To overcome the restrictions of only investigating linear relationships with linear discriminant analysis as variable selection method, a neural network wavelength selection method was used. A fully connected Kohonen neural network with three layers was trained and used for classification. The input layer consisted of the reflection variables, the hidden layer consisted of one, two or three hidden neurons, and the output layer consisted of two neurons, one for volunteer potato class and one for sugar beet class. The wavelength selection procedure was as follows: fifty percent of the dataset was used for training of the neural network and fifty percent was used for classification. A forward inclusion method of input variables was used. More specifically, the first variable separates the two output classes the best with a net including one input variable. Each next step the net is expanded with one input variable and a wavelength variable is included that separates the two output classes better compared to the remaining wavelength variables and increases classification accuracy. The forward inclusion approach for choosing discriminating wavelengths reduced the number of wavelengths needed for classification from 167 to a maximum of ten wavelengths needed for 100.0% correct classification of the wavelength spectra. This is true even for the simpler neural networks with only one or two hidden neurons in the net, which is nice because less hidden neurons means that the relations described by the neural net can be easily understood. The results show that both in the visual and in the near infrared reflection region some wavelengths are responsible for discrimination between sugar beet and volunteer potato plants. A neural network approach was successfully used to identify discriminating wavelength variables. These wavelengths can now be used to choose a sensor system for detection of volunteer potato plants.

Automated weed detection in winter wheat by using artificial neural networks

Kluge, A. and Nordmeyer, H., Julius Kühn-Institute, Institute for Plant Protection in Field Crops and Grassland, Messeweg 11/12, 38104 Braunschweig, Germany

Weeds are distributed heterogeneous within agricultural fields and in many places weed densities are below threshold values where weed control is not necessary. Site specific weed control is a strategy based on threshold values for herbicide use (i.e. the spray or no spray approach). The use of variable dosages is an additional concept (i.e. the low dose or high dose approach). Site-specific weed control requires detailed information about the spatial distribution of weed species and densities. A problem is how to map the spatial weed distribution in a short time and at low cost. Since manual weed estimation seems therefore not applicable, methods for automatic weed recognition are needed. In this regard, weed detection is a key stage in developing decision support systems. An image analysis system was developed to identify weed species (*Galium aparine, Matricaria chamomilla*) in winter wheat. The system consists of two firewire cameras using CMOS and CCD technology. For image analysis an algorithm was developed by calculation and classification of different plant characteristics. A method measures geometric attributes of plants and put them in relation. From the resultant characteristic vector an assignment to a class of plants is made including a high error ratio. A false classification of the weed plant leads to different undesirable effects depending on the weed and the crop. The error in weed recognition results basically in overlapping parts of plants and their morphology. Another essential factor is the localisation and segmentation of the objects to be investigated. In this paper more methods of image recognition to extract features will be analysed. It considers texture, structure and geometry of the analysed plants. Furthermore it is attempted to establish a more robust localisation for the objects. The retrieved information will be evaluated by neural networks and they will be used for classification. Single plants, now called objects, are automatically segmented by operations like k-mean-algorithm for coloured pictures, watershed-transformation for grey-scaled pictures and crass-fire-algorithm for binary pictures. With the Canny-algorithm the edges of objects are located and eight measure points are placed with the same distance. Vectors of distance between points are calculated and the neural network is trained by Euclidean distances. After training neural networks different pictures are used to analyse the correctness and to proof the recognition rate. If the recognition is not satisfactorily a new network is generated. A lot of different networks are generated that will be proofed on its capability. The model for the neural network is a backpropagation net. Classification of weed plants by object criteria can be used to distinguish between dicotyledonous weed species like *Galium aparine* and *Matricaria chamomilla* at growth stage BBCH 10-12 and crop plants. For *Galium aparine*, a mean recognition rate of 84.6% could be achieved. With neural network compared to the geometrical approach the same performance and nearly same computing time can be achieved, but additional morphological effects can be included. Therefore, by neural networks the classification of overlapping plants is possible. However, a method separating overlapping objects is part of the research, generally.

Spectral signatures of diseased sugar-beet leaves
Mahlein, A.-K., Steiner, U., Dehne, H.-W. and Oerke, E.-C., University Bonn, INRES-Phytomedicine, Nussallee 9, 53115 Bonn, Germany

Crop diseases are often heterogeneously distributed in agricultural fields. An understanding of the spatial and temporal distribution and spread of pathogens is necessary for site-specific fungicide applications. As shown in previous studies, remote sensing can be a useful device to detect and observe spectral anomalies in the field. Conventional remote-sensing techniques use vegetation indices to describe the vitality of vegetation cover. Remote sensing is used for the assessment of the effect on yield formation of crops, however there are limitations of these approaches for disease detection. In this study hyper-spectral near-range sensing was used to examine its potential to detect and identify the diseases incidences on sugar-beet leaves. The spectral properties of sugar-beet plants infested with *Cercospora beticola* (Sacc), *Erysiphe betae* (Vanha) Weltzien, and *Uromyces betae* (Persoon) Léveillé have been assessed. Using a handheld spectrometer, reflectance spectra from diseased plants were investigated from 450-1,050 nm and compared on the leaf and plant level. Significant differences between the disease-specific spectra were detected in the NIR-range. Structural and pigmentation indices obtained from specific wavebands gave significant differences in plant vitality depending on the disease. The hyper-spectral vegetation indices calculated from leaf spectra showed disease-specific decreases in the chlorophyll content of plants. Infestations of *C. beticola* and *U. betae* caused increased values of the anthocyanin-reflectance-index. There was a strong relationship between disease severity and the values of vegetation indices.

Development of an integrated approach for weed detection in cotton, for site specific weed management

Efron, R.[1], Alchanatis, V.[1], Cohen, Y.[1], Levi, A.[1], Eizenberg, H.[2], Yehuda, Z.[3] and Shani, U.[3], [1]ARO - The Volcani Center, Institute of Agricultural Engineering, P.O. Box 6, Bet Dagan, Israel, [2]ARO - The Volcani Center, Dept. of Phytopathology and Weed Research, Newe Ya'ar Research Center, P.O. Box 1021, Ramat Yishay, 30095, Israel, [3]Faculty of Agricultural, Food and Environmental Sciences, The Hebrew University of Jerusalem, Dept. of Soil and Water Sciences, P.O. Box 12, Rehovot 76100, Israel

The ability to properly control and eliminate weed infestation in cotton field can be a crucial factor for the crop's profitability. Site specific weed management has long been marked as the next step in precision agriculture and the tools of remote sensing and image processing stands in the front line of this discipline. Several researches that have being conducted in this field showed encouraging results for distinguishing weeds from the crop, but no mechanism has been implemented yet in the agriculture routine. This fact has been driven us to develop a low cost, yet effective, method for weed map preparation which later on will be used for on-the-go herbicide application. A new approach is being suggested here which includes combining information from a hydro-thermal model into image processing algorithms. This information consists of the expected number of plants per m^2, leaves per plant, etc. The suggested system is constructed from a simple RGB camera, mounted on a tractor's sprayer boom and connected to an on-board computer. The information coming from the camera is a video file with a rate of 7.5 frames per second. The pictures are then being extracted from this file and analyzed in a MatLab™ environment. The process involves applying different vegetation indices using the RGB channels, object separation based on morphological features compared with incoming information from the hydro-thermal model and data from previous years. The result of this process is a 'weed map' calibrated with dGPS data that had been collected during the image acquisition stage. The information achieved each year will also be used for continuous update and amendment of the hydro-thermal model.

Selectivity of weed harrowing in cereals with sensor technology in Germany
Rueda Ayala, V.P. and Gerhards, R., University of Hohenheim, Weed Science, Otto-Sander-Str. 5, 70599, Germany

In four field experiments it was investigated whether intensity, timing, and direction of post-emergence weed harrowing in winter and spring cereals influenced the selectivity. Selectivity was defined as the relationship between crop burial in soil immediately after treatment and weed control, same as defined in Denmark. Each experiment was designed to create a series of intensities: by increasing the number of passes and the penetration of the harrow tines in the soil, both at varying crop growth stages. Objective estimation of soil cover through differential image analysis was used to compare with danish experiences from color digital images. Special attention is being given to statistical analysis procedure, mainly to determine the effect of treatments at different crop growth stages. Image analysis also was used to discriminate weed species and contrast with the manual counting; classification of images will be finished after the final date of data collection. Selectivity has been influenced by timing of harrowing application. It seems to be improved at later crop growth stages (22 to 27) by using stronger harrow intensities, although weed control appears to be reduced. Harrowing across crop rows seemed to have no influence on selectivity, nevertheless percentage of soil cover and yield are still under analysis and measurement, respectively. Additional data from a soil sensor have been acquired to support the computed soil coverage; it was intended to obtain information to automatically control the set up of the harrow tines according to soil compaction and weed coverage.

Automatic derivation of weed densities from images for site-specific weed management
Weis, M. and Gerhards, R., University of Hohenheim, Department of Weed Science, Otto-Sander-Straße 5, 70599 Stuttgart, Germany

Site-specific herbicide applications can save large amounts of herbicides and improve management practices. One crucial part of a system for site-specific weed management is the measurement of the spatial variability of weed densities. A system was developed to identify different weed species from images taken in the field. The automation has the potential to increase the spatial density of weed sampling points. A manual sampling with high density of points is unfeasible due to the costs. Image processing algorithms are used to generate a shape description for each plant in the image. A classifier can be constructed that assigns weed and crop classes to the plants based on the shape features. Weed density maps are generated using the results of the classification. The weed maps are transformed to application maps, which are used for the site-specific herbicide application. The shape of the plants vary with their growth stage and may be segmented into parts, e.g. single leaves, in the image processing. Therefore different classes for each species may be introduced in the process. To avoid over- or underestimation of the actual number of weeds some of the classes are aggregated using weight factors. To derive transformation functions the results of the automatically derived weed counts are compared to manual measurements of weed densities, which were derived from the images and in the field. The results show that the raw classification results are linearly related to the actual number of manually counted weeds and the results can be used as input for a site-specific decision component.

Evaluation of a ground-based weed mapping system in Maize

Andujar, D.[1], Dorado, J.[1], Ribeiro, A.[2] and Fernandez-Quintanilla, C.[1], [1]ICA-CSIC, Serrano 115B, 28006 Madrid, Spain, [2]IAI-CSIC, Carretera de Campo Real km 0.2, 28500 Arganda del Rey, Spain

Satellite and airborne methods have been developed in the past to detect weed patches in the fields. However, these methods are strongly dependent on sky cloudiness, they have a relatively low resolution, the cost of the images can be high and time is lost in processing the images, causing delays in herbicide application. Ground-based, machine-mounted sensors offer numerous advantages for practical real-time applications. Although various image analysis systems have been developed and tested to identify weeds in different crops, their operational complexity makes them slow and expensive. This paper assesses the relative accuracy of a simple system based on optoelectronic sensors in comparison with direct measurements made on digital images. Although this system does not allow to discriminate between weed species or groups of species, it may be well fitted to the type of herbicide applications currently applied in maize in Spain. A ground-based weed mapping system was constructed using three passive optoelectronic sensors (WeedSeeker®), a DGPS receiver, a data acquisition card and processing system and a system operation software. The three sensor units were mounted on the front of a tractor at a 0.75 cm distance between them and at 60 cm height above ground level. Consequently, the system was able to explore a 2.25 m band, corresponding to three row crop, detecting the vegetation present in 0.34 strips in the middle of the inter-row maize area. The working capacity of the system was higher than 1 ha h^{-1}. Weed populations present in two maize fields were mapped with this system when the crop was in a 4-6 leaves stage. In order to validate the system with highly reliable data, digital images were obtained in numerous geo-referenced points distributed throughout the two fields, assessing weed cover present in each of the images using a software specifically designed for that use. The comparison between the data obtained with the ground-weed mapping system and those obtained from the digital images indicated an excellent agreement (>90%) between the two sets of data. Fifty percent of the 'failures' corresponded to weedy areas not detected by the autonomous system and the remaining 50% of failures were clean areas assessed as weedy. The weed cover level present in each area did not influence the reliability of the system.

Evolution of agricultural machinery: the third way
Berducat, M., Debain, C., Lenain, R. and Cariou, C., Cemagref, Ecotechnologies et Agrosystèmes, 24 Avenue des Landais, 63172 Aubiere, France

The agro equipments have always been an important vector for the development of agriculture. For several years, we have been witnessed to an irremediable increase in the size of agricultural machines. If this 'first way', encouraged by farm machinery industry, is synonymous of high outputs, a lot of disadvantages can nevertheless be identified in term of soil compaction, high fuel consumption, difficulty to control large width implements on irregular soils, or to integrate the traffic on campaign roads. A 'second way' have recently been proposed by several research laboratories (North of Europe, Japan), based on light weight robots for a small scale farming at the plant level (detection of position of each plant during sowing operation and localized interventions after during all the production process (hoeing, fertilizing…)). This approach is well suited to high added value product such as market gardening or flower productions. However, for crops like cereals which represent the first productive sector in Europe (260 millions tons in 2006 for 51 millions of hectares), smart previous robot machines even in swarm working configuration would certainly not be able to assume harvest operations in large production areas. Thereby, another scenario, called the 'third way', could consist to propose medium machines in size and power but always integrating a high degree of technologies. Medium power could mean the possibility for agricultural machines to easier benefit of future car component developments in term of transmission (engine-wheel pack…) or energy motorization (electric fuel cell…). For preservation of economic competitiveness, these machines could compensate the decreasing of the work width by an increase of the forward speed. A high degree of modularity could also authorize the cooperation of machines, with a leader machine always driven by an operator, and one or several other following. The paper will underline main advantages and limitations of each way. Some first technological robotic developments in link with the third way will be presented.

Mass flow sensor for combines with bucket conveyors

Zandonadi, R.S.[1], Stombaugh, T.S.[1], Queiroz, D.M.[2] and Shearer, S.A.[1], [1]University of Kentucky, Biosystems And Agricultural Engineering, 128 Barnhart Bldg., 40546, USA, [2]Universidade Federal de Vicosa, Engenharia Agricola, Universidade Federal de Vicosa, Vicosa, MG, 36570000, Brazil

Most commercially available yield monitoring technology for grain harvesting equipment is designed to work with paddle-type clean grain elevators. There are some harvesting machines such as peanut and dry edible bean harvesters that use bucket conveyors to move grain from the cleaning shoe to the clean grain tank. This conveyor design prevents the use of currently available impact plate, optical, or radiometric mass flow sensors. Researchers at the University of Kentucky have developed a torque-measurement technique for sensing mass flowrate through a bucket conveyor. One significant advantage of this technology especially for developing countries is that in-field calibration can be performed with static weights thus eliminating the need for weigh carts or immediate weigh scale receipts. This paper will describe the sensing technology and its potential for in-field static calibration.

Automated, low-cost yield mapping of wild blueberry fruit

Swain, K.C., Zaman, Q.U., Schumann, A.W., Percival, D.C., Dainis, N. and Esau, T.J., Nova Scotia Agricultural College, Engineering Department, 39 Cox road, B2N5E3, Canada; swainkc@yahoo.com

Wild blueberry (Vaccinium angustifolium Ait.) is a unique crop native to Northeastern North America with annual production of 82 million kg of fruit (market price around $352 million) covering an area of 79,000 ha. Wild blueberry fields are developed from native stands on deforested farmland by removing competing vegetation. The majority of fields are situated in naturally acidic soils that are low in nutrients and have high proportions of bare spots, weed patches and gentle to severe topography. This crop is perennial in nature, having a vegetative growth season (sprout year) followed by a productive season (fruit year). Considering, the uniqueness of wild blueberry, it is a challenge to apply site-specific management practices to maximize the profitability and minimize environmental risks. Yield maps along with topography and soil nutrient maps could be used to develop site-specific nutrition programs for wild blueberry production. The objective of this research was to develop an automated yield monitoring system for real-time wild blueberry fruit yield mapping. An automated yield monitoring system was developed and mounted on a specially designed Farm Motorized Vehicle (FMV). The system consists of a 10-mega pixel, 24-bit digital color camera (Canon Canada Inc., Mississauga, ON, Canada), Trimble Ag GPS 332 (Trimble Navigation Limited, Sunnyvale, CA) for geo-referencing and a laptop computer. The camera was mounted at the front of the vehicle, aiming downwards at a height of 1.5 m with a clear view of the ground. The DGPS antenna was mounted above the camera. Custom image processing software was developed in 'C' and 'Delphi' programming languages to estimate the blue pixels representing ripe fruit in the field of view of each image and express them as a percentage of total image pixels. Two wild blueberry fields were selected in central Nova Scotia to evaluate the yield monitoring system. Calibration was carried out at randomly selected 40 data points (20 each) in the two wild blueberry fields. The ripe fruit was hand-harvested out of a 0.5 x 0.5 m steel-framed quadrant using a commercial blueberry rake. Linear regression was used to calibrate the actual fruit yield with percentage blue pixels in both fields. A linear regression model through the origin (y=bx) was highly significant in field 1 (R^2=0.90) and field 2 (R^2=0.97). Real time yield mapping was then carried out by acquiring images on the moving FMV at a spacing of 1.25 m and a ground speed of 0.5 m/sec. The software also processed the image to extract the percentage blue pixels for the blueberry field in real-time. The estimated yield per image field of view along with geo-referenced coordinates was imported into ArcView 3.2 GIS (ESRI, Redlands, Calif., USA) to map fruit yield. The results indicated that the automated yield monitoring system could be used for fruit yield mapping on a large scale in wild blueberry fields. This information could be used to implement site-specific management practices within the wild blueberry fields to optimize productivity while minimizing the environmental impact of farming operations. An automated yield monitoring system could be incorporated into wild blueberry harvesters for real-time fruit yield estimation and mapping.

Development and first tests of a farm monitoring system based on a client-server technology
Mazzetto, F., Calcante, A. and Salomoni, F., University of Milan, Institute of Agricultural Engineering, Via Celoria 2, 20133 Milan, Italy

It's a common knowledge that farm operations monitoring must be considered the starting point to realize a complete and integrated Precision Agriculture system. Therefore, it is necessary to collect the actors taking part to the event and physical effects related to it. For each field process, it is essential to know: a) what's happened; b) where the event has been carried out; c) who is involved in the event; d) what and how many productive factors have been employed. The Institute of Agricultural Engineering of Milan is involved in the development of farm monitoring systems from many years. In particular it faces both hardware and software aspects, such as the identification of the most suitable technological solutions (sensors, communication bus, positioning systems), and the realization of a database based on the technology of the so called 'Field Datalogger'. In this work, last version of the farm monitoring system is presented: it is characterized by client-server logic and wireless data download. This system is installed in the experimental farm of the University of Milan, and it is based on a mixed network (ZigBee + TCP/IP) connected in real time to a central server. Actually, clients connected to the net are represented by all farm tractors and the two devices for the measurement of slurry level in storage tank. In detail, each tractor is equipped with a datalogger which collects: a) the vehicle position and speed through a GPS receiver, b) the engine speed, c) the temperature of exhausted gas (for the estimation of the fuel consumption), d) the actuation of the PTO, e) the numerical code of the coupled operating machine (O.M.). The identification of the O.M. is obtained through ZigBee technology. To this aim, every O.M. is equipped with a transmitter that is activated by an accelerometer: the machine movement enables the transmission of an univocal identification code. Each tractor is connected to a TCP/IP net by GPRS modem in order to allow the data upload to the central server. In order to monitor farm slurry flows a system has been realized; it implements an ultrasonic sensor to measure the storage level every 1 minute. Again, data transmission is realized through a GPRS modem with real time connection. Data have been collected by a software installed on the farm server which is endowed of a component that allows the creation of multiple connections on the TCP/IP stack. The software verifies - on a dedicated socket - if any device is calling and, when the vehicle/device is connected, the server qualifies the logon and allows data download. Data flow coming from every connection is managed by the server and stored on a local database. So structured data can be visualized on a digital map using the client software - FleetMonitoring ver. 2.1.0 - and they can be used to compile reports, concerning single mechanized operations, through query submissions on the database. About the monitoring of slurry storage level, StorEyes ver. 1.02 software allows the analysis of filling and emptying events through the measurement of any slurry flows. This system contributes to realize a farm historical archives, and it also offers a useful instrument for traceability and environmental safeguard. To this end is on progress the Metamorfosi project, whose main target is to implement a geografical territorial information system, using described technologies, to realize an objective and complete control of distribuited nitrogen amounts.

Agri yield management: practical solutions for profitable and sustainable agriculture based on advanced technology

Hadders, J., Hadders, J.W.M. and Raatjes, P., Dacom B.V., P.O. Box 2243, 7801 CE Emmen, Netherlands

Introduction The globally increasing demand for food and crops for bio fuels is outpacing the agricultural production capacity. Farm input costs are related to the ever rising costs of energy, forcing the economic need to minimize inputs for crop production. Environmental concerns and the availability of usable irrigation water put pressure on sustainable crop production. Growing more food with fewer resources poses a great challenge for the agricultural sector. Agri Yield Management Agri Yield Management (AYM) systems provide the farmer with the necessary tools to achieve this goal. The AYM system as developed by Dacom in The Netherlands makes use of advanced technologies like intelligent sensors, internet and GPS. Software combines the collected information into practical solutions that supports the farmer in his daily operations. This AYM system is unique because it integrates multiple aspects of the crop production such as water, chemicals and fertilizer in relation to the yield. Precision timing and using the right product has proven to be profitable through the years. The system for disease control was evaluated in field trials around the world and consistently demonstrated a reduction of chemical inputs of around 25% as compared to traditional practices while maintaining excellent disease control. Also the use of intelligent soil moisture sensors in combination with weather forecast proved to lead to a much more efficient use of water. Around the world reductions of 49% of water and 31% yield increase have been shown. Sensor Technology Sensor measurements from local fields are important input in real time management systems. The sensors are put in the field and exposed to all kinds of climatic conditions and possible mechanical maltreatment. Hourly collection of reliable and accurate sensor information is a laborious undertaking. Constant validation of the produced information is a necessity. The Dacom AYM system handles data of thousands of data collection points with multiple sensors from around the world. Raw data is collected through telemetry systems and internet in a central databank. In the control room, a quality check is applied both automatically and manually by a team of technicians to ensure reliable data before it is distributed to farmers and advisors. Automated collection of real time crop data and the development of more intelligent sensors are some of the challenges that lay ahead. Software Technology The software provides the linkage between the farmer and all the gathered information. Models calculate practical, real time advice. Traditional these models are based on local conditions with mathematical algorithms. The new generation AYM models are biological models and can be applied worldwide. For example the Dacom Phytophthora model has been successfully applied over more than a decade under conditions ranging from desert climate to sea climate. Though, it is only still the forerunner farmers that make use of AYM systems. The challenge is to develop software that makes more farmers turn away from their traditional practices. Conclusion The challenge for agriculture is to produce more with less. AYM provides an important contribution towards the future demand for food. Challenges are the real-time collection of reliable crop environment information and methods for adaptation of these innovative systems by farmers to change from traditional practices to new AYM methods.

Flow behavior analysis of a rfid-tracer for traceability of grain
Steinmeier, U., Von Hörsten, D. and Lücke, W., Agricultural Engineering Section, Department of Crop Science, Gutenbergstraße 33, 37075, Germany

For quality management, product liability and costumer satisfaction the knowledge of the origin of products is a crucial factor. The Agriculture Engineering Section developed a new system for traceability of grain based on Radio Frequency Identification technology (RFID). The RFID data carrier is encapsulated in epoxy resin and adapted to the grain. This transponder unit (tracer) remains in the grain from the time of harvest until shortly before processing. In order to investigate the application of the tracer it is important to know how the tracer has been constructed for marking grain. In different experiments the flow characteristics of the tracer are analysed. Therefore tracers with different form and density were designed. These tracers had been inserted in the grain and tested in different transport devices and a bunker. A preliminary experiment showed a low demixing rate of the tracer which has similar form and density of the grain. In this experiment the density had the highest effect beside the form. Based on these results further experiments are made: The experiment set-up is a bunker (500 mm width, 200 mm depth and 2,000 mm height) with a variable outlet hopper. The angel of the hopper is changeable in three steps for simulate different flows e.g. mass flow. The bunker can be filled by sample containers which have a size of 2 dm^3. When the sample containers filled with grain and tracers then they have to be fixed on a depot which is on top of the bunker. By pulling a plate slide the mixture of grain and tracer flows into the bunker. One after another the content of the sample containers are filling the bunker. The same sample containers are used to discharge the bunker at the outlet hopper. The simulation based on two several grain batches, one is coloured and the other one non-coloured. This is important to get information about the mixing of two different batches. These batches are also marked with individually numbered tracers. With image analyse the mixture rate is determined of non-coloured and coloured grain. Then a metal detector is helping to detect the tracers. In the tracer a steel ball is included for a safe detection. By separating the tracer out of the grain, the number of the tracer is connected with the number of the sample container. The tracers which have been marked the non-coloured batches have to be found by the non-coloured grain batches. And in the case of coloured one it is vice versa. The prospect is that the tracers with similar flow behaviour to the grain are the right one for a practical use for a traceability application. About this bunker experiment there will be final results at the end of October 2008.

Gamma ray sensor for top soil mapping: THE MOLE
Loonstra, E.H., The Soil Company, Kadijk 7b, 9747 AT Groningen, Netherlands

Gamma ray sensor for top soil mapping: The Mole In the '90's the Physics Department of the State University of Groningen (NL) developed an in situ on-the-go gamma ray sensor. Initially, the sensor was used for mapping of minerals on the seafloor. Later on, The Soil Company further developed the sensor for high resolution mapping of soil characteristics, like clay, grain size, loam and organic matter, on land. The Molehas been put into practice in Europe since 2003 and showed promising results. Principle of The Mole The sensor the Mole of The Soil Company measures in situ the gamma ray decay of the radio active trace elements K^{40}, Th^{232}, U^{238} and Cs^{137}. The sensor is optimised for measurements under dry circumstances and measures top soil (30-40 centimetres). The measured activity concentrations of the trace elements are used to unravel the composition of soil. Each soil type can be considered as a unique combination of radio active trace elements; a fingerprint. The Mole in practice The Mole consists of a metal tube containing the sensor in combination with an electronic read out device, a GPS, a computer with data logging software and data analysis software. Field surveys are done driving the sensor at a speed of 6 km/hr. Contact with soil is not required. Field data (spectrum) is transformed into individual gamma ray concentrations of the trace elements. In order to transfer the field data into soil properties a calibration curve is build. The Soil Company has built an unique database for this purpose, which is constantly growing with sample results from new projects. After calibration curves are drawn up GIS software is used to produce the actual soil maps. Applications The Mole has proven to be benificiary in many ways. The soil maps made with readings The Mole from have proven to be valuable for farmers that conduct precison farming. Especially, good financial results were made with the applications in variable planting distances of seed potatoes, variable fertilising and the use of granular product for nematode control. The sensor is also useful in locating soil disturbances such as old rivers or archeological spots. The mapped physical soil properties are used for matters of irrigation and nature development. Comparison other soil sensors The Mole is different compared to other well known sensors: (1) The gamma ray spectrum remains constant over time which makes calibration very easy. A huge step towards universal soil calibration. (2) Restrictions on sensing days are only controlled by workability of the field and rain. Sensing can therefore be done throughout the year. (3) Crop (residue) or frost does not effect sensor readings. (4) Sensing is non-destructive as the sensor is positioned above the soil. (5) The gamma ray spectrum represents the top 30-40 cm of the soil; the most important part for crop growth.

Combined sensor system for mapping soil properties

Mahmood, H.S., Hoogmoed, W.B. and Van Henten, E.J., Agrotechnology and Food Sciences Group, Wageningen University, Farm Technology Group, Bornsesteeg 59, Technotron 118, 6708 PD Wageningen, Netherlands

There is a growing demand for rapid, accurate and inexpensive methods to measure and map soil properties to be used in precision agriculture. Because the acquisition of fine scale information on the variation in soil properties by manual soil sampling and conventional laboratory analysis is time consuming, labour intensive and cost prohibitive. Moreover, simultaneous measurement of several soil physical and chemical properties has been a desire for many years. The proximal soil sensors such as gamma ray, electromagnetic induction and visible-near infrared are able to measure soil properties in real-time through either direct contact or at a close range to the soil. Gamma ray sensor can measure clay content, potassium, organic carbon, iron content, soil depth and parent material. Electromagnetic induction sensor (EM38) can measure total ground conductivity, organic matter content and moisture content. Visible-near infrared reflectance spectroscopy has been used to characterize soil minerals (Ca, Mg, Fe, K, P etc.), organic matter and molecular compositions. However, in certain soil conditions like presence of salinity or gravels or in moist soils, a single sensor does not give reliable response for a specific soil characteristic. For instance, it is difficult using γ-ray sensor alone to distinguish between clay content and gravels as both result in similar strong signals. EM38 cannot discriminate between sandy soils and gravels as both show similar and low apparent electrical conductivity (EC_a) and salinity also affects data interpretation. This limitation could be overcome by the combine use of different sensors underlying different physical principles that often produce complementary information to eliminate this ambiguity and unreliability. This paper focuses on a preparatory literature study to a (PhD) project aimed at developing a combined sensor system which may contain a gamma ray sensor, an EM38 and a visible-near infrared soil sensor to get complementary information on surface soil properties such as soil texture, organic matter, soil pH, soil electrical conductivity (EC), phosphorus, and potassium. It is hypothesized that the combined responses of these sensors enables one to collect high resolution field soil data for making real-time soil maps of different soil properties. This high resolution soil information may be used for site-specific crop management and provides as a research tool to help in understanding field scale soil variability. It is expected that results will demonstrate higher effectiveness of combined sensor systems as compared with the single sensor in its application to investigate various soil physical and chemical properties in precision agriculture.

Reproducibility of different composite sampling schemes for soil phosphorus

Schirrmann, M.[1], Domsch, H.[1], Von Wulffen, U.[2], Zauer, O.[3] and Nieter, J.[3], [1]Leibniz-Institute for Agricultural Engineering, Engineering for Crop Production, Max-Eyth-Allee 100, 14469 Potsdam, Germany, [2]State Office of Agriculture, Forestry and Horticulture (LLFG), Strenzfelder Allee 22, 06406 Bernburg (Saale), Germany, [3]Dawa Agrar, Am Plan 3, 39326 Dahlenwarsleben, Germany

Soil phosphorus exhibits a high spatial variability in fields due to numerous influences like chemical processes, soil erosion or fertilization history. To estimate its spatial distribution it is known to sample with small sample spacing and high sample size. However, the problem is that mass sampling strategies are not cost effective and not feasible. Therefore, a common alternative is to physically mix sampling cores gathered over a distinct area to reduce laboratory analysis costs. But is the composited sample a reliable estimate for the sampled area? To answer this question we took sampling cores in grid cells on different positions and compared the outcome distributions from many cells. The objective was to evaluate the reproducibility by examining the correlation and variation among each composite design. In this case study we divided a field in 59 regular grid cells with an area of 100 m x 100 m. Three composite layouts were used with different core positions aligned on lines in each grid cell: First, a circle line in the middle of the cell with 5 m radius, second, a line in the middle of each quarter of the cell and, third, a core sample in the middle of each ninth of the cell. We repeated the third design by sampling 10 m apart from the core positions to account for fine scale differences. Thus, positions were most concentrated in the first design and most uniformly distributed in the third design. Additionally, we implemented a mass sampling design with a sample spacing of 25 m to approximate the real phosphorus distribution for 16 grid cells. The mean among the different composite designs were almost constant while the variance was lowered when the sample cores were more distributed. Also, nugget variance was smaller and the spatial structure appeared stronger when examining standardized semivariogram values. Spatial autocorrelation observed with the Moran's I was weaker when using the more concentrated first design. These effects result from averaging the underlying spatial process when the sample support is raised. When observing correlation and mean error the first design was most different. However, there were also differences seen comparing the second and third design with each other. In accordance, the comparison with the estimated spatial distribution derived from the mass sample design showed highest errors for the first design, medium for the second and lowest for the third design. These results show that core distribution for a composite sample design does have an implication on the estimated phosphorus distribution. The more concentrated the cores are laid out in a grid cell the higher is the resulting deviation from the real distribution. Thus, farmers should orientate sampling lines in grid cells in a way that most of the area is covered by sampling cores homogeneously.

The capability of non-destructive geophysical methods in precision agriculture to capture subsoil mechanical strength

Hoefer, G. and Bachmann, J., Leibniz University of Hannover, Institute of Soil Science, Herrenhaeuser Str. 2, 30419 Hannover, Germany; hoefer@ifbk.uni-hannover.de

To date an easy access to regionalise subsoil compaction on the field scale has not been realised with the common accessible soil physical methods. Caused by their own methodical limitation, results can only be produced in the laboratory but with a limited significance for the probed soil pit and therefore presenting only the local state of mechanical strength. Existing field methods to access this problem allow simple punctual measurements but are not adequate for a larger spatial resolution. Non-destructive probes, on the other hand, provide a better spatial resolution but in general do not show the state of mechanical stress. A solution for this problem could be an approach to the state of mechanical strength focusing on the stress-at-rest-coefficient, K0, in combination with various independently measured geophysical values like the apparent electrical conductivity (ECa), the electrical conductivity (EC) and the electromagnetic reflection (EMR). Firstly, we tested the validity of the underlying basic assumption that the horizontal stress component can be used to characterise the compaction state of the soil. This procedure consists of assigning penetration resistance (PR) values to the principal stress (sigma x) as a function of depth, normalised by a relation defined by the PR value (measured with a Penetrologger) of the greatest accessible depth (0.8 m, limited by the used probe). Secondly, we analysed the significance of ECa (with an EM38 probe), EC (done with a Veris 3100 sensor and the new developed 'Geophilus Electricus' sensor) and EMR (assessed with a ground penetrating radar) as non-destructive methods to regionalise inhomogeneities in soils. We took approximately 10,000 measurements at two test fields located in the loess belt of Northern Germany. The measurements were taken over a time period of 4 years at different times of the year, mainly at spring and autumn time. We detected various subareas according to the status of the mechanical strength in the subsoil, which affects e.g. bulk density and hydraulic properties of the soil. The results from the experimental sites showed a strong correlation between penetration resistance logger and the signal of the various geophysical methods, especially in the areas with higher PR values. This is mainly caused by a good response of the lateral water content in the significantly compacted subsoil between 0.30-0.40 m. The results lead us to the conclusion that in precision agriculture nearly every of the applied geophysically based methods and sensors can be used as detectors for the capture of subsoil mechanical strength in loess derived or comparable homogenous soils. Therefore the methodology introduced can be a useful tool for the regionalisation of subsoil compaction.

Comparison of different EC-mapping sensors

Lueck, E.[1] and Ruehlmann, J.[2], [1]University of Potsdam, Geosciences, Karl-Liebknecht-Str. 24, 14476 Golm/ Potsdam, Germany, [2]Institute of Vegetable and Ornamental Crops, Theodor-Echtermeyer-Weg 1, 14979 Großbeeren, Germany

Digital soil mapping becomes more and more important not only in Precision Farming. In the past miscellaneous technologies and sensors, among others geohysical techniques were tested to get information about the spatial variability of the soil properties. Especially the measurement of the elctrical conductivitiy (EC) may be useful for delineating heterogeneity in soil parameters like texture, water content as well as fluid conductivity, compaction and organic matter. Conductivity measurements can be performed either with galvanic or capacitive coupling electrodes or with help of electromagnetic method. In this context, the VERIS 3100 (veris technologies, USA), the ARP (Geocarta, France) and the EM38 (Geonics, Canada) are the most popular devices. All existing sensors differ not only concerning the working principle but also concerning the depth range and the depth sensitivity. We want to present a comparison of these instruments with our new soil mapping system – Geophilus electricus which was developed during the last few years. Measurements were performed at different scales and landscapes and were compared among each other as well as either with direct push data (measured directly within the soil) or with conventional geophysical investigation with fixed electrodes in standard arrays. The conductivity maps were quite similar. Because Geophilus electricus is capable to measure amplitude and phase for five channels simultaneously we get more detailed information also about the vertical structure and about layers within the investigated depth range.

Interaction of tillage direction and lag distances for directional semi-variograms of implement draught
Mclaughlin, N.B., Agriculture and Agri-Food Canada, Research Branch, 960 Carling Ave., Ottawa, ON, K1A 0C6, Canada

Zones of high soil strength can be identified from spatial tillage implement draught data collected in conjunction with normal tillage operations. High density draught data are typically noisy and require filtering with filter parameters adjusted according to the scale of variability. Previous research has shown significant differences in semi-variogram parameters for tillage implement draught data parallel to and perpendicular to the direction of travel. Some of the differences are likely due to natural anisotropy in soil parameters within the field, while others are due to the effects of micro-topography on tillage implement operating depth and narrow compaction zones from previous wheel traffic. Spatial data on primary and secondary tillage implement draught were collected in a large field over several years using different tillage directions. Data were normalized, averaged in a one meter grid, and directional semi-variograms were calculated. The lag distances of the semi-variograms parallel to the direction of travel representing the inherent large scale variability in soil strength were approximately the same for the different tillage directions, and were consistently much larger than lag distances of semi-variograms perpendicular to the direction of travel. The results clearly show that apparent anisotropy in draught data is strongly influenced by tillage direction changes. These results indicate that multi-directional tillage implement draught data collected over several years are required to obtain a true assessment of the scale of soil strength variability in different directions.

Mapping traffic patterns for soil compaction studies using GIS

Meijer, A.D.[1], Heiniger, R.W.[2] and Crozier, C.R.[1], [1]North Carolina State University, Soil Science, 207 Research Station Rd, Plymouth, NC 27860, USA, [2]North Carolina State University, Crop Science, 207 Research Station Rd, Plymouth, NC 27860, USA

Soil compaction is a common problem in the southeastern United States, and can lead to a host of soil and plant growth problems. Since one of the main causes of soil compaction is equipment traffic in fields, the idea of limiting traffic to certain rows can help alleviate the onset of soil compaction. The objective of this study is to determine the amount of traffic occurring in North Carolina fields and the effect that the level of traffic had on soil compaction as measured by soil bulk density. GPS was used to map all traffic on these fields in 2006. Using measurements of tread widths and wheel spacing, a series of processes in a GIS was performed to generate a map indicating the level of traffic that occurred in each area of the field. After all field operations were complete, fields were sampled for bulk density. Sample locations were based on the number of tire passes that had occurred. Samples were taken where there had been 0, 1, 2, and 4 passes of equipment tires. Initial results showed that 65-85% of the field's area was tracked at least once. Bulk density ranged from 0.5 to 0.8 g/cm3 in the organic soil, and from 1.6 to 1.8 g/cm3 in the sandy soils. Initial results show that in the organic soil, areas of the field that were tracked at least four times had significantly higher bulk density in the 0-10 cm depth than the areas that received no tracks. The study is continuing on research stations in 2008 and 2009 to better look at the influence of tire traffic on soil bulk density.

Searching for the cause of variety

Jukema, J.N., Wijnholds, K.H. and Berg, Van Den, W., Applied Plant Research, Arable farming and field production of vegetables, Edelhertweg 1, 8219 PH Lelystad, Netherlands

For many years research has been done for the use of remote sensing in nitrogen application rates. Worldwide but also in The Netherlands a lot of research has been done towards this topic. Good relations have been found regarding the need of nitrogen and various calculated indices. With that in mind it should be possible to save nitrogen, increase yield and improve the homogeneity of the harvested crops. Food chains are asking for more homogeneous primary products. With heterogeneous products adjustments have to be made to the production process and the process will be less efficient. In addition, the effects on our environment are also becoming more and more important. These put increasing limitations on modern agriculture. Site-specific farm management can optimize the input of nutrients to the need of the plants. With this in mind it seems a matter of time before these systems will be common practice. The fact that their have been found good relations between the need of nitrogen and various calculated indices you can't say that all the variation within the field is caused by nitrogen. Within the project 'Perceel Centraal', a project in association with Agrifirm, HLB, IRS, and around 30 farmers, research has been don to the causes of variety, charted by areal imaging, within the field. The inspected crops are spring barley, potatoes and sugar beets. The project is located in and around Valthermond, The Netherlands on sandy- and peaty soils. Research has been done at the experimental location PPO 't Kompas. This area is a part of the Netherlands with a lot of difference within the field. Discussing the Yara LORIS® 'biomass maps' with farmers pretty soon became clear that causes of the variation not only is induced by nitrogen, they also can be found in drought, structure, nematodes, organic matter, minerals, etcetera. This scale of causes is one of the reasons why farmers don't adopt nitrogen applications based on remote sensing immediately. Although it is clear that Nitrogen wasn't the only reason causing the variation within the field it obviously is one of the main reasons. To make the right decisions the biomass maps should be interpreted first and then a decision should be made how the handle the field. It turned out to be very difficult and time taking for farmers to find out what the reason is of the variety within their fields. Because of that it occurred that valuable information ended up in the bookcase. To get the maximum value out of the biomass maps the project 'Perceel Centraal' developed a checklist for cereals, potatoes and sugar beets with which the cause of the infield variation can be found easily. After using the checklist the right choices can be made concerning the optimization of the farmers fields.

Delineation of site-specific management zones using geostatistics and fuzzy clustering analysis
Castrignano, A.[1], Fiorentino, C.[1], Basso, B.[2], Troccoli, A.[3] and Pisante, M.[4], [1]CRA, IS AGRO, Via Ulpiani, 70125, Italy, [2]University of Basilicata, Crop Systems, Forestry, Environmental Sciences, Via Ateneo Lucano, 10, 85100 Potenza, Italy, [3]CRA, Cereal Research Centre, SS 16, 75100 Foggia, Italy, [4]University of Teramo, Food Science, Via C.R. Lerici 1, 64023 Mosciano S.Angelo TE, Italy

Rationale Site specific management enhances natural resources, minimizing environmental impact and increasing economic returns for farmers. Different criteria for delineating management zone (MZ) have been proposed. Such delineation involves some sort of clustering, and previous study demonstrated that there is not a single or widely accepted method. In most cultivated fields, soil and crop may vary gradually rather than abruptly, therefore in such conditions it is very difficult to unambiguously draw the borders of MZ. The application of fuzzy set theory to clustering algorithm has allowed researchers to better account for the continuous variability in natural phenomena. Similarly, Geostatistics treats multivariate indices of soil and crop variation as continua in a joint attribute and geographical space. The objective of this paper is to compare different procedures for creating MZ and for testing the question of the number of MZ to create. Methodology One hundred georeferenced measurements of soil texture, vegetation hyperspectral data and crop (LAI, biomass at harvest, gluten content of grain) properties were collected on a 12-ha field cropped to durum wheat in three seasons (2005-2008). The trial was carried out in the experimental farm of CRA-CER (Foggia, south-east Italy) and the locations of the samples were chosen so that they were evenly distributed on the field. The techniques compared were: the iterative Self-Organising Data Analysis Technique (ISODATA), one of the most widely used unsupervised clustering algorithms; the fuzzy c-means algorithm, which uses a weighting exponent to control the degree to which membership sharing occurs between classes; a multivariate geostatistical approach, referred as factor kriging, according to which any soil and crop attribute is considered as a random regionalised variable, varying continuously and its gradual geographical application is described by a covariance function and finally, a combined approach based on multivariate geostatistics and a non-parametric density algorithm of clustering. Results All the methods produced consistent results, creating the subdivision of the field into 2 or 3 distinct classes of suitable size for uniform management. The clusters differed mainly in the textural properties and the southern part of the field was more productive. Each method shows advantages and disadvantages: fuzzy c-means algorithm provides the user with two performance indices useful for deciding how many clusters are most appropriate for creating MZ. However, it does not take into account spatial dependence between samples and in some cases the two performance indices may be dissimilar. Therefore, the final decision of how many clusters to use for creating MZ may require a combined use of different clustering algorithms.

Mapping spatial variation in growing willow using small UAS

Rydberg, A.[1], Hagner, O.[2] and Aronsson, P.[3], [1]JTI-Swedish institute of agricultural and environmental engineering, Box 7033, S-750 07 Uppsala, Sweden, [2]Swedish university of agricultural sciences, Department of Forest Soils, Box 7001, S-750 07 Uppsala, Sweden, [3]Swedish university of agricultural sciences, Department of Crop Production Ecology, Box 7043, S-750 07 Uppsala, Sweden; anna.rydberg@jti.se

The attempt to reduce the usage of fossil fuels has lead to an increased usage of biofuels. Even though willow is a crop that generates a high energy exchange, and that is profitable in relation to many other dedicated energy crops, there has been a relatively low interest among famers to grow willows. This has to do with factors related to the longer time between planting and harvest for willow compared to annual crops. This means longer time before any income can be secured and lower flexibility to adjust cropping to fluctuating market prices. The low interest is expected to remain if not the profitability of willow growing is considerably increased. In order for the profitability to increase, either the expenses have to be significantly reduced or the price of the product has to increase considerably. This would be possible to obtain if the grower could guarantee a product of a certain amount and quality, and at a certain time. This requires a possibility to estimate standing biomass in willow plantations, among other factors. The objective of this study is to evaluate if new remote sensing technique using a small Unmanned Aircraft System (UAS), can be used as a cheap and reliable method for estimating standing biomass in willow plantations in an accurate and efficient way. This study, which is part of a larger project evaluating different ways of estimating standing biomass in willow plantations using different remote sensing techniques, evaluates if spatial and spectral information from an UAS can describe and quantify spatial variations of standing biomass in willow plantations. A small hand-launched UAS - Personal Aerial Mapping System (PAMS) equipped with an autopilot was used to systematically acquire imagery of six willow fields from 200 m altitude at two different occasions. The system software was subsequently used to automatically produce orthophoto mosaics with high spatial resolution. The fields were selected to have different willow varieties and different degrees of homogeneity. The occasions were selected according to when the green intensity of spectral reflectance from willow leaves is expected to be the most (May) and when the leaf area is expected to have reached maximum (August). The aerial images were analysed manually and automatically, and compared with ground measurements from field inspections. It is possible to differentiate between the different varieties and to map areas with different homogeneity with high precision using the orthophoto mosaics images generated with the UAS.

iSOIL: exploring the soil as the basis for sustainable crop production and precision farming

Van Egmond, F.M.[1], Nüsch, A.-K.[2], Werban, U.[2], Sauer, U.[2] and Dietrich, P.[2], [1]The Soil Company, Kadijk 7b, 9747 AT Groningen, Netherlands, [2]Helmholtz Centre for Environmental Research – UFZ, Dept. Monitoring & Exploration Technologies, Permoserstrasse 15, 04318 Leipzig, Germany; egmond@soilcompany.com

Precision farming is based on the concept of optimizing crop production and increasing profit margins by taking into account the spatial within-field variation in soil, soil moisture and other environmental conditions that influence crop growth. To be able to adjust crop and soil management to this natural variation, the type and extent of the variation should be known to the farmer. Ideally in a format that enables him and his machinery to adapt his crop management to the relevant variation in such a way that he can optimize his crop production process in an easy and semi-automated way. This requires reliable and coherent high resolution digital maps of relevant variation such as soil texture, soil organic matter, soil nutrients, soil moisture, slope, aspect, compaction (risk), erosion risk. The focus of the iSOIL project is on improving fast and reliable mapping of soil properties, soil functions and soil degradation threats. This requires the improvement as well as the integration of geophysical and spectroscopic measurement techniques in combination with advanced soil sampling approaches, digital soil mapping and pedophysical approaches. The outputs of iSOIL encompass methodologies, reliabilities, standards and possibilities for mapping all the above mentioned and can therefore be of great value to precision agriculture. The project aims to supply tools and standards for mapping e.g. to precision agriculture end-users to enable fast, easy and reliable mapping of all relevant soil, hydrological and environmental parameters. The resulting soil property maps can e.g. be used for precision agriculture applications such as variable planting distance of potatoes, variable liming and variable fertilizer application. The soil degradation threats studies, e.g. erosion, compaction and soil organic matter decline can be used to improve the sustainability of current farming practices to ensure fertile and sufficient quality lands for crop production in the future.

Comparision of yield mapping methods and an analysis of rainfall effects on temporal variation

Rogge, C.B.E.[1,2], [1]Technical University of Munich - Weihenstephan, Department of Life Science Engineering, Am Staudengarten 2, 85354 Freising, Germany, [2]Cranfield University at Silsoe, Cranfield University at Silsoe, MK45 4DT Silsoe, Bedfordshire, United Kingdom

The aim of this paper is to determine robust statistical methods for providing management data for the spatial and temporal variability in yield maps. Therefore alternative methods for the analysis of yield data were reviewed. The study was based on previous doctoral thesis by Blackmore (Cranfield University at Silsoe) and Steinmayr (TUM Weihenstephan), which has its origins in the work of Blackmore. The common procedure in yield mapping is a two step process. In the first step the data are corrected, i.e. erroneous data strings are either corrected or removed from the data set. As the methods usually only differ in the data correction process, a closer study is made of the sources of errors. In the second step the actual yield map is usually produced by interpolating. Here the most commonly used methods are Kriging and Inverse-distance. Three different correction algorithms were compared; namely the expert filters of both Blackmore and Steinmayr and Noack's H-method. These were investigated to determine the amount of data points that are simply removed and how many are corrected. To analyse spatial and temporal variability, data sets from both TUM and Cranfield University were used. In his first analysis Blackmore found temporal stability in the spatial variation of yield with time, however as a result of further analysis he found that there was greater temporal instability, suspecting temporal variation in rainfall to be the cause. Steinmayr, who further developed Blackmore's technique, also found stability in the German data. Different analyses from many different countries indicate that water is the relevant factor. Stability is high for example in dry regions such as Australia and under irrigation conditions such as centre pivot irrigation systems in the US, but is more variable in temperate regions with variable weather patterns. It has therefore been decided to add an intermediate step to the process described above. The concept formulated in this study was to separate data sets from 'dry' and 'wet' years in order to analyse the temporal trends. The results of this showed that the standard deviation of yield on average decreases after the separation.

The economic potential of precision farming: an interim report with regard to nitrogen fertilization
Wagner, P., University of Halle, Farm Management group, Luisenstrasse 12, 06099 Halle, Germany

The idea of managing the in field heterogeneity is not new. Nowadays, different sensors and actuators are available to record and to react according to the sub field circumstances. Furthermore, some decision rules, which give the recommendations based on the sensor information, are also available. For example, for the site specific nitrogen fertilization, different approaches can be used for a better management of the grain production management. Despite the fulfillment of these prerequisites, the practical utilization of these site specific management tools is quite low. The paper tries to answer the question why acceptance is such low and what has to be done to increase acceptance. Materials and methods: On the basis of two different locations in Germany the paper will show the actual economic potential which can be achieved by the farmer using existing precision management tools. The example of site specific nitrogen fertilization was chosen. Various field trials during the last three years were conducted to compare different site specific decision rules (mapping approach, sensor approach and decision rules generated by means of an artificial neural net) with the uniform field treatment. For this purpose, strip trial designs on field scale were used to take all concerned factors (e.g. the accuracy of practical farming technique) into account which are influencing the economic potential. To examine only the influences caused due to the site specific management, the field trial data were exploited with a geostatistical approach. Results and conclusions: The results of the field trials were very differing. On the on hand Precision Farming advantages are to be observed, on the other hand even Precision Farming losses. The outcome depends on the a) used decision rules (quality of the decision rules) and b) in-season weather conditions. While the utilization of the 'mapping-strategy' shows almost no positive results, the sensor approach shows sometimes negative, sometimes positive results. Only the 'neural-net-approach' always shows positive results (in every year and in all locations). In fact, the last mentioned approach needs the most spatial high resoluted information in order to generate its decision rules. Obviously, collecting and processing information is worthwhile. Especially the (most successful) neural-net-approach will be described in detail. On the basis of these field trial results and additional historical weather data, calculations of the risk will be carried out to examine the economic potential of this site specific technology under dynamic aspects. Furthermore, some reflections will be made with regard to the requirements of the further development concerning the decision rules for Precision Farming.

Optimal path planning for field operations

Hofstee, J.W.[1], Spätjens, L.E.E.M.[2] and IJken, H.[1], [1]Wageningen University, Farm Technology Group, Bornsesteeg 59, 6708 PD Wageningen, Netherlands, [2]Agrovision, Postbus 755, 7400 AT Deventer, Netherlands

Many farmers in the Netherlands rent each year new fields to grow potatoes or flower bulbs. With these fields they have no experience from the past and have to do a path planning before starting the work. Most fields are not rectangular but have non-parallel sides, are L-shaped, or have obstacles within the field and it is not always easy to make plan. Contractors face the same problem when they arrive at the field of the customer and have to do path planning immediately. In the future we have to deal with the same problem with autonomous vehicles for field operations. They need to have a plan; there is no driver to guide the vehicle over the field. So there is a need for a tool for path planning of agricultural operations. The objective of the work described in the paper is the development of a tool for path planning of field operations. This tool has to consider the shape of the field and the required operations and then has to determine what the most optimal path will be. The tool is developed with Matlab. It reads the coordinates of the border of the field. These coordinates are processed and only the coordinates of real corners are kept. For this field an optimal path has to be found. In general fields that have a trapezoidal shape are the most the efficient to farm. The developed tool is based on that only convex fields are farmed. Concave fields are split into smaller convex fields. For each concave corner split lines are projected parallel to each side of the field. All these split lines are ordered and, starting with the field limited by the first split line, the field grows by adding subfields as long as the shape of the new field stays convex. This results in a convex field and a remaining subfield. If this subfield is concave the procedure has to be repeated until all subfields are convex. Operations for each subfield are only planned parallel to each side of the subfield; random directions are not considered. For each direction the costs for the operation are calculated. The costs are based on working speed, working width, a fixed time for turning at the headlands, and fixed hourly costs for the operation. Working speed and width can be set and depend on the operation. The costs base makes it possible to add costs for overlap of operations, additional travel time, or the decision to not farm small inconvenient subfields. In our case we have chosen to use spraying a potato crop as operation since farmers have to spray their potato crop several times during the growing season and an optimal path for this operation is expected to be the most profitable. This also means that other operations as planting have to be derived from the spraying operation. The developed tool is tested for several real fields. For regular fields the resulting optimal path is as expected, parallel to the longest side. For more irregular fields the calculated optimal path seems to be not always the optimal path. This is also a basis for rethinking the way field operations are executed nowadays; they are not necessarily optimal. In some cases the non-optimal path is caused by that at this moment no costs are accounted for overlap or for additional headlands with no crop required when the working direction is not the same for parallel subfields, causing turning in a cropped field. While working on the tool it became clear that there are a large number of factors and boundary conditions that have to be taken into account for a final tool.

Combined coverage path planning for field operations

Bochtis, D.D.[1] and Oksanen, T.[2], [1]AUTH, Agriculture Engineering, Kath. Rossidi 37, 54655 Thessaloniki, Greece, [2]TKK Helsinki University of Technology, Automation and Systems Technology, P.O. Box. 5500, Otaniementie 17, 02015 Hut, Finland

Searching an optimal path to cover the whole field to fulfill certain operation has been found to be extremely difficult problem to solve mathematically or algoritmically. However, a need for such algorithm or method has been seen during last years as the parallel tracking devices has become more common in tractors and other agricultural machines. Even if mathematically the most optimal path is not found, an automatic algorithm that gives feasible routes can improve efficiency of agricultural field operations. Area coverage planning determines a path that guarantees that a machine will pass over every point in a given field without overlaps or missed areas and all obstacles must be avoided. This includes the next three procedures 1) Decomposition of the coverage region into sub-regions, 2) Selection of a sequence of those sub-regions and finally, 3) Generation of a path that covers each sub-region. During the last year, two researches regarding the field coverage planning for agricultural machines have been independently presented, which by coincidence are complementary to each other. The first approach results the optimum decomposition of a complex-geometry field into sub-fields and the optimum driving direction in each field. The second one presents an algorithmic approach towards computing traversal sequences for parallel field tracks, which improve the machine's field efficiency by minimizing the total non-working distance travelled. In this paper, a first attempt to connect the two approaches is presented, in order to provide a complete method for field area coverage planning that is directly applicable to autonomous agricultural machines. Examples of optimal planning for a number of given fields are given. In a first stage, using 'prediction' and 'exhaustive search' methods, each field is divided into sub-fields and for each one of them the driving direction is determined. In a second stage, the optimal sequence that the machine visits the sub-fields is determined and the optimal field-work pattern is produced for each sub-field. For this stage, a combinatorial optimization algorithm was used. This algorithm is based on 'heuristics operations' and has been developed for the special case of agricultural operations. In a third stage, the planning is completed by using a new method that provides a point to point motion planning for the most common agricultural maneuvers.

From sensor values to a map: accuracy of spatial modelling methods in agricultural machinery works
Kaivosoja, J., MTT Agrifood Research Finland, Plant Production Research, Vakolantie 55, 03400 Vihti, Finland

Precision agriculture (PA) is based on the knowledge of in-field variability. In PA actions measured field data go always through spatial interpolation and modelling procedures. Errors, inaccuracies and generalisations in these procedures can cause systematic errors and can lead to cumulative mistreatments in PA. The main error sources at a spatial modelling perspective are positioning accuracy and the temporal linkage between sensor value and the corresponding specific work area at a specific moment. Factors such as posture determination, system and sensor delays, spread pattern changes and a lack of differential gear functioning effects mostly on the correspondence. Also the division between sensor filtering and spatial filtering can be unclear. The aim of this study was to determine and to evaluate the extent of these error factors. By knowing the errors caused by spatial interpolation and modelling, the development requirements for modelling techniques and their relevance can be determined and analysed. Most common field farming machines including harvesters, seeding machines, harrows, sprayers and spreaders were examined and analysed from the spatial modelling perspective. Interacting factors were determined and classified. Near optimal spatial interpolation and modelling methods were constructed for exemplary machine types. The effects of the factors were simulated and determined using simulated or previously driven and collected field data. Results were accomplished by incorporating or leaving out different factors, and also focusing on present common spatial modelling practises. In-depth studies focused on a combine harvester and a fertilizer spreader. A combine harvester was chosen because of its popularity in PA. The fertilizer spreader was chosen due to most complicated spatial modelling requirements. This study gives new information about mobile machinery's spatial modelling and its accuracy. The information can be compared to other machines input or output error sources such as measuring technology and sensor accuracy. Also proportion with biological knowledge can be made. Positioning errors may emerge as overlaps and gaps and for that matter is a crucial source of error for in-field variability information. There are also external variables such as wind, soil structure, soil erosion and water management, which change the machine's spatial input over a time or during work. Providing faultless information for PA usage is still far away, but knowing better the effect of the spatial modelling is a step forward.

Field-scale model of the spatio-temporal vine water status in a viticulture system
Taylor, J.A.[1], Tisseyre, B.[2] and Lagacherie, P.[1], [1]INRA, UMR LISAH, Bat 21, 2 Place Pierre Viala, Montpellier 34080, France, [2]Cemegref-SupAgro Montpellier, UMR ITAP, Bat 21, 2 Place Pierre Viala, Montpellier 34080, France

Monitoring vine water status both temporally (through the season) and spatially is of importance for growers and winemakers. The vine water status is critical at all stages of production as it impacts on vine vigour, yield and harvest quality. Correct and timely knowledge of vine water status should be the basis for any within season management aimed at optimising grape yield and quality. The opportunity to manipulate management is particularly relevant in irrigated vineyards where growers have some control over soil moisture conditions.However, the plant water status is a parameter which varies significantly over time depending on the climate of the year. It also shows significant spatial variability within vineyards and vineyard blocks depending on soil (landscape) spatial variability. It has therefore traditionally been a very difficult and costly parameter for growers to measure and manage. As a consequence, studies in this area have tended to be conducted at either very broad scales (regional) or very fine scales (plant-specific).The lack of vine water status data for decision support at the vineyard level is the key driver for a new project at INRA Montpellier (begun July 2008). The project aims to create a spatio-temporal model that is able to merge the existing fine-scale and broad-scale vine water status models/data with other external ancillary and model data to assess the temporal and spatial changes in vine water status at scales comparable with production (from vineyard blocks to winery cooperative areas).This paper will present the first stage of this work addressing identification of a suitable methodology for the spatio-temporal model at the field-scale. For two vineyard blocks, near Narbonne, France, vine water status has been measured at 49 sites per field across two seasons (5-6 times per season). A simple linear model has been developed to interpolate results of the measurements between sites, however this requires the initial measurement of vine water status at a large number of sites within the field. Ancillary information on apparent soil electrical conductivity, multi-spectral and thermal imagery of the canopy, terrain attributes and existing soil maps are available. The main hypothesis of the work is that these data can be used to minimize the number of samples required to create surfaces (maps) for field management. The suitability of regionalised autoregressive exogenous variable analysis, Bayesian maximum entropy analysis and spatio-temporal kriging for mapping the data will be assessed to identify a) which is the best model, b) what are the best predictors within the model and c) what is the relationship between output quality and sampling density.

Hydropedology and pedotransfer functions (PTF)

Zacharias, S., Helmholtz Centre for Environmental Research – UFZ, Dept. Monitoring & Exploration Technologies, Permoserstrasse 15, 04318 Leipzig, Germany

It is an always growing insight in the research and science community that interdisciplinarity and a perspective related to the landscape scale is necessary if cross-cutting issues like watershed-management or different agricultural management practices will be examined. To meet the challenges which result from questions related to soil and water at varying spatial and temporal scales – bridging traditional pedology, hydrology and soil physics as well as other related disciplines is absolutely necessary. In the last years the new concept of hydropedology- the combination of knowledge and expertise in the fields of pedology, soil physics and hydrology - attracts a wide interest. The 'Critical Zone'-Concept as a system concept which covers the subsurface environment (root zone, vadose zone, groundwater zone) and the land surface including vegetation as well as surface water bodies provides a sustaining framework for interdisciplinary research on the terrestrial environment. Simulation models are an important tool in evaluating environmental effects. In soil science related models the required information are often derived through expensive laboratory measurements. Often large numbers of samples are needed to cover the temporal and spatial variability of these properties. Using the relationship between soil hydraulic parameters and properties, which are easier to measure, pedotransfer functions (PTF) represent an effective way for model parameterization. Apart from the advantages of PTFs (easy to derive, easy to use, inexpensive) there are also some problems. So often the representation of soil structure induced effects is poor. In the presentation advantages and disadvantages of PTF will be discussed. Examples from the own work represent the use of PTFs in the context of hydrological modelling. A careful selection or construction of PTFs offers an effective way towards improved model parameterization and model prediction.

A multi agent simulation approach to assess agronomic income sources in the North China Plain: case study for Quzhou county

Roth, A. and Doluschitz, R., Computer Applications and Business Management in Agriculture, Farm Management, Universitaet Hohenheim, Schloss, Osthof-Sued, 70599 Stuttgart, Germany

The North China Plain (NCP) covers an area of around 328,000 km^2 and is one of the most important regions of cereal crop production in China. Wheat and maize rotations and one season cotton are the most common cropping systems. The region contributes at an amount of about 50% to the countries wheat production and about one third of maize yields. Crop production in the NCP was focused in the last decades on increasing yields to meet the growing food demand accompanied by the limitation of arable land as a result of urbanization rate i.e. of the Beijing District. Food production needs can nowadays only be achieved by the optimization of agricultural management, i.e. fertilizer input, irrigation, improved crop rotations. The focus on increasing yields raised serious environmental problems, like water shortage and pollution, air pollution and soil contamination. Hence the development of future land use system approaches improving these conditions is essentially. This may provide both a high production level as well as a protection of resources. The multidisciplinary collaborative International Research Training Group project (IRTG) 'Modelling Material Flows and Production Systems for Sustainable Resource Use in Intensified Crop Production in the North China Plain', funded by the Deutsche Forschungsgemeinschaft (DFG) and the Chinese Ministry of Education, was launched to detect the potential of adjustments in cropping systems and to further develop management practices for sustainable resource use and protection of environmental conditions while assuring a high yield level. The here presented subprojects work concentrates on the construction of a modelling framework of different spatial-temporal scales in order to regionalize the detected effects of changing land use patterns. A huge number of software and tools are used in socio-economics and ecosystem modelling approaches. Hence actually the author evaluates different approaches on their applicability and the reliability of expected outcomes. Primarily we want to prove on province level the concept of Cellular Automata (CA). In order to parameterize transition matrices field based statistical analysis of household survey data is used. Simulations, i.e. of farmer income evolution are planned to be obtained by transition rules for different cell stages. The probabilistic infinity to create cells even in hierarchical order provides the opportunity to regionalize simulations. Next we evaluate the applicability of the crop model DSSAT (Decision Support System for Agricultural Transfer) to simulate and regionalize water use efficiency, fertilizer efficiency and yield under iterative management scenarios. With its GIS extension AEGIS/WIN for the evaluation of spatial probabilities of future development chains for the above mentioned scope areas research it provides an ideal analysis tool. We present primarily analyses of spatial dependences of farmer income on province level in the North China Plain. These outcomes are planned to be constructed by a spatially GIS sourced Cellular Automata based on the project GIS database Agro-Environmental Information System (AEIS).

Sensitivity to climate change with respect to agriculture production in Hungary

Erdélyi, É., Corvinus University of Budapest, Department of Mathematics and Informatics, Villányi út 29-43, 1118 Budapest, Hungary; eva.erdelyi@uni-corvinus.hu

It is evident that global warming is one of the most serious problems we face in the 21st century. Using geographical analogues Gaál and Horváth showed that the possible future climate in Hungary - predicted by the scenarios - would be similar to the present climate of South-Southeast Europe. We have proved that the risk of maize and winter wheat production in Hungary have increased independently to the risk aversion of the decision maker. In search of the reasons, we analyzed the temperature and precipitation needs of the plant in each phenological phase. In this paper we investigated the effects of climate change on the growing period and production of maize and winter wheat, as the two most cultivated plants in Hungary. Debrecen, the basic object of our calculations is an important centre of agricultural production in Hungary, so we would like to interpret the results in this aspect. Climate scenarios can be defined as relevant and adequate pictures of how the climate may look in the future. During our research, we applied the principles defined by IPCC and we used some of the most commonly accepted scenarios: scenario BASE with the parameters of our days; scenarios created by Geophysical Fluid Dynamics Laboratory, GFDL2534 and GFDL5564; UKHI, UKLO and UKTR scenarios worked out by United Kingdom Meteorological Office. For comparison we used the historical data of the climate scenarios reference period. In our crop model research we used the simulation method by the 4M model. It has been developed by the Hungarian Agricultural Model Designer Group and based on the CERES model, adapted to hungarian circumstances. The simulations were run for meteorological data forecasted by climate scenarios and downscaled to Debrecen, as weather inputs. Climate change affects agriculture in many direct and indirect ways. We examined the frequencies of extreme temperature values during the growing season. We also analyzed the effects of changing climate on the growing periods of corn and winter wheat. We can expect increasing accumulated heat in the vegetation period in the future, so we investigated how the lengths and the starting points of the growing periods of the plants change. It can be said that phenological phases of corn and winter wheat have shortened and happened earlier as a result of temperature increase, especially in the first periods of growing. Harvesting is predicted to be a week earlier in the future. We show the results by comparing the historical data and the climate scenarios. Finally we used the model for finding an adaptive strategy for increasing the yield. For both biomass and grain mass quantity the simulation results are very promisin. for both plants. Modeling is a great tool for investigating the future circumstances without having expensive and long experiments. It can help us find a good strategy in preparing for the future. There is a wide scientific consensus that if these changes continue, significant damage to global ecosystems, food production and economies will ensue. Interdisciplinary, collaborative research projects are very much needed all over the world.

Rotation and the temporal yield stability of landscape defined management zones: a time series analysis

Grove, J.H.[1] and Pena-Yewtukhiw, E.M.[2], [1]Univ. of Kentucky, Plant and Soil Sciences Dep., N122 ASCN, Lexington, KY 40546-0091, USA, [2]West Virginia Univ., Div. of Plant and Soil Sciences, P.O. Box 6108, Morgantown, WV 26506, USA; Eugenia.Pena-Yewtukhiw@mail.wvu.edu

Yield is often used to delineate management zones, but this approach is confounded by yield's weather dependence, causing yield to evidence temporal variability. The term 'yield stability' is used to describe this variability. Agronomic management (genetics, nutrition, irrigation, etc.) options also influence yield stability, and among these, crop rotation is one of importance. Our objective was to describe the influence of crop rotation on the temporal yield stability of landscape defined management zones. The 1987 to 2007 harvests (21 years) from a large rotation study established at the University of Kentucky research farm near Lexington were used. Yield data for two corn (*Zea mays* L.) rotations, monoculture corn (C-C) and corn alternating yearly (W/S-C) with winter wheat (*Triticum aestivum* L.)/double-crop soybean (*Glycine max* L. Merr.), were taken over four landscape position-management zones (shoulder, upper backslope, lower backslope and footslope). Yield data were analyzed: by ANOVA to partition yield variance due to time, space and zone; by Spearman rank correlation to find the relative stability of management zones, by rotation; and by time series analysis, after linear detrending, using Box-Jenkins methodology to give an autoregressive/ moving average model. The 21-year average yields were lower for C-C (8.32 ± 2.56 Mg ha^{-1}) than for W/S-C (9.62 ± 2.68 Mg ha^{-1}). Yields for W/S-C were greater in all zones. The shoulder exhibited lowest yields (8.35 ± 2.90 Mg ha^{-1}), while the footslope exhibited greatest yields (9.31 ± 2.44 Mg ha^{-1}). Spearman rank correlations by management zone, within a given rotation, were high (above 0.80). With C-C, the shoulder and lower backslope exhibited lower rank correlations with other zones, being most dissimilar, while the upper backslope and footslope were most similar. In the W/S-C rotation, rank correlation analysis found the shoulder most dissimilar, while other zones were most similar. Choice of rotation impacted the relative relationships among landscape zones as regards yield stability. A plot of yield versus time for individual landscape zones exhibited slightly linear upward trend (average slope of 0.13 Mg ha^{-1} yr^{-1}) in both rotations. There was a greater slope in the W/S-C trend yield for all but the shoulder zone. After removing linear yield trend, Box –Jenkins time series analysis found that present year C-C detrended yield, in three of four zones, was positively related to the yield observed three years before. However, for the upper backslope, detrended yield was positively related to the yield observed five years before. The greater temporal 'lag' indicates that the upper backslope was least stable, in this rotation. Detrended W/S-C yield was related to yield observed only one to three years prior, indicating this rotation has been more 'yield-stable' during the 21-year experiment, though this was not equally true across the four zones. Shoulder and lower backslopes exhibited similar temporal yield behavior with yield autocorrelation at a three-year lag, regardless of rotation, but upper backslope and footslope were significantly more stable in the W/S-C rotation, with a yield autocorrelation at only one- or two-year lags. These 21-year models indicate that similar management zones would require a minimum of three years of previous yield data in order to forecast management zone yield response behavior.

Bioclimatic modelling of the future ranges of crop species: an analysis of propagated uncertainty

Marinelli, M.A., Corner, R.J. and Wright, G., Curtin University, Department of Spatial Sciences, Western Australia; R.Corner@curtin.edu.au

The agriculturally significant region of south west Western Australia, has a 'Mediterranean' climate and its agriculture is currently characterised by the rainfed broad-acre cultivation of cereals and pulses. The rainfall in this region has been trending downwards over the last 30 years and much agriculture is already practised close to climatic limits. It is inevitable that under predicted future trends the range in which crops are grown will need to be reconsidered. One way of simulating this is with the use of bioclimatic models such as BIOCLIM. BIOCLIM is a probabilistic model that can be used to investigate and predict species distributions for both native and agricultural species. Its results have greatest validity when studying relatively large (subcontinental) areas. The version of BIOCLIM used in this study uses three basic spatial climatic inputs (monthly maximum and minimum temperature and precipitation layers) and a dataset describing the current spatial distribution of the species of interest. Our work has investigated how uncertainty in the input data propagates through to the estimated spatial distribution for Field Peas (Pisum sativum), using as current species distribution a dataset of successful field plot tests carried out across the study area. In the case of future distribution modelling the input climate data layers obviously contain uncertainty, difficult to quantify, which will vary depending on the climatic modelling and future emission scenarios used. However the uncertainty in 'current condition' is more readily quantifiable and we have identified two components to it. Such spatial data sets are frequently the result of both spatial and temporal averaging. In the study area this averaging has been carried out using a sparse network of climate stations over a period during which climatic variables have been showing distinct trends. The maximum possible extent of this uncertainty has been quantified as follows. Firstly the difference between the decadal and 50 years means for all the valid data points used in the generation of the climate data surfaces was determined by examination of the source data (the Global Historical Climatology Network data base). Secondly the errors in spatial interpolation were quantified. Uncertainty propagation was carried out using Monte Carlo methods with up to 4500 simulations being run for any scenario. In addition we have investigated the effect of different frequency distributions for the input uncertainty Our results clearly show the effect of uncertainty in the input layers on the predicted specie's distribution map. In places the uncertainty significantly influences the final validity of the result. Work continues with the intention of incorporating potential uncertainties in future climate predictions.

A framework for motion coordination of small teams of agricultural robots

Vougioukas, S., Aristotle University of Thessaloniki, Agricultural Engineering, Aristotle Univ. Campus, Faculty of Agriculture, 54124 Thessaloniki, Greece

The current trend in agricultural production is to use larger, heavier, high-efficiency agricultural machines. For example in large-scale grain harvesting, teams of large harvesters, unloading carts and transport trucks are used to carry out the operation in reasonable time. An alternative approach for agricultural production is to use teams of smaller autonomous machines. The prospect is that with proper operations planning and coordination such teams could achieve high collective work rates with greater redundancy and competitive cost. Consider a team of R tractor-robots operating in the same field. Each robot has a desired sequence of waypoints to visit.and is equipped with its own tracking controller which computes an optimal M-step-ahead control sequence based on Non Linear Model Predictive Control (NMPC). The rth vehicle's planned optimal trajectory is computed by using the optimal control sequence in the state equation. Motion coordination is necessary to avoid collisions between robots with intersecting trajectories, but it is also necessary for cooperation between robots. In this work a distributed control approach is adopted. Each robot trajectory is broadcasted to all other vehicles at each time instant over a high speed Wireless Local Area Network (WLAN). Motion coordination is achieved by changing the NMPC cost function and constraints of the corresponding tracking controllers. In 'master-slave' mode the robot-master can change directly components of the slave's tracking controller such as the desired trajectory of the slave robot, or its cost function, or constraints. In 'peer-to-peer' mode each robot has a 'rank' and may affect another robot's motion by introducing weighted terms in its cost function. The weights are proportional to the rank. For example, if two robots are in a head-on collision course, each robot will introduce a proximity penalty term in each other's NMPC cost function. However, the bigger machine would have a higher rank and thus would cause the smaller machine to deviate much more from its original path. In cases of equally ranked robots, oscillations could occur and hence a random temporary re-assignment of ranks is used. Due to numerical performance constraints, as well as practical communication constraints such an approach could only be practical for small teams of robots. Simulation experiments in C++ will be presented which support the validity of this approach. Collision avoidance will be shown for small robot teams turning at the headland of a field and motion coordination will be demonstrated for a robotic harvester and loading truck case.

Model-based loading of agricultural trailers

Happich, G., Lang, T. and Harms, H.-H., Technische Universität Braunschweig, Institute of Agricultural Machinery and Fluid Power, Langer Kamp 19a, 38106 Braunschweig, Germany

There is a trend in agricultural engineering towards high-performance harvesting machines with growing operating width and throughput. As much as performance and throughput are rising, the transportation units, usually tractor-pulled trailers, are characterized by increasing transportation volume. If harvesting and transport are combined in parallel operation (e.g. self-propelled forage harvester), the driver of the harvesting machine as well as the driver of the transport unit has to pay a high degree of attention to the loading process. But losses of harvesting goods caused by missing the trailer have to be kept at a minimum, the complete transport volume should be utilised and collisions between the involved machines have to be avoided. Overloading processes with large-scaled machinery mostly imply that the visibility into the transportation unit is severely limited. The main aim of this research project is to develop and analyse several model based loading strategies exemplified on a forage harvester and a corresponding transport unit. In a former project the forage harvester had been used as the prototype for developing a GPS-based position control of the spout. The model based loading means an enhancement of the automation of the loading process. First objective of this research project is that a software model of the heap of bulk goods during the overloading process has to be developed. Basal analysis on heaps of agricultural goods like grass and maize silage are essential. By combining the software model, the space model of the transportation unit and the throughput, the current status of loading is predictable and different loading strategies can be spotted, tested and scrutinized with regards to efficiency and the facilitation of work. The research project started in January 2007 and will run for 3 years. The general Setup of the project, the principles as well as the latest results concerning the softwaremodel will be presented.

Procedures of soil farming allowing to reduce compaction

Kroulik, M.[1], Loch, T.[1], Kviz, Z.[1] and Prosek, V.[2], [1]CULS Prague, Department of agricultural machines, Kamycka 129, 16521 Prague, Czech Republic, [2]CULS Prague, Department of Machinery Utilization, Kamycka 129, 16521 Prague, Czech Republic; kroulik@tf.czu.cz

Introduction Evaluation of new technologies in agricultural machinery guidance is very important and can help producers choose the right equipment for their applications. The GPS guidance systems are used almost in all field operations. In relation to the soil protection against an erosion and a compaction others possibilities are looked for guidance systems utilisation in this time. The passages across field are necessary but an organisation of them is mainly random. Compacting the soil by traffic reduces the soil infiltrability, hydraulic conductivity, porosity, and aeration and increases bulk density and impedance for root exploration. In Czech Republic, around 40% of the arable soil stands in danger of soil compaction. The use of a technology named controlled traffic may minimize or eliminate the need for deep tillage or subsoiling, since CT is based on maintaining the same wheel lane for several years. The CT was primarily suggested since soil compaction caused by heavy agricultural machinery has become an acknowledged problem in reducing agricultural yields. The guidance systems accuracy promotes controlled traffic farming system expansion. It is one of the equipments which can be used in the soil protection again the compaction. Objectives In order to gain enter data for further CT systems observation, several measurements concerning frequency and total area of machinery passages in a field were done. A frequent problem is missing or overlaps between two adjacent tracks with broad implements. Material and methods DGPS receivers were placed into a machine for monitoring of all machinery passages across observed fields with 2 s logging time for position data saving. All field operations in the particular field were observed for 3 variants of tillage systems during one year. Conventional system with ploughing, conservation tillage and direct seeding systems were evaluated. The total area covered/run-over by all machine's tyres were calculated with help of the software ArcGIS 9.2. Different guidance systems were evaluated in this study as well. It was based on the amount of guidance error (i.e., the deviation of the vehicle from the desired path). Conclusion The results showed that 96% of the total field area was run over with a machine at least ones during a year, when using conventional tillage, and 65% and 43% of the total field area were run-over when using conservation tillage and direct seeding respectively. It was calculated that 144% of covered area was run-over repeatedly for conventional tillage, 31% for conservation tillage and only 9% for direct seeding. The results show considerable high number of tyre's contacts with soil.

Parallel guidance system for tractor-trailer system with active joint
Backman, J., Oksanen, T. and Visala, A., TKK Helsinki University of Technology, Automation and Systems Technology, Otaniementie 17, 02015 TKK, Finland; juha.backman@tkk.fi

Parallel tracking or auto guidance systems are becoming common in tractors. Auto guidance system with accurate positioning system allows driving very accurately in straight driving lines. If the driving lines are not straight but curves, it is mathematically much harder to lay driving lines so that in every point of curve, the shortest distance to adjacent driving line is constant. And it is even harder if the vehicle to be driven along the path is not just tractor but tractor-trailer and certain point in trailer has to follow the curve. If the field has slopes, the trailer necessarily does not follow kinematic route and more measurements are required to compensate the error. In this paper a developed path tracking system is presented. In a system an ISO 11783 compatible tractor was used together with towed combine seed drill. Tow bar of the seed drill was customized; one joint was added to that. That joint is controlled with hydraulic cylinders and in that way there are two active controllable variables in the system: angle of front wheels in the tractor and an active joint in the tow bar. The measurements used in navigation are: RTK-GPS receiver in tractor, laser scanner in seed drill to detect edge of previous swath and attitude estimation from inertial and magnetometer measurements. A focus in this research was to develop a path tracking algorithm where the trailer follows the adjacent swath. Two different algorithms were developed: the simple one is based on just tractor navigation and curve bending with direct laser scanner based tow bar control; and the advanced one which is based on nonlinear model predictive control. In model predictive control approach a full kinematic model of the tractor-trailer system with active joint is utilized and the laser scanner measurement is an auxiliary state. For testing purposes a simulator was developed where kinematic model is utilized and noises are modeled. The first field tests were giving promising results with the simple algorithm. The active joint in trailer tow bar was found to be essential as the response from control to error is much quicker than from front wheel control. Laser scanner was found to be reliable to detect an edge of the previous swath when the produced small furrow at the edge is clear. In the simulator the model predictive control worked nicely. Final tuning of the model predictive control approach in a field is under way and the conclusions of that will be available in the final paper.

A vision-guided mobile robot for precision agriculture

Ericson, S.[1] and Åstrand, B.[2], [1]University of Skövde, Box 408, 54128 Skövde, Sweden, [2]Halmstad University, Box 823, 30118 Halmstad, Sweden

In this paper we have developed a mobile robot which is able to perform crop-scale operations using vision as only sensor. The advantage of using vision as sensor is that it is scaleable, i. e. resolution can be selected, and it provides a measurement of the position relative the crops on the field. The vision system is also not sensitive to wheel slippage, which can be expected on an agricultural field. The GPS system provides a global position which can be used to know where on the field the robot is located. However, this position is of limited value for crop-scale operations since the position has to relate to an old map. The objective with this paper is to show a method to provide a good local position relatively the crops. This position can be used for performing tasks as mechanical weeding and other operations close to the crops. The robot is a modified electric wheel chair with a row-following system and a visual odometry system. There are in total three cameras mounted on the robot, one for the row-following and two for the visual odometry. The cameras used are MV BlueFOX-120aC, which are USB cameras with a frame rate up to 100 frames/sec. These are connected to two image processing computers. A third computer integrated in an outdoor touch screen is used for controlling the robot and to provide a user interface. All computers are running Ubuntu Linux and Player. The row-following system uses Hough transform to extract lines from the front looking camera. The crop rows on the field are identified and a control system is controlling the steering to follow one row ahead of the robot. This is done with high accuracy. A second control system controls the velocity based on data from wheel encoders. In these test the velocity reference is set to a constant value, which means the wheel will rotate at constant speed. However, this does not mean that the speed of the robot is constant. Due to the environment on the field where wheel slippage is expected, this will not be a good measurement for the robots motion. Therefore a visual odometry system is used, providing a non-contact measurement of the actual motion in both x- and y-direction. Experiments are performed on an agricultural field with a clear row structure to follow. The robot is placed in the beginning of a row and it is programmed to go a specified distance. The final position is marked and the error is calculated. The result shows that the precision of the system is enough for crop scale operations.

An imaging system to characterise the mechanical behaviour of fertilisers in the context of centrifugal spreading

Villette, S.[1], Piron, E.[2], Martin, R.[1] and Gee, C.[1], [1]ENESAD, DSI - UP GAP, 26 bd Dr Petitjean - BP 87999, 21079 Dijon cedex, France, [2]Cemagref, Les Palaquins, 03150 Montoldre, France

In Europe, the application of mineral fertilisers is mainly performed by centrifugal spreaders because of their robustness, their simplicity and their low cost. The quality of the deposit pattern is known to be sensitive to many parameters including fertiliser physical properties and machine characteristics. Thus, the setting of the spreader with respect to the mechanical behaviour of the fertiliser is crucial for the quality of the spreading. Because of difficulties in adjusting the machines, major differences in the quality of the spread pattern deposition are observed in the field and numerous studies have reported the lack of fertiliser uniformity. Nowadays, economic and environmental considerations call for improvements in the quality of fertiliser spreading. Moreover, in the context of precision farming, variable rate applications require a perfect control of the spread pattern to avoid inadequate overlapping. Since autonomous feedback systems for automatic uniformity control are complex to design and are not available yet, there is a need to better characterize and model the process of centrifugal spreading. This process consists of two main stages: the motion of the fertiliser particles on the spreader and the ballistic flight of the particles. In this study, we address the problem of characterising the fertiliser behaviour on the vane by analysing the fertiliser ejection. Numerous mechanical models demonstrate that the acceleration on the spinning disc can be characterised by the friction coefficient of the fertiliser against the vane. Consequently, the ejection parameters depend on this friction coefficient. Nevertheless, when the friction coefficient is measured with traditional methods, the values do not provide good results to predict the motion of fertiliser particles by using mechanical models. In this study, the 'equivalent' friction coefficient EFC of the fertiliser is deduced from a mechanical model of the motion of the particles on a spinning disc and from the measurement of the outlet angle of the particles (i.e. when they leave the disc). The outlet angle is extracted from an image of the fertiliser particle trajectories in the vicinity of the spinning disc. An automatic image processing has been developed to extract information from the location of the trajectories with respect to the location of the disc axle. This 'equivalent' value corresponds to the friction coefficient value of a fictive single particle which would have a pure-sliding motion on the vane and which would have the same behaviour that the real fertiliser particles regarding the ejection parameters (outlet angle or outlet velocity). Experimental values have been obtained for two fertilisers (KCl and NPK), for various disc and vane configurations and for various rotational speeds. Results demonstrate that the method can be used to obtain a relative classification of fertilisers. On the other hand we demonstrate that it is difficult to obtain an absolute value for the EFC considering available mechanical models. This paves the way to established a normalized method in order to perform a fertiliser typology with respect to real spreading behaviour. This typology is a prerequisite to the efficient use of experimental models in order to: help operators in the setting of their machines, reduce the number of calibration tests and develop on-the-go uniformity control sensors and feedback loops.

New high speed image acquisition system and image processing techniques for fertilizer granule trajectory determination

Hijazi, B.[1], Cointault, F.[1], Dubois, J.[2], Villette, S.[1], Vangeyte, J.[3], Yang, F.[2] and Paindavoine, M.[2], [1]ENESAD, UP GAP, 26, Bd Dr Petitjean, 21000 Dijon, France, [2]LE2I Laboratory, University of Burgundy BP 47870, 21078 Dijon, France, [3]ILVO, T&V, Burg. Van Gansberghelaan 115, 9820 Merelbeke, Belgium

To better manage the fertilizer granule input in the field for environment respect, the whole centrifugal spreading process must be studied to determine the fertilizer distribution on the ground. Since several years we develop image acquisition systems to determine the granule trajectories at ejection. Particularly, the motion of the granules (40 m/s) required a specific imaging system based on the combination of a high-resolution digital camera and stroboscopic flashes. The multi-exposure images were analyzed with motion estimation mehtod using the combination of the Markov Random Fields (MRFs) method and an initialization of the motion based on the modelling of the projection. Even if the results were good (90% of correct recognition), this method needs 1) invariance of luminosity between two successive images, 2) a good initialization of the motion depending on the determination of spreader parameters directly in the images. Moreover this method is limited to the detection of 2D-motion that is to the detection of motion with a flat disc, not representing commercial spreading systems. In this paper we first present the newly developped high-speed imaging system based on the use of an illumination system with power-leds and an electronic card to combine the taking of photographs with the vane run. In parallel, we have investigated new motion estimation methods based on block matching methods and Gabor Filters, to propose an alternative to the previous method. The scientific interest concerns the development of motion estimation algorithms for large displacement (70 pixels/image) of small objects which constitutes a niche not currently exploited. The first results obtained on simple and synthetic images are quite good with both methods (90% of correct detection) but the first tests on real images of ejected granules give for the moment no significant information due to the particular motion of the fertilizer: size of the granules (5 mm), motion discontinuity, large displacement of the granules, effect of the centrifugal force. Therefore, the combination of the MRFs technique and motion initialization appears as the best solution and we currently improve this global technique with the automatic determination of the fundamental spreader parameters.

Simulating the dynamic changes in winter wheat after grazing

Harrison, M.T.[1,2], Evans, J.R.[2] and Moore, A.D.[1], [1]CSIRO, Plant Industry, GPO Box 1600, Canberra ACT 2601, Australia, [2]The Australian National University, Research School of Biological Sciences, GPO Box 475, Canberra ACT 2601, Australia; Matthew.Harrison@csiro.au

Winter wheat can be grown for the dual-purpose of livestock grazing and grain production. The introduction of cultivars capable of producing large quantities of vegetative and grain biomass has recently increased the adoption rate of this farming system in Australia. To gain a better understanding of the interactions controlling crop growth rates, a modelling approach was required. The objective of this study is to simulate the growth response and grain production of winter wheat subjected to sheep grazing. An existing crop model, SUCROS2, was extended with a defoliation subroutine. Dry matter accumulation was modified from leaf photosynthesis to a radiation-use efficiency approach. Water stress was calculated as the ratio of actual to potential transpiration rate, with increasing stress reducing growth rates, but increasing the daily fraction of carbon allocated to roots. Grazing was simulated as a reduction in the current shoot biomass and leaf area index. The proportion of daily shoot removed was weighted towards leaf biomass. Carbon allocation patterns following defoliation were shifted in favour of shoots and leaves, tending to restore genetically-determined root-shoot and leaf-shoot ratios. Model calibration was carried out using measurements of soil water, leaf, stem and grain biomass. Measurements were taken from experiments conducted near Canberra, Australia, over two years using two wheat cultivars and a range of grazing intensities. Validation was performed with driving variables and grain yields measured in previous experiments at the same site. The model produced reliable estimates of soil water-use, shoot biomass and grain yield. Simulations revealed two main insights. First, moderate grazing during prolonged periods of limited rainfall caused a delay in soil water-use. This was the result of a reduction in leaf area, even though the rate of root elongation was increased. Second, moderate grazing intensities could be conducted after canopy closure without adverse affects on growth rates, due to the counteracting effects of reduced light interception and increased soil moisture. In both cases, greater availabilities of soil moisture during grain filling were conducive to greater grain yields than would have occurred under no grazing.

Modeling competition for below-ground resources and light within a winter pea (*Pisum sativum* L.) – wheat (*Triticum aestivum* L.) intercrop (Azodyn-InterCrop): towards a decision making oriented-tool

Malagoli, P.J.[1], Naudin, C.[1], Goulevant, G.[1], Sester, M.[1], Corre-Hellou, G.[1] and Jeuffroy, M.-H.[2], [1]Groupe ESA, Plant Ecophysiology and Agroecology, 55, rue Rabelais BP 30748, 49007 Angers cedex 01, France, [2]INRA, Agronomie, UMR 211 Agronomie INRA AgroParisTech, BP 01, 78850 Thiverval-Grignon, France

Background and objective Grain legume-cereal intercrops allow a gain of productivity for each species grown along the growth cycle on the same piece of land under low input (of which nitrogen (N) fertilizers) levels. This is partly due to a better use of soil nitrogen (larger available soil N per plant for cereal N uptake and an increased contribution of N fixation for pea) resulting in LER (Land Equivalent Ratio) values higher than 1. Higher index values are reached when crossed combinations (species and crop management systems) are fully optimized within a given soil and climate environment. Modeling is a powerful tool to save time and explore a wide range of combinations. It can be further used as a decision making oriented-tool provided below-ground (water and nitrogen) and light sharing is satisfactorily simulated. Our work aimed at building a new dynamic intercrop growth model (Azodyn-InterCrop (IC)) based upon Azodyn for wheat and Afisol for pea. A special attention was brought to rules of soil water, nitrogen and light sharing between the two species. Model outputs (dry matter accumulation, N taken up, leaf area index (LAI) and related species ratios, light interception by intercrop, soil N) were then tested against independent datasets from experiments conducted in Pays de la Loire (France) under various nitrogen fertilizer applications. Final performances simulated by Azodyn-IC (grain yield and protein content, nitrogen taken up by intercrop) were also confronted to extra independent datasets. Model description Nitrogen and water partitioning between pea and wheat is firstly driven by nitrogen and water demand of each species. When intercrop demand is larger than soil supply then water and N acquisition is limited by root exploration, soil nutrient supply and water and N taken up by the companion species as it concurrently depletes available below-ground resources. The 'functional' root layer concept developed for the model allows to account for (1) root exploration of each species, (2) demand based-competition for below-ground resources between species and (3) advantage towards species with a faster root penetration rate. Leaf area expansion is driven by nitrogen daily taken up, itself computed through an adapted version of N dilution curve to intercrop growth. Light sharing depends on LAI growth and leaf properties (reflectance, leaf angle) of each species so a nitrogen deficiency results in an unbalanced competition for light, a slower growth rate ultimately a lower yield of one or both species. Conclusions Model outputs show Azodyn-IC can satisfactorily simulate N taken up, LAI, light interception efficiency and crop growth of each sole- and intercropped species along the growth cycle leading to realistic yields for the applied N fertilizer rates. It thus provides a validation of rules chosen to share the studied resources between the two species. It also emphasizes competitions for light and below-ground resources are tightly and dynamically linked within intercrops. Azodyn-IC simulations crossing crop managements (N fertilizer applications, sowing density and sowing date) are currently under progress.

Conductance model evaluation for predicting yield and protein content of lupin-cereal forage crops in organic farming
Azo, W.M., Davies, W.P., Lane, G.P.F. and Cannon, N.D., Royal Agricultural College, Cirencester, GL7 6JS, United Kingdom

The Conductance model simulates plant growth using an electrical analogy; describing competition by considering space occupied by the crown (zone of the smallest circle to include all foliage viewed from above) and consequently light interception. It is based on individual plant data, where leaf area is presumed to be a time independent function of dry weight and growth is the product of intercepted light converted into dry matter. The model has been successfully used to predict growth of various crops. Forage potential of a crop is determined by dry matter yield and protein content, as they influence ruminant nutrition. Maximising cost-effective production of nutritious, traceable. home-grown feed of quality protein and energy is increasingly important for farmers seeking alternatives to grass and clovers. Lupins offer a highly digestible protein crop opportunity, but are inconsistent in performance on UK farms. Combined with appropriate cereals, however, lupins could offer an attractive option for good quality protein and energy rich forage – subject to a better understanding. Field trials occurred on land certified for organic production at Harnhill Manor Farm Cirencester (51°42'N latitude 01°59'W longitude) in 2005 and 2006. The site was 135 m above sea level and soil pH 6.0 to 6.8. Randomised block experiments were used with four replicates. Bi-crops were established with three lupins; *Lupinus albus*, cv Dieta, *Lupinus angustifolius*, cv Bora and *Lupinus luteus*, cv Amber and three cereals; wheat (*Triticum aestivum*) cv Paragon, millet (*Pennesitum americanum*), cv Mammoth and triticale (Triticosecale Wittmark) cv Logo. Crops were drilled late April and wholecrop harvested on 20 July, 3, 17 and 31 August. Replicate spaced plants were sown 1m apart in 1m rows on 25 April in 2005 and 2006, and 24 March in 2005. The first harvests were made 20 June 2005 and 5 May 2006 and then weekly. Prior to harvest, individual plants were measured using calibrated card and photographed. Leaf areas were measured using Aequitas software (Dynamic Data Link Ltd, Cambridge). Digital images were also studied using Aequitas, to measure ground area covered by the plant, and the diameter of the smallest circle embracing the leaves. Areas of the smallest circle representing the crown zone of each plant were calculated. Fresh weights determined total biomass, and samples dried to determine dry matter yield and subsequent chemical analysis. Kjeldhal was used to determine nitrogen content. Key relationships between leaf area and plant dry weight, and between the crown zone area and plant dry weight were strong and enabled good predictions of dry matter yields for both mono-crops and bi-crops. Crude protein content and plant dry weight also showed a good new correlation for the lupin cultivars and spring wheat cv Paragon. The main limitation using the Conductance model being its poorer predicative capacity during plant maturity and crop canopy senescence. Performance prediction was also poorer where weed competition was more intense in this organic situation.

Crop modelling based on the principle of maximum plant productivity

Kadaja, J. and Saue, T., Estonian Research Institute of Agriculture, Teaduse 13, 75501 Saku, Estonia

Complex influence of environmental conditions on crop yield, its variability and changes is essential from several aspects, such as seasonal and long-term yield forecasting, assessment of past conditions and region's suitability for production, selection of alternative crops and varieties, and so on. The methodology based on the principle of maximum plant productivity introduced by H. Tooming (1967) – stating that adaptation processes in a plant and plant community are directed towards providing the maximum possible net photosynthesis productivity in the existing environmental conditions – allows to observe and model the maximum production and yields under different limiting factors divided into agroecological groups: in general, into biological, meteorological, soil, and agrotechnical groups. According to the concept of reference or model yields, these groups are included in the model separately, step by step, starting from optimal conditions for the plant community. The main categories of reference yields are, in descending order, potential yield (PY), meteorologically possible yield (MPY), practically possible yield (PPY), and commercial yield (CY). This set of yield categories provides us with an ecologly based reference system for comparison and analysis of different yield values obtained from field trials as well as from model experiments. Additionally, each of these categories represents certain kind of ecological resources for plant growth expressed in yield units. PY, expressing the solar radiation resources in yield units, is the maximum yield of the given species or variety possible under the existing solar radiation, with all the other environmental factors considered to be not limiting. MPY is the maximum yield conceivable under the existing irradiance and meteorological conditions, with optimal soil fertility and agrotechnology. MPY expresses agrometeorological resources (for prolonged periods – agroclimatic resources) in yield units. PPY, the maximum yield achievable under the existing meteorological and soil conditions, considers soil tilling to be optimal and the influence of plant diseases, pests and weeds to be absent. CY takes into account all the factors limiting the production process and the crop yield. This approach is realized in a potato crop model POMOD, in present state allowing to compute PY and MPY. The analysis of variability of agrometeorological resources was carried out by a 107 year long data series in Tartu, Estonia. Significant trends of MPY were not observed, but there exists noticeable increase in variability. For early variety 'Maret', the standard deviation of the MPY increased from 6.4 t ha^{-1} in 1901–1980 to 9.5 t ha^{-1} in 1981–2006 ($P<0.01$ by F-test); for 'Anti', the change from 10.0 to 13.5 t ha^{-1} ($P<0.05$) was recorded. Based on the MPY category, a method for its climatic and seasonal probabilistic forecast, i.e. forecast of agrometeorological resources, is proposed. In Tartu, the climatic probabilistic forecast is not symmetric, average of the MPY distribution being 55.6 t ha^{-1} and median 58.6 t ha^{-1}. The MPY values slightly above average are prevalent, but are more sparsely set below median. The lowest yields in the series, below 30 t ha^{-1}, are related to excessively wet years, whereas the MPY values between 30 and 40 Mg ha^{-1} are mostly affected by dry conditions.

Automatic working depth control for seed drill using ISO 11783 compatible tractor
Suomi, P.[1], Ojanne, A.[1], Oksanen, T.[2], Kalmari, J.[2], Linkolehto, R.[1] and Teye, F.[1], [1]MTT Agrifood Research Finland, Plant Production Research, Vakolantie 55, 03400 VIHTI, Finland, [2]TKK Helsinki University of Technology, Automation and Systems Technology, Otaniementie 17, 02015 TKK, Finland

Reliability of field work documentation and the functionality of PA systems set critical requirements in the instrumentation of implements. Precision agricultural machinery should be easy to use and be able to operate without the need for automatic online controls that are map-based. In this paper a measuring system, a control system and the instrumentation of a prototype combine driller's working depth is described. A preliminary survey was carried out to choose sensors for the instrumentation of working depth control in Junkkari Maestro 3000 combine driller. In the first phase of the research, a reliable measurement system based on multiple sensors was developed. A model was then developed for automatic control of the working depth. The measurements used in model were from: two ultrasonic rangers and two wheels for measuring the height of the drill body from the soil, three sensors for measuring the angle of coulter links and one angular position sensor located at the supporting wheel joints. In the first phase, all the sensors were calibrated indoors on a concrete floor. In this primary calibration both the tractor and the drill were elevated 100 mm from concrete floor to enable the calibration of all the sensors within a range of 0-100 mm working depth, which is more than required for crop farming. A RTK-GPS receiver was used for positioning in the field. For validating the theoretical model, the real depth of seeds was obtained from the soil profile by slicing after the germination of seeds. ISO 11783 based control unit prototype in the drill was developed earlier in the project. The developed static model is used to calculate the working depth online from 8 sensors. The working depth is controlled by changing the height of drill from the supporting wheels. A hydraulic cylinder controlled the height of the drill, which was connected directly to the ISO 11783 Class 3 compatible tractor's hydraulic valve. The control commands to the hydraulic valves of the tractor are transmitted using ISO 11783 network. In the control system architecture cascade control is used. The height of drill is controlled (PD) an inner loop algorithm, and in the outer loop the set point of height controller based on depth measurement (PID). In the field tests automatic working depth control was evaluated. The target working depth varied from 10 to 50 mm. Preliminary tests showed that the constant error in measured working depth was 10-12 mm.

Robotic control of broad-leaved dock

Van Evert, F.K.[1], Samsom, J.[2], Polder, G.[1], Vijn, M.[3], Van Dooren, H.J.[4], Lamaker, A.[5], Van Der Heijden, G.W.A.M.[1], Kempenaar, C.[1], Van Der Zalm, A.[1] and Lotz, L.A.P.[1], [1]Plant Research International, P.O. Box 16, 6700 AA Wageningen, Netherlands, [2]Mts Samsom-Visser, Gagelweg 1, 3648 AV Wilnis, Netherlands, [3]Applied Plant Research, P.O. Box 167, 6700 AD Wageningen, Netherlands, [4]Animal Sciences Group, P.O. Box 65, 8200 AB Lelystad, Netherlands, [5]MARIN, P.O. Box 28, 6700 AA Wageningen, Netherlands

Broad-leaved dock (*Rumex obtusifolius* L.) is a common and troublesome weed with a wide geographic distribution. In organic farming, the best option to control the weed is manual removal of the plants. Manual removal may require up to 1600 hours year^{-1}. In this report we describe the development and first tests of a robot to detect and control broad-leaved dock. The robot consists of a rigid frame of 1.25 x 1.11 m to which four independently driven wheels are attached. Tight maneuvering is not required, thus we implemented skid steering in order to keep construction light and to keep costs down. Power is provided by a diesel engine. Weeds are detected with a downward-looking camera that provides full-colour images with a resolution of 1.5 mm per pixel. Image sections with grass contain more colour and intensity transitions than image sections with broad-leaved weed. Colour and intensity transitions are quantified by dividing each image into sub-images (tiles) of 8x8 pixels and subjecting each tile to two-dimensional Fourier analysis. The power of the Fourier spectrum for all spatial frequencies above zero is a measure for number of colour and intensity transitions. Following Fourier analysis, a combination of thresholding and morphological processing is used to determine whether the image contains one or more weeds and to determine the location of each detected weed. Once detected, weeds are controlled using the method proposed by Austrian farmer Ferdinand Riesenhuber. This method consists of a chopper with a single 0.18 m blade that rotates about a vertical axis and is pushed into the ground at the location of the weed. The chopper is powered by a high-speed hydrostatic motor capable of rotating at 1,500 rpm. The chopper is raised and lowered by a hydraulic cylinder. The chopper assembly can be moved along a rail that is fastened to the front of the vehicle. The rail can be folded for transport; when extended, it allows the chopper to move over a distance of 2 m. Construction of the robot has been completed. In field tests the robot was run at 0.5 m s^{-1}. At this speed, more than 90% of weeds were detected and positioning of the chopper occurred with a precision on the order of 0.01 m. The time required to position and operate the chopper was determined to be 12 s. We conclude that our robot to detect and control broad-leaved dock provides an attractive alternative to manual removal of weeds.

First prototype of an automated rotary hoe for mechanical weeding of the intra-row area in row crops
Gobor, Z.[1], Schulze Lammers, P.[2] and Wendl, G.[1], [1]Bavarian State Research Center for Agriculture, Institute for Agricultural Engineering and Animal Husbandry, Vöttinger Str. 36, 85354 Freising, Germany, [2]Institute of Agricultural Engineering, Technology of Crop Farming, Nussallee 5, 53115 Bonn, Germany; zoltan. gobor@lfl.bayern.de

As a component of successful non-chemical weed control intra-row weeding shall be considered as a final weed elimination procedure and not as a primary method. Conventional methods for inter-row weed control can handle with approximately 80% of the field area in row planted crops. However, the weeds occur in the remaining area between (intra-row) and around the crop plants (close-to-crop) have a much bigger impact on the development and yield of the plants. Online detection of the single plant position and the plant/weed distinction are the bottlenecks of automated intra-row weeding, but concerning the expeditious development in this field it is to expect that appropriate systems would be available on the market in near future. In the meantime, construction and adjustment possibilities of implements considering the role of soil properties and mechanics need to be optimised toward universal intra-row weeding tools, which can be used in different plant spacing systems, different plant intra-row distances and growth stages. A virtual prototype of a system for intra-row weeding imitating the manual hoeing motions under the soil surface was designed. The hoeing tool consists of an arm holder and three or more integrated arms rotating around the horizontal axis above the crop row. Tests and simulations of the hoeing trajectories carried out with the virtual prototype have increasingly facilitated the design process and significantly shortened the path from the idea to the prototype. A simplified methodology and system for plant position detection based on the spectral characteristics of crop plants combined with the context information of the planting pattern was developed and tested. The experimental results showed that the combination of the RGB sensor and laser sensor can be used for accurate detection of the plant centre position independently from illumination conditions. The servo system built in the physical prototype was operated in a mode with direct software control providing rotational speed adjustment according to the forward speed of the carrier, intra-row distance between successive crop plants and the observed angular position of the arms. The controlling algorithm and software solution were developed in the Labview® environment. The main task of the controlling software was permanent calculation, checking and change of the recent rotational speed of the hoeing tool in real time. Tests have proved that depending on the angular adjustment of the duckfoot knives an uncultivated area big enough to avoid damaging of the plants can be left around the plants during the intra-row weeding with the developed system.

Auto boom control to avoid spraying pre-defined areas
Mickelåker, J. and Svensson, S.A., Swedish University of Agricultural Sciences, Dept. of Agriculture - Farming Systems, Technology and Product Quality, Box 104, 230 53 Alnarp, Sweden

Non-sprayed buffer zones are widely used methods to protect sensitive objects within or adjacent to fields treated with pesticides. These buffer zones should protect the objects from pesticide residues caused by spray drift and leakage. The use of untreated buffer zones has been mandatory for Swedish farmers since 1997. E.g. recommended minimum safe distances to sensitive areas for spray situations are 12 m to wells, 6 m to streams, ponds or lakes, and 1 m to ditches and drainage wells. Studies have shown that buffer zones are not always respected and that there is a need to improve the protection of sensitive objects. On a typical sprayer, all nozzles on a section of the boom are either on or off. A higher resolution in sprayer control would reduce the untreated area but still be able to respect the buffer zones. The objectives of this study were to develop and evaluate a system that automatically could shut off single nozzles along the spray boom to avoid pre-defined areas in the field. Such a system could improve management of buffer zones, reduce the risk for mistakes and improve record keeping from spraying activities. A commercial spray controller, Legacy 6000 (Teejet), could consider pre-defined areas by using the field boundary features. During 2007, the system was used in practice on a conventional trailed sprayer with a 24 m boom and 7 boom sections. Experiences from mapping and spraying were documented. Also accuracy of the system was studied in an experiment using video recorder and analysis of the video frame by frame. The following season, 2008, a small sprayer with a 6 m boom and individual nozzle control was constructed. This year, the study of accuracy focused on small circular buffer zones, e.g. close to drainage wells. Spraying was done on an asphalt surface during a sunny day, leaving wet and dark marks when nozzles were activated. The whole buffer zone was photographed directly after spraying, and the pictures were later analysed manually to determine the accuracy of the sprayer. Different controller settings, spray nozzles and GPS receivers were used. The control system made it possible to achieve an automatic deactivation of the sprayer at pre-defined areas. Both methods, video analysed frame by frame and photographs on an asphalt surface, could be used to determine the accuracy of the auto boom function. The difference between experiment field boundary and boom activation/deactivation for the conventional sprayer had a standard deviation of 0.51 and 0.55 m, respectively. The average differences could be adjusted by the controller settings. The sprayer with individual nozzle control had an average standard deviation of 0.35 m for the different settings. Precise mapping was achieved by the use of RTK-GPS on an ATV, but map management was time consuming and involved a knowledge threshold for the farmer. The application error could not be explained only by sprayer construction and positioning error. Delays and variations in the electronic control system also influenced the accuracy of the sprayer. Accuracy was found to be sufficient for the task and the lack of precision could be handled by an acceptable increase of the buffer zone distance.

Image processing algorithms for a selective vineyard robotic sprayer

Berenstein, R., Edan, Y., Ben-Shachar, O., Shapiro, A. and Bechar, A., Ben-Gurion University of the Negev,
Departments of Industrial & Management, zela 21, 84965 Omer, Israel; ron2468@gmail.com

This paper presents image processing algorithms for a selective robotic sprayer for vineyards. Two types of image processing algorithms were developed: direct spraying to grape clusters; controlling the amount of spraying material according to foliage density and location. Direct spraying toward grape clusters is a common way of spraying pesticides and fertilizers in vineyards. Two types of machine vision algorithms were developed to detect grape clusters in vineyards. The first algorithm originated by examining vineyard images. In vineyard images the difference in the amount of edges between the foliage and the grape clusters is well shown. According to this assumption the first algorithm was developed. All non green pixels were filtered from the main image. Edge detection using Sobel method was then performed on each of the RGB channels. The sum of edges from the three RGB images was combined into one image. A threshold of the image enabled to separate the grape clusters from the background. The second algorithm for specific grape spraying uses a decision tree for separating the grape clusters from the background. A data set was created from vineyard images sampled along the grape season. Using the data set, a decision tree was created. The decision tree was used for separating the grapes from the foliage. Spraying the foliage is the most common way of spraying vineyards. Separating the foliage from the background was achieved by isolating the green color from the image. This was achieved by comparing the different RGB channels with each other. A pixel was defined as GREEN in the following case: If the GREEN channel is larger than the RED channel & the GREEN channel is larger than the BLUE channel & the GREEN channel is larger than a specific threshold. All image processing algorithms were tested on real-data acquired using movies acquired in vineyards along the growing season of 2008. Foliage detection results indicate high ability in separating the foliage from the background. As well as the foliage separation algorithm, the grape clusters detection algorithm, shown high ability to detect the grape clusters within the vineyard row. The paper will detail the different algorithms and present detailed results.

WURking: a small sized autonomous robot for the farm of the future

Van Henten, E.J.[1,2], Van Asselt, C.J.[3], Bakker, T.[3], Blaauw, S.K.[1], Govers, M.H.A.M.[1], Hofstee, J.W.[1], Jansen, R.[1], Nieuwenhuizen, A.T.[1], Speetjens, S.L.[3], Stigter, J.D.[3], Van Straten, G.[3] and Van Willigenburg, L.G.[3], [1]Wageningen University, Farm Technology Group, P.O. Box 17, 6700 AA Wageningen, Netherlands, [2]Wageningen UR Greenhouse Horticulture, P.O. Box 16, 6700 AA Wageningen, Netherlands, [3]Wageningen University, Systems and Control Group, P.O. Box 17, 6700 AA Wageningen, Netherlands

Autonomous robots for agricultural practices will become reality soon. These mobile robots could take over regular task such as scouting for weeds and diseases, plant specific applications, yield and field mapping and for instance the release of info-chemicals for attracting predators of pests. This paper presents WURking, a small sized autonomous robot that can be used for a wide range of tasks on the Farm of the Future. Small sized light weight robots fit into the small scale farming approach of precision farming. They induce less soil compaction and offer the opportunity of a more weather independent access to the fields. And due to their small size, safety issues will be less critical under autonomous operation than with large machines. Given the current and ever growing size of today's farm machinery, deployment of small scale machines would mean a paradigm shift in agriculture. The potential of small scale possibly collaborating robots is investigated in the EU FutureFarm project. WURking was designed for navigating within row crops like corn. It consists of a mobile platform with three independently driven steerable wheels. The platform carries ultrasound sensors and a gyroscope used for navigation between the crop rows, end of row detection and headland turning. A camera is mounted for detection of objects like weeds. High level control of this robot is based on LabView. A data fusion technique is used to extract from the redundant set of sensor data, the position and orientation of the robot relative to the crop rows. Feedback linearization of the non-linear system dynamics yields a simple linear controller structure which is fed by state estimates generated by a Kalman filter using the raw ultrasound sensor data. Head land turning is based on a proportional controller using data from the gyroscope. The performance of WURking was evaluated in two ways. First of all, WURking participated in the Field Robot Event competition in 3 consecutive years during which it was tested in a variety of tasks. Ending at the last place in 2006, performance improved the next years and a 4th and 3rd place were obtained in 2007 and 2008, respectively. Secondly, in 2008 the performance of WURking was evaluated in a corn field using independent measurements of the position and orientation of the robot in the field. Besides sketching the potential of small sized machines and describing the WURking platform, this paper will present results of the performance evaluation of this autonomous mobile platform.

A supply/demand, single-organ crop growth model

Seginer, I.[1] and Gent, M.[2], [1]Technion, IIT, Civil and Environmental Engineering, Technion City, 32000 Haifa, Israel, [2]Connecticut Agricultural Experiment Station, Forestry & Horticulture, New Haven CT 06504-1106, USA; segineri@techunix.technion.ac.il

Traditional growth models are of the form $S = dM/dt = e\,[p\{I,C,T\} - m\{T\}]\,f\{M\}$ where M is crop structural mass; I, C and T are light flux, carbon dioxide concentration and air temperature of the shoot environment; p is specific gross photosynthesis, m is specific maintenance respiration, f is light interception function, and e is a conversion factor. Often this carbohydrates (CH) supply (source) term (hence S in the formula above) is considered to be the only growth limiting factor. Sometimes, however, the potential for growth is smaller than the supply of CH. The limiting factor may then be the size of the growing organ, or low temperature, or the scarcity of any one of various nutrients. This is expressed in terms of a demand (sink) term, such as $D = [s\{T\}+g\{T\}]\,f\{M\}$ where s is potential structural growth and g is the corresponding growth respiration. S and D have been used to determine the partitioning (allocation) of CH to the various organs of a crop, using either $A = S{\times}D$ or $A = \min\{S,D\}$ to quantify assimilation into structure. In either case $A\{T\}$ has an optimum at an intermediate temperature. Nicolet is a dynamic growth model with supply and demand functions similar to the above, where a CH/nitrate pool is used as a daily buffer to overcome supply and demand mismatch. By reducing to zero the CH pool size, the model simplifies considerably and becomes quasi-steady-state. In this form there is an unambiguous optimum temperature which produces maximum growth. Growth rate decreases with increasing distance from that temperature, similar to observations. Furthermore, the optimal temperature increases as the supply term increases, again similar to observations, meaning that a higher light level justifies (e.g., in greenhouses) a higher temperature. At low temperatures, the demand (sink) term is the limiting growth factor, while at high temperatures it is the supply (source) term. Simulations with Nicolet show that often the dynamic model drifts to one of the extreme states of the CH pool, justifying the use of the quasi-steady-state approach. Additional restricting factors may be treated by the model. Suppose that nitrogen, N, is in short supply. Assimilation, then, may be expressed as $A = S{\times}D{\times}N$ or $A = \min\{S,D,N\}$. Following the experience with Nicolet, the simplified model can not only predict structural growth, but also non-structural carbon-growth, which is the difference between supply and demand, namely $\max\{S\text{-}D,\,0\}$. If $D > S$ then structural material is consumed for maintenance respiration at a rate $D - S$. The simplified, quasi-steady-state growth model, based on Liebig's law of the minimum, may serve usefully in studies of season-long control problems (e.g., using optimal control methodology) as well as producing short term results for cases where consistently $S > D$ or $D > S$.

f.

Prediction of within field cotton yield losses caused by the southern root-knot nematode with cropping system model-CROPGRO-cotton

Ortiz, B.[1], Hoogenboom, G.[2], Vellidis, G.[3], Boote, K.[4] and Perry, C.[3], [1]Auburn University, Agronomy and Soils, 201 Funches Hall, 36849, USA, [2]University of Georgia, Biological and Agricultural Engineering, 1109 Experiment St, 30223, USA, [3]University of Georgia, Biological and Agricultural Engineering, 2329 Rainwater Road, 31793, USA, [4]University of Florida, Agronomy, 404 Newell Hall, 32611, USA; bortiz@auburn.edu

Southern root-knot nematode [*Meloidogine incognita* (Kofoid & White) Chitwood] (RKN). This pathogen is responsible for twice as much cotton (*Gossypium hirsutum* L.) yield losses as to all other nematodes across the U.S. Cotton Belt. Yield losses, expressed in cotton lint, by plant-parasitic nematodes have increase from 1% to 2% in the 1950s to 4.39% in 2000. Understanding the impact of RKN due to local environmental conditions, such as soil texture, on cotton growth and development is important for spatial management. The objective of this study was to predict yield losses on a producer's field having three management zones with different risk levels for RKN damage. The spatial variability of cotton yield as consequence of RKN parasitism was simulated with the CSM-CROPGRO-Cotton Model. The model was adapted to simulate RKN damage throughout the implementation of two strategies: (1) RKN acting as sink of soluble assimilate, and (2) RKN inducing a reduction of root length per root mass and root density. These factors were adjusted with data collected in 2007 from an experiment that studied the interaction of RKN infection and drought stress. The population of RKN population density was collected three times during the growing season on a 50 x 50 m m grid (0.25 ha cell size) and the average population by zone was entered into the model to remove assimilates on a daily basis. The reduction of root length per unit mass was implemented by modifying the root length to weight crop factor. After calibration the model was applied to predict differences in yield (seed plus lint) losses for a producer's field in southern Georgia, USA, on the three management zones for the 2006 season. The model simulation indicated that cotton yield was highly impacted by RKN population. Cotton yield decreased by an average of 1,165 Kg ha^{-1} (30% reduction) in the zone that had the highest RKN population compared to the zone that had the lowest RKN population. In general, the simulated values followed the same trend as the observed values. The results showed that CSM-CROPGRO-Cotton model, after adaptation for RKN specific effects on the rate of assimilate consumption and the root length to weight ratio, can be used to simulate yield losses due to nematodes under the conditions of a commercial field.

Use of geographic information systems in crop protection warning service
Zeuner, T. and Kleinhenz, B., ZEPP, Rüdesheimer Str. 68, 55545 Bad Kreuznach, Germany

One of the important aims of the German Crop Protection Services (GCPS) is to reduce spraying intensity and to guarantee an environmentally friendly crop protection strategy. ZEPP is the central institution in Germany responsible for the development of methods in order to give the best control of plant diseases, so far more than 20 met. data -based models were developed and introduced into practice. This study shows that it is possible to obtain results with higher accuracy for the models by using Geographic Information Systems (GIS). The influence of elevation, slope and aspect on met. data were interpolated with GIS and the results were used as input for forecasting models. The results of interpolation are saved in a grid over Germany. At the moment the area of Germany ($357,050$ km^2) is represented by ca. 570 met. stations, that is one met. station each 626 km^2. A grid cell is 1 km wide which means that after interpolation each square kilometre of Germany is represented by a virtual met. station (= grid cell). Multiple Regression method is used to interpolate temperature and relative humidity. A comparison between real temperatures and interpolated temperatures showed results with high accuracy. The coefficient of determination in all cases ranged between 96 and 99%. Interpolated met. data are made available for disease forecasting models. Absolute differences of forecasted and recorded dates for Late Blight first occurrence were three days, which must be regarded as a highly accurate result. Currently additional disease forecast models are adapted and validated to this method.

Plant-specific and canopy density spraying to control fungal diseases in bed-grown crops

Van De Zande, J.C.[1], Achten, V.T.J.M.[1], Schepers, H.T.A.M.[2], Van Der Lans, A.[3], Michielsen, J.M.G.P.[1], Stallinga, H.[1] and Van Velde, P.[1], [1]Wageningen UR - Plant Research International, P.O. Box 16, 6700 AA Wageningen, Netherlands, [2]Wageningen UR –Applied Plant Research (AGV), P.O. Box 420, 8200AK Lelystad, Netherlands, [3]Wageningen UR – Applied Plant Research, Research Unit Flower Bulbs, P.O. Box 85, 2160 AB Lisse, Netherlands

Matching spray volume to crop canopy sizes and shapes can reduce the use of plant protection products, thus reducing operational costs and environmental pollution. Developments on crop adapted spraying for fungal control are highlighted in arable crop spraying. A plant-specific variable volume precision sprayer, guided by foliage shape and volume (canopy density sprayer; CDS) was developed for bed-grown crops to apply fungicides. Sensor selection to quantify crop canopy and spray techniques to apply variable dose rates are evaluated based on laboratory measurements. Sensor-nozzle combination delay time was determined for the different nozzle settings and combinations. Optimal sensor-nozzle distances could be determined to specify sprayer design. Potential volume rate savings are evaluated based on crop canopy structure development evaluations during the growing season of bed-grown flower bulbs (tulip, lily) and potatoes. Based on the laboratory experience a prototype CDS sprayer was built using either a Weed-It or a GreenSeeker sensor to detect plant place (fluorescence) or size (reflectance). Variable rate application was either done with a pulse width modulation nozzle or a Lechler VarioSelect switchable four-nozzle body. Spray volume could be changed from 50-550 l/ha in 16 steps. Spray deposition, biological efficacy and agrochemical use reduction were evaluated in a flower bulb and a potato crop during field measurements using a prototype CDS sprayer. Spray volume savings of a prototype plant-specific sprayer are shown to be more than 75% in early late blight control spraying in potatoes. In flower bulbs (lily) it was shown that in fire blight control on average spray volume could be reduced by 25% and at early crop development stage even by more than 90%. In both crops biological efficacy was maintained at the same good level as of a conventional spraying.

Data collection and two-way communication to support decision making by pest scouts

Hetzroni, A.[1], Meron, M.[2], Frier, I.[3] and Magrisso, Y.[3], [1]ARO, Agricultural Engineering, P.O. Box 6, 50250 Bet Dagan, Israel, [2]Migal, P.O. Box 831, 11016 Kiryat Shmone, Israel, [3]ScanTask, 8 Hamaayan St., 58811 Holon, Israel; amots@volcani.agri.gov.il

Knowledge and information are key factors in pest management decision making. This work is part of an endeavor to establish a collection, storage and dissemination system for pest scouting data, as part of a decision support system for pest management. A data collection system was developed based on pocket-pc's and was deployed in three regions. The concept was proven to be feasible and was accepted and used by the pest scouts. Yet, it was used only to collect data and transfer it to the central repository; it lacked two-way communication to provide the feedback required in order to support decision making in the field. Objectives were to provide pest scouts with a feedback mechanism to support decision on pest management's measures. The handheld pocket-pc type data collection units were replaced by cellular phones linked to the repository data server. A flexible data collection protocol based on a cellular workflow engine (ScanTask Ltd.) was implemented on the phone which serves as a data collection unit. Upon login, local data tables are updated from the main server, data such as a list of plots and crops pertinent to user. The scouting records, including time and location are transferred, upon communication availability to the server. The repository was designed for internet interface to respond to spatial queries. Predefined queries are available for the end user from the cellular terminal, such as history of infestations. Experimenting with cellular units confirmed the advantages of the units compared with the pocket-pc units. No hardware failures or communication difficulties were reported. GPS data was available with almost all records.

Prototype system of monitoring farm operation with wearable device and field server

Fukatsu, T.[1], Sugahara, K.[1,2], Nanseki, T.[2] and Ninomiya, S.[1], [1]National Agriculture and Food Research Organization, National Agricultural Research Center, 3-1-1, Kannondai, Tsukuba, Ibaraki, 305-8666, Japan, [2]Kyushu University, Faculty of Agriculture, 6-10-1, Hakozaki, Fukuoka, 812-8581, Japan; fukatsu@ affrc.go.jp

It is important for a farmer to monitor crop growing, field environment and farm operation automatically. We have already developed a field monitoring system with Field Servers which provide high-speed communication and collected data from various sensors including movable cameras. In order to monitor detailed information about various farm operations, we propose a monitoring system of farmer's operation with Field Servers and a wearable device which equips a RFID reader and motion sensors. In this system, we try to recognize farmer's operation based on sensor data and scanned RFID tags which are posted on all farming material, facility and machinery. This method enables to apply various situations without changing traditional system. By analyzing scanned sequence with pattern matching, this system estimates farm operation and feeds back adequate action and useful information to a farmer. Field Servers are also controlled with the cooperation of farmer's operation in order to collect data both from fixed point and perspective of a farmer. In this paper, we construct a prototype system which is assumed a situation of pesticide preparation. In this system, RFID tags are attached to some points of a shelf and some pesticide bottles, and farmer's operation is monitored by a wearable RFID reader automatically. When the system recognizes an operation of taking a certain bottle, it controls neighbor cameras of Field Servers to record target process and it provides appropriate pesticide information to wearable computer display. The prototype system works well in the experimental situation and shows some future tasks such as recognition problem, hardware performance, response of complicated operation and so on. By improving the system to monitor not only farmer's operation but what he feels and notices, the system will be more useful and effective to study farmer's behavior. Through the result and discussion about prototype system, we propose a new concept of user-based monitoring system which identifies user's perception and reaction as a sensor node in a sensor network.

Information requirements and data sources for automated irrigation control and supervision in tree crops

Casadesus, J., IRTA, Irrigation Technology, Av. Alcalde Rovira Roure, 191, 25198 Lleida, Spain

Advances in sensors, communications and data acquisition have increased the potential for supervision and control of remote field processes in agriculture. However, increasing the capacity for remote data acquisition does not necessarily lead to an improved capacity to manage such processes, as the gap between the raw data and the information required for controlling the process can be relevant. Then, it makes sense that the design of applications for control and supervision first considers which are the information requirements for the involved algorithms and then revise what data sources –i.e. sensors or external data- and processing methods can be used to support them. In the specific context of tree-crop drip irrigation we designed, implemented and tested a system for automated remote control and supervision of irrigation. The main functionalities of the system were to optimize the daily water dose and to detect eventual malfunctions. The characterization of the information requirements in an abstract way, that could be implemented using different data sources, facilitated a common interpretation and use of four information pieces among different scenarios: 1. Quantitative prediction of the water loss imposed by the environmental conditions. Basically, evapotranspiration (ETo), which can either be calculated from a set of local meteorological sensors, or estimated from a simplified subset of these sensors, or queried to an external meteorological network. 2. Quantitative indicators of vegetation activity, which can be either based on crop coefficients (Kc) published by irrigation support organizations, or be locally estimated from sensor data, such as vegetation indices, imaging, etc. 3. Quantitative indicators of soil water status. As sensors in soil or in plants use to show a high variability both in space and time, deriving reliable information from them is not trivial. Usually, the data recorded by these types of sensors show a distinct daily pattern, where the instantaneous absolute values can be meaningless and the required information has to be derived from the day to day evolution of the pattern. Here is where the gap between data and information use to be larger. 4.Occurrence of malfunctions in the irrigation system, detectable with the analysis of data from water counters and pressure detectors in the hydraulic system. In the implemented system, different sources of data –that is, different types of sensors and some external data sources- have been tested for deriving the required information, while keeping unaltered the full functionality of the system. This shows that focusing the application design on the information requirements, rather than on the physical sensors, may allow applications for field supervision and control to be more robust and adaptable to each particular context.

Hands free registration of tillage activities in agriculture

Janssen, H.[1], Van Der Wal, T.[2], Beek, A.[3], Van Rossum, J.[3] and Uyterlinde, M.[3], [1]WUR-Alterra, Centre for Geo Information, Postbus 47, 6700 AA Wageningen, Netherlands, [2]Portolis, Spoorbaanweg 23, 3911 CA Rhenen, Netherlands, [3]Food Process Innovations, Van Ledenberchstraat 10, 2334 AT Leiden, Netherlands

European arable farmers increasingly suffer from an administrative burden caused by intense governmental legislation and regulation and information demand of the agri-chain. The geo-ICT domain provides useful tools to their increasing call for simplifying the rules and administration. Global Navigation Satellite Systems (GNSS) combined with Geographical Information Systems (GIS) and wireless communication (GPRS, UMTS) make up a perfect toolkit to develop hands free registration systems for crop husbandry activities on parcel level, and thus assisting the farmer in his time consuming administrative tasks. The presentation will set the scene of geo-enabled farming in the context of the technological development of satellite navigation, GIS-datasets and precision and development of sensors for the registration of machinery activities. The results of an experiment will be presented in which innovative arable farmers together with geo-researchers and hardware developers built a hard- and software demonstrator to register 'where', 'when' and 'what' on parcel level without human interaction. The collected data are used for farm management and to deliver information to the agri-industry chain and retailers. Future developments and research challenges will be presented based on literature review and own experiments. The benefits of integration of geo-technology and workflow management will be elaborated.

A multi-level modelling approach for food supply chains using the Unified Modeling Language (UML)
Lehmann, R., Fritz, M. and Schiefer, G., Institute of Food and Resource Economics, Chair for Business Management, Organization and Information Management, Meckenheimer Allee 174, 53115 Bonn, Germany

Food supply chains differ from supply chains in other sectors by their complexity which makes modelling in this field an exceedingly difficult task. Even though supply chain management in food industry is gaining increasingly in importance, most of the existing modelling approaches still focus on intra-enterprise business and information processes and only include other actors of the supply chain as a secondary issue. But also the few approaches particularly developed for supply chains show weaknesses since they are just considering single views on the supply chain. Consequently, in order to meet the increasing supply chain orientation, existing modelling approaches have to be aligned to tasks of food supply chain management by including different views into one model. However, the complexity and the resulting multitude of interacting aspects require models which allow analysing all these different aspects in single but interrelated modelling levels. Hence, the present paper gives a multi-level modelling approach considering a requirement, process and information view using the Unified Modeling Language (UML). On a first level, all chain-relevant business transactions are identified, therewith providing a general overview of the most important chain-actor requirements. These requirements serve as the basis for a second level, where the requirements can be further specified by implying detailed processes and resources as well as their interdependencies. Finally, by knowing the detailed activities, a third level can be modelled focusing on information availability and flow. In this paper, the modelling levels as well as their interrelations are presented using the example of the pork chain. Aim of this deterministic approach is to build a modelling framework which can be used as a basis for following probabilistic food supply chain simulation models.

BoniRob: an autonomous field robot platform for individual plant phenotyping

Ruckelshausen, A.[1], Biber, P.[2], Dorna, M.[2], Gremmes, H.[3], Klose, R.[1], Linz, A.[1], Rahe, F.[4], Resch, R.[4], Thiel, M.[1], Trautz, D.[3] and Weiss, U.[2], [1]University of Applied Sciences Osnabrück, Faculty of Engineering and Computer Science, Albrechtstr. 30, 49076 Osnabrück, Germany, [2]Robert Bosch GmbH, Corporate Sector Research and Advance Engineering (CR-AE32), Postfach 30 02 40, 70442 Stuttgart, Germany, [3]University of Applied Sciences Osnabrück, Faculty of Agricultural Science and Landscape Architecture, Oldenburger Landstr. 24, 49090 Osnabrück, Germany, [4]Amazonen-Werke H.Dreyer GmbH & Co. KG, Postfach 53, 49202 Hasbergen-Gaste, Germany; ruckelshausen.os@t-online.de

Electronics, communications and sensor technologies are strongly pushing innovations in agriculture and have become key technologies in this field, offering options for economical as well as ecological benefits. Recently, increased research activities have been started to develop autonomous field robots for future applications in agriculture. However, field tests of first experimental platforms demonstrate the high challenge for the related technologies under complex and dynamic field conditions. The low robustness of the single robot is a major hindrance for the introduction of first products of an autonomous field robot or robot swarms. In order to have a robust platform available, the authors have identified the phenotyping of plants as a task, where interpreted sensor information – such as plant height or spectral signatures - are already the actual result of the robot application. Moreover, a continuous complete characterisation of all plants in a field would be a revolutionary method for agricultural field trials. The realisation of such a system strongly depends on interdisciplinary competences. As a consequence industrial partners from electronics and agricultural equipment as well as research institutes (engineering and computer science as well as agriculture) are working together (funded by the Federal Ministry of Food, Agriculture and Consumer Protection, Germany), experiences in the development of autonomous systems inside and outside agriculture are available. Moreover several cooperative partners as future users of the autonomous field robot are included. The new vehicle 'BoniRob' has been designed thereby combining the mechanical robustness with flexibilities due to measurement heights, row widths and field navigation. Advanced technologies are used for navigation and agro-sensors, including 3D and spectral imaging as well as sensor fusion and probabilistic robotics, also a safety concept is included. The internal system communication (control units, sensors, documentation, user interface) is based on Ethernet and thus allows a real-time implementation. The plant phenotyping is related to a single plant, thus a RTK-DGPS system is integrated; the authors have recently demonstrated the feasibility of an individual plant detection based on sensor fusion and high-accuracy GPS. The concept and results of first field tests of the flexible and modular system will be presented. The intended robustness of the autonomous field robot BoniRob will offer future options for other applications (including actuator developments) and robot swarms.

Preparing a team to field robot event: educational and technological aspects

Oksanen, T.[1], Tiusanen, J.[2] and Kostamo, J.[3], [1]TKK Helsinki University of Technology, Automation and Systems Technology, Otaniementie 17, 02015 TKK, Finland, [2]University of Helsinki, Agrotechnology, Koetilantie 3, 00014 Helsinki, Finland, [3]TKK Helsinki University of Technology, Engineering Design and Production, Sähkömiehentie 4, 02015 TKK, Finland; timo.oksanen@tkk.fi

The Field Robot Event is a yearly competition for small field robots, and it is mainly meant for student teams. A joint student team from Finland has participated the Field Robot Event four times (2005-2008), every year the robot and the student group has changed. A robot could be built among just one technology branch (eg. robotics or agrotechnology) but we have set high interdisciplinary education goals. During the four years 29 students have got experience to build a small agricultural field robot. This paper describes the evolution of the education. The goals of the projects have been: possibility to apply adapted theoretical knowledge; learn team working in a technologically heterogeneous group; build a robot from scratch and to get acquainted with mobile robot subsystems; build something that must actually work; and finally put the result into test in the competition. Several technologies have been tried out in mechanics, mechatronics, software and control implementation. Only the machine vision has been quite similar every time. Choosing suitable RC model car parts has been one educational aspect, but decreasingly since there have been backlashes with adequacy. Robot price is one competition factor and it is hard to find parts with both high quality and low price. Therefore e.g. sensors have been very similar throughout the competition history, which has strongly directed education. On the other hand, low level electronics is one essential part of building a mobile robot and therefore it is necessary to include also those parts in education. Our robots have always had a webcam for row-, weed-, and docking station detection. OpenCV image processing library has been found to have a good computing performance and supporting webcam. More advanced image processing libraries or software tools are available, but those have been considered to be too advanced in sense of to educate machine vision. Signal processing, navigation and position/state estimation has been increasingly done in Matlab/Simulink. In the first year only some controllers were tuned with Simulink, later a kinematic robot model and a 2D simulator were developed. 2008 all position estimation and navigation along with all calculations (excluding machine vision) were done in Simulink. Focusing on one good tool keeps the education concise. In product development more advanced tools with rapid prototyping capabilities are more and more important in the future. C++ code generated from the Simulink model is used for real time computing. For tuning and developing logics etc., the simulation model of robot together with control system runs in Simulink and after tuning it is easy to deploy online code just by pressing a button and connecting signals to real signals in the runtime computer. As this phase is more fluent in the development process compared to traditional software development, there is more time to concentrate developing algorithms and tuning the parameters. The last team developed also a visual simulator that was connected to Simulink kinematic environmental model and in that way it was also possible to simulate the camera image. Simulating robot behavior enabled evaluation of chosen technologies before building the hardware in the short project timeline.

Safe and reliable: further development of a field robot

Griepentrog, H.W.[1], Andersen, N.A.[2], Andersen, J.C.[2], Blanke, M.[2], Heinemann, O.[3], Nielsen, J.[1], Pedersen, S.M.[1], Ravn, O.[2], Madsen, T.[4] and Wulfsohn, D.[1], [1]University of Copenhagen, Hojbakkegaard Alle 30, 2630 Taastrup, Denmark, [2]Technical University of Denmark, Ørsteds Plads, 2800 Lyngby, Denmark, [3]Hako Werke, Hamburger Straße 209-239, 23843 Bad Oldesloe, Germany, [4]Agrocom Vision, Boegeskovvej 6, 3490 Kvistgaard, Denmark; hwg@life.ku.dk

High levels of reliability and safety are necessary and can be achieved by novel developments within automated perception, diagnosis and decision making and fault tolerant operation. The project builds upon recent developments of outdoor robots for autonomous field operations at Copenhagen University and Technical University of Denmark. The aim is to add functionality to the existing robot prototype so that it will behave in a safe, reliable and effective manner under unmanned operation. The specific project objectives are to (a) specify and identify descriptions of reliable and safe behaviors and strategies under outdoor unmanned conditions, that meet legislative as well as task-determined requirements, (b) develop sensor-based perception system for relevant description of the machine's behavior as well as it's environment, (c) develop internal generic robot fault handling and decision support routines, (d) add hardware and software components to the existing field robot, (e) upgrade user interface and usability routines, (f) implement and test the improved robot and (g) conduct an economic and technology assessment. A failure modes and effects analysis (FMEA) was conducted for the existing machine and the intended autonomous operations. The identified and ranked risks and hazards were used to select safety procedures, components and sensor technologies in order to fulfil the requirements. Furthermore the controller system was redesigned and signal monitoring software was developed to achieve higher levels of reliable and fault tolerant control. Additionally the existing machine system and the proposed operation conditions were assessed by a machine safety consultant to check the compliance of existing legal requirements. Based on FMEA and legal safety consultancy a redesign of the machine system was completed as follows: The supposed operation is spraying and mowing in semi-public orchards, the machine will be unmanned but not unattended, an operator prepares, checks and starts the machine in the operation area, during machine operation the operator is around and is supposed to complete other working tasks, an additional stereo vision camera and a laser scanner will provide more information about the machine environment to avoid collision with obstacles and improves the machine navigation, this will contribute to reduce the dependency on GPS information, additional bumper switches allow the machine to stop when other collision sensors fail and fault tolerant software routines allow reliable machine controls. Conclusions are to reduce the identified risks that (1) additional hardware and software monitoring is necessary to optimise the systems behaviour and (2) the operations have to be reduced to particular applications under semi-public conditions to achieve a reasonable economic viability. Some safety risks had to be excluded or tried to be met by defining rules e.g. for the operator behavior. It was identified that due to its complexity the most difficult and challenging task is to increase the reliability and fault tolerant behaviour of the installed computer control system.

Simple tunable control for automatic guidance of four-wheel steered vehicles

Bakker, T.[1], Van Asselt, C.J.[2], Bontsema, J.[3], Müller, J.[4] and Van Straten, G.[2], [1]Tyker Technology, Papiermakersbeek 16, 3772 SV, Netherlands, [2]Wageningen University, Systems and Control Group, P.O. Box 17, 6700 AA Wageningen, Netherlands, [3]WUR Greenhouse Horticulture, P.O. Box 17, 6700 AA Wageningen, Netherlands, [4]University of Hohenheim, Institute for Agricultural Engineering, 70593, Stuttgart, Germany

Automation of agricultural machinery reduces driver fatigue and improves precision of the work performed and could finally exclude the need for a driver. Automatic guidance solutions exist for two-wheel steered and tracked tractors. The kinematics of these machines are systems with two degree of freedom: driving straight (in x-direction, with straight front wheels or same speed of tracks) and rotating around an instantaneous centre of rotation (when steering). With the tendency of increasing sizes of agricultural machines, more machinery is equipped with four wheel steering to keep some degree of maneuverability with these big machines. Four wheel steered vehicles have in principle three degrees of freedom (drive in x-direction, rotate, and drive in y-direction), but in practice the steering angles of front and rear wheels of these machines are normally coupled. The way of coupling can be different 'steering modes' like: front and rear wheels steer in opposite direction, only front-wheel steering is used, or the front and rear wheels steer in parallel and sometimes the rear wheel is set to a fixed offset to obtain 'dog-like' moving. Because these steering modes predefine the relation between front and rear wheel angles at any time, these steering modes limit the vehicles' degrees of freedom: in case of the opposite steering to rotating around a point and driving straight (two degrees), and in case parallel steering to moving sideways or straight (two degrees). It is obvious that using the degrees of freedom of a four-wheel steered vehicle by not having a fixed relation between front and rear wheels could potentially improve its maneuverability, and improves headland turning performance. The objective of this research was to develop a practical method for path following control applicable for any four-wheel-steered vehicle using its full degrees of freedom. It should be possible to define the vehicle control point as a position in the coordinate system attached to the vehicle, which can be different depending on the application. The controller should be simple to tune and the controller should be able to handle wheel steering angle constraints. The path following control consists of a method that inverts the kinematic model of the four-wheel steered vehicle, into a very simple model. For this simple model two PI controllers are designed. The tuning of the PI-controller is simple and straightforward. The inversion method always results in one unique solution for the actual control which satisfies the constraints. For the implementation, the kinematic model inversion can just be seen as a general applicable software module that can be set according to the vehicles' parameters. The method is implemented on the Intelligent Autonomous Weeder platform of Wageningen University. Path following results presented in figures and video show its performance and practical applicability.

Development of a small agricultural field inspection vehicle

Gottschalk, R., Burgos-Artizzu, X.P. and Ribeiro, A., Instituto de Automatica Industrial - CSIC, Systems, Crta. Campo Real, km. 0.2. La Poveda, 28500 Arganda del Ray - Madrid, Spain

Weed mapping is a valuable tool when it comes to optimising the use of resources in managing the attempts that are made to control weeds. Weed mapping has traditionally been done using random sampling techniques amongst others, using a team of experts working in the field. One alternative to this tedious task, which requires moving the experts with the consequent associated cost, is the use of photographic samples, with and without controlled lighting. Once in the laboratory, an expert can estimate the amount of weeds and the growth state of the crop, on the basis of the set of photos that have been taken at the georeferenced sampling points. This method can be used to observe the photos several times and to correct errors in estimation, in view of the fact that the visual assessment of the photos is a subjective process in which it is easy to adapt the observation to an overall situation; for example, an average density can be estimated as high if the set of photos is not very dense and as low if the opposite is the case. This paper presents the first vehicle prototype developed at the IAI-CSIC that is able to move autonomously across agricultural fields, taking high-resolution photos, in sampling points defined by the user, for the generation of a weed distribution map. The used vehicle was the Traxxas E-MAXX electric driven model truck. It is approximately 52cm long, 42 cm wide and has a ground clearance of 10 cm. The vehicle is powered by two 7.2 V battery packs and has two main driving motors and two steering motors. It is four-wheel driven which enables the vehicle to move through rouge terrain. All the important steering and drive chain parts which need to withstand aggressive movement are made. The only industrial component used for the entire vehicle was the on-board PC/104 computer system including its power supply. The variety of connections of the PC/104 was a big advantage for connecting peripheral devices such as the camera, the motor controller, an external hard drive and a wireless access point used for external vehicle supervision and remote control. The developed vehicle uses the webcam for the navigation between two lines of agricultural crop. The relative vehicle position is calculated by segmentation and classification of the images and by then extracting geometrical lines corresponding to the crop rows. Webcams typically have a viewing angle of less than 50 which would make it necessary to mount the camera at an approximate height of 150 cm. The 'Live! Cam Notebook Ultra' developed by Creative was chosen to overcome previous problem, since it has a viewing angle of approximately 80. The autonomous vehicle was tested successfully in an agricultural environment without controlled lighting.

Assessing the potential of automatic section control

Stombaugh, T.S., Zandonadi, R.S., Dowdy, T. and Shearer, S.A., University of Kentucky, Biosystems and Agricultural Engineering, 128 Barnhart Bldg., 40546, USA

One of the newest innovations in Precision Agriculture is automatic section control for application equipment. Automatic section control systems will continuously record the areas of a field that have been covered during a field operation and then automatically turn on and off sub-sections of a machine to prevent double coverage of previously treated areas. Research on a cooperator farm in Kentucky, USA has shown potential savings of as much as 25% in very oddly-shaped fields using automatic section control. On the other hand, potential savings in rectangular fields would be almost zero. Likewise, larger machine or section widths will cause overlapped areas to increase. To make an informed decision about purchasing and implementing the technology, producers need to know the potential savings for a given field shape. This paper describes a software tool that can be used to evaluate the percentage of a field that would be overlapped for different boom section sizes.

Performance of auto-boom control for agricultural sprayers

Molin, J.P., Reynaldo, E.F. and Povh, F.P., University of Sao Paulo, Rural Engineering, Av. Pádua Dias, 11, 13418-900 Piracicaba, SP, Brazil

Auxiliary guidance systems like light bars and the auto guidance systems have been intensively adopted by farmers. One of these devices is the auto-boom control, dedicated to sprayers to provide both guidance and to control individual sprayer sections or individual nozzles. The use of those devices is based on a field boundary and internal overlapping georeferenced by a GPS receiver. If its accuracy is limited, the equipment will not offer the amount of savings in chemicals and in environmental mitigation and based on that we developed and used a procedure for measuring the impact of different GPS signals and compared the effect of vehicle speed and field boundary shapes. We also compared the use of a commercial auto-boom control versus manual control by operators with low and high training abilities, based on no auxiliary devices and with light bar showing the borders on the screen for day and night operation. The GPS signals tested were RTK, Omnistar VBS and no correction with internal filtering on the firmware. The vehicle was a self-propelled spray machine with a 24 m spray boom width separated in four or eight sections and the speeds were 7, 18 and 30 km h^{-1}. The shapes of field boundary were simulated by entrances or exits at 90^0, 45^0, and 30^0. The borders were saved on the auto-boom control memory and as the vehicle crossed the border at each treatment and each of the tree replication the time of turning on or off the boom sections was detected by a pressure sensor on the nozzle. An infrared sensor detected the exact moment of the boom crossing the border. The data is under analysis, indicating that the use of low accuracy GPS signal like the use of internal firmware only, is not sufficient for a minimum boom control accuracy that may provide the benefits that this kind of technology offers, with results similar to the manual operation. The information is crucial, especially for countries where public augmentation systems are not available, indicating that users may have to upgrade the GPS signal with extra costs.

The development of a computer vision based and real-time plant tracking system for dot spraying
Nørremark, M., Olsen, H.J. and Lund, I., University of Aarhus, Department of Agricultural Engineering, Blichers Allé 20, DK-8830 Tjele, Denmark

Deformable object tracking in real time is essential for many applications such as control of autonomous vehicles, surveillance, agriculture automation, and so on. It is desirable that an autonomous vehicle for plant nursing perform guidance and executing high precision tasks e.g. to crops and weeds. Spraying only a single droplet on a leaf of a weed (dot sprayer) that eliminate the leaching of herbicides to the environment is one such high precision task, where the computer vision system should maintain a dynamic visual model of the external world and positioning the spray target in real time. The objectives of the presented research were (1) to develop a computer vision system for tracking green objects of the size of seedling leaves, (2) to estimate position of spray objects in real time while the vision system moves over the object, and (3) to test the system with a dot sprayer on a test stand in the laboratory. The main components of the dot sprayer system was a digital colour camera equipped with an embedded computer (eXcite exA640-120c, Basler Vision Tech., GER). Images covered a 180x135 mm area on the soil surface. The spraying unit consisted of 16 small solenoid valves (Willett 800, Videojet Tech. Inc., US), the camera and a microprocessor. The microprocessor received data packets for which valves to open from the camera's serial port. Plants were detected as blobs and positions were determined by the blob's centroid coordinate in the image. These coordinate values were stored in the memory and for a new image the position of each detected blob was compared to the previous ones. If a blob in the recent image was found to be at the same x-position as a blob in the previous image and a defined distance behind it in the y-direction, it is taken as a new position for the blob. In this way the blobs was tracked through the image from top to bottom and eventually enter a zone in the bottom part of the image. This zone was divided in 16 squares each having a size of approximately 10 by 10 mm. When a blob was located within such a square the corresponding valve was ejecting a droplet of 0.5 microliter to the centre position of the square. The dot spraying system was tested with simulated plants placed on a plane surface. All simulated plants were sprayed with a single droplet. However, the research has realized several problems to be solved before the computer vision based and real-time plant tracking system for dot spraying can be operational in the field.

SensiSpray: site-specific precise dosing of pesticides by on-line sensing
Van De Zande, J.C., Achten, V.T.J.M., Kempenaar, C., Michielsen, J.M.G.P., Van Der Schans, D., Stallinga, H. and Van Velde, P., Wageningen UR, P.O. Box 16, 6700AA Wageningen, Netherlands

Crop protection products (CPP) are used in agriculture to protect the crop against pests, diseases and weeds. The use of CPP assures high yields at harvest and high quality agricultural products. Several sprayings of CPP are done per crop per season. The used amount of CPP per spraying is generally based on the advised dose and generally does not take into account what is to be sprayed, the target. During the growing season generally the dose is not adapted to changes in crop canopy structure. More often the whole field is treated uniformly although spraying some patches in the field would have been sufficient. In order to deal with these variations in crop development and site-specific variations in the field, a sensor based spray technology was developed; SensiSpray. The system can be built on a boom sprayer and consists of sensors to detect crop variation and a spray system to automatically change spray volume depending on the sensor signal. The sensors used were Ntech Greenseeker sensors measuring crop reflection from which the NDVI output signal was used. Electronics and software were developed to use the output signal of the Greenseeker sensor to adapt spray volume. For the variation of spray volume Lechler VarioSelect nozzle bodies were used fitted with four different low-drift venturi flat fan nozzles (Lechler IDKN 12001, 120015, 12002, 120025). On a 27 m working width boom sprayer 7 sensors were placed for each boom section to control spray volume per boom section of 3-4.5m wide. Spray deposition measurements were performed to test the sprayer accuracy in adapting spray volume based on the reflection signal per section. A grassland field was prepared in 20 m wide bands which were next to each other extra fertilized, mowed and treated with a herbicide to create difference in vegetation colour and therefore reflection. The sprayer was driven [speed ?] over the field in an angle of 45° to show the individual section changes on the sprayer boom. Measuring spray deposit was done using a fluorescent dye (Brilliant Sulpho Flavine) added to the water in the spray tank. The individual section sensor reacts on the change in reflection and spray volume was adapted per boom section. The spray volume and therefore dose adaptation was according to the changes in measured reflection (NDVI) of the different bands of grassland. The accuracy of the system to adapt spray volume was within 1-2 m of the borderline of the different grass bands. To demonstrate the potential use of the SensiSpray system it was used in potato haulm killing spraying. The variation in the field of the greenness of the potato canopy at desiccation spraying before harvest was used to vary spray volume of the herbicide used. The spray volume and dose adaptation was based on calculation rules of the Minimum Lethal Herbicide Dose system (MLHD) relating reflection measurements with minimum dose needed to kill of potato canopy. In the 2007 and 2008 season different potato fields were sprayed and a general use reduction in CPP for potato haulm killing was circa 50%. During the 2008 season first tests were done in late blight (*Phytophthora infestans*) control in potatoes adapting spray volume to the crop development during the growing season. First results show no difference in disease development between conventional spraying and canopy adapted dose spraying with the SensiSpray system with good protection against late blight.

Comparing companies and strategies: a genetic algorithms approach

Hennen, W.H.G.J., LEI, Plants Systems, Alexanderveld 5, 2585 DB The Hague, Netherlands

Different types of reference values or standards can be used to compare the results of a company, e.g. data from previous years (historical), data based on planning or simulation models, normative standards based on scientific research and average performance of a comparable group (external comparison within the same year). The last type is appealing as entrepreneurs are eager to compare their results with other comparable companies. The problem is to find a suitable comparable group. The method and tool Face-IT (Farm Accounts Compared by Evolutional Improving to Top-combination) has been developed to compare a company with the average of a comparable group of companies and explore the consequences of strategies for the company. An application in the dairy sector with accounting data is successfully used in several projects. To find a suitable or even optimal group of companies out of numerous different combinations, a genetic algorithm was developed and implemented as a search mechanism. The objective is to find companies that are not only individually comparable, but as a group also have the same average values of the selected classification factors. The method as well as applications are described in this paper. As a very generic comparison tool, Face-IT can also be applied in surveys, or as a marketing instrument or for exploring success factors.

Prioritizing agriculture research projects emphasizing on analytical hierarchical process (AHP)

Mortazavi, M., Rajabbeigi, M. and Mokhtari, H., Institute of Technical & Vocational Higher Education of Agriculture, P.O. Box 13145-1757, Tehran, Iran

The ultimate goal of an agriculture research system is on-time, correct and clear response to the problems and expectations of agriculture household and stakeholders. In this respect, though, due to variation and frequency of the problems and expectations and as well as many limitations such as financial deficit, short time and shortage in work force and equipments etc, the system can not be thoroughly responsive. Therefore, the necessity for optimizing the system to response through prioritizing the research projects has been a major challenge before the responsible managers and authorities. In this paper, the analytical hierarchical process (AHP) has been introduced as a well known multi criteria decision making method that combines qualitative and quantitative criteria for prioritizing the research projects of the Institute for Modification and Provision of Sugar Beet Seed. For the implementation of the mentioned methodology, principals and methods of prioritizing the research projects have been studies and then by determining the final decision making criteria the priority of the projects in the Institute have been determined by drawing decision hierarchy tree. Required data was gathered through pair wise comparison questionnaires filled by the experts and researchers. Finally we used expert choice software to analyze and determine the priorities.

Using formal concept analysis to calculate similarities among corn diseases

Souza, K.X.S., Massruhá, S.M.F.S. and Fernandes, V.M., Embrapa Agricultural Informatics, Av. André Tosello, 209 - Barão Geraldo Caixa Postal 6041, 13083-886 - Campinas, SP, Brazil

Diagnostic reasoning either automated or conducted by diagnosticians relies on having a set of symptoms and trying to match them against a (sparse) matrix containing the complete set of diseases and their corresponding symptoms. Whenever a symptom manifests itself, the diagnostician activates in this matrix several possible diseases which could manifest that symptom. The reasoning proceeds up to the point in which a very small subset of diseases (ideally one) are known to cause those symptoms. The reasoning process that goes from the causes to the consequences has received increasing attention since the proposition of Parsimonious Covering Theory by Peng & Reggia in early 90's, in opposition to the one commonly found in expert systems that were based from the consequences to the causes. As the number of symptoms is usually large it is necessary to group similar diseases together in such a way that the most frequent symptoms are asked first. In this way, the reasoning further reduces the space of possible diseases by excluding those that do not manifest that symptom. This paper evaluates similarities among diseases considering the set of common and distinct symptoms and proposes a method for structuring the space of search. The similarity measure used in this paper was based on Formal Concept Analysis (FCA), a data analysis technique based on Lattice Theory and Propositional Calculus. FCA is especially suitable for exploration of symbolic knowledge (concepts) contained in a formal context, such as a corpus, a database, or an ontology. For the application of FCA, the mathematical relation <diseases, symptoms> expressed in the above matrix is mapped to FCA's. As a result, the ordering algorithm produces a mathematical structure called Concept Lattice, which shows on the top the most common symptom and in the bottom the least frequent one. The diseases appear linked to the point which encompasses, in the lattice hierarchy, all respective symptoms. Using the lattice, the similarity measure evaluates how close diseases are by counting the number of structuring elements they have in common. The application of FCA to this problem gave interesting results, as for instance, sets of symptoms that never occur in isolation, which indicate that perhaps the system should ask for only one of them since the other is implied. Another result was that diseases with greater similarity value coincided with those grouped together by a human expert (phyto-pathologist), an indication of the quality of the similarity measure.

Probability distribution on the causal bacteria of clinical mastitis in dairy cows using naive Bayesian networks

Steeneveld, W.[1], Van Der Gaag, L.C.[2], Barkema, H.W.[3] and Hogeveen, H.[1,4], [1]Utrecht University, Department of farm animal health, marburglaan 2, 3584 CN Utrecht, Netherlands, [2]Utrecht University, Department of Information and Computing Sciences, Padualaan 14, 3584 CH Utrecht, Netherlands, [3]University of Calgary, Department of Production Animal Health, T2N 4N1 Calgary, Canada, [4]Wageningen University, Business Economics, Hollandseweg 1, 6706 KN Wageningen, Netherlands

Clinical mastitis (CM), an inflammation of the udder in dairy cows, needs an effective treatment to eliminate the infection that causes it. CM can be caused by a wide variety of bacteria. Knowing the causal bacteria and, subsequently, using appropriate treatments, will increase the cure rate of CM. Bacteriological culture will provide the information about the causal bacteria. However, because CM needs a treatment immediately after diagnosis, culture information comes too late. In the absence of bacteriological culture results, several other sources of information are available on a dairy farm that could aid in the diagnosis of the causal bacteria of a CM case. The objective was to provide a posterior probability distribution on the Gram-status (Gram-negative vs. Gram-positive bacteria) for CM cases based on information on clinical signs, several cow factors and season of the year, and to examine the accuracy of these probabilities. The posterior probability distribution on the Gram-status of the bacteria was developed using Naive Bayesian Networks (NBN). A NBN will give a probability for the outcome variable being in a certain state. These networks can handle missing values and are known for their powerful classification performance. Data were used from 274 Dutch dairy herds that recorded CM over an 18-month period. The final dataset contained information on 3,534 CM cases, all cases were classified into Gram-negative and Gram-positive bacteria. The NBN was constructed from data using a wrapper method with forward selection on variables. Several information sources of a cow that are usually available at a dairy farm were included as variables in the NBN. These information sources include cow factors (parity, month in lactation and quarter position), historical data (on somatic cell count and CM), clinical signs (being sick or not, and color and texture of the milk) and season of the year. For each CM case a posterior probability distribution on the Gram-status was determined. Using the most likely bacteria (with the highest posterior probability) as the predicted value, the accuracy of classifying CM cases into Gram-positive or Gram-negative pathogens was 73%. Since only CM cases with a high probability on a single causal pathogen will be considered for pathogen-specific treatment, accuracies based on only classifying CM cases above a particular probability were determined. For instance, for 30% of the CM cases, classification according to the Gram-status reached an accuracy of 97%. The presented NBN can be used as a guideline for decision support on choice and duration of treatment of CM cases. For CM cases with high posterior probabilities of CM being caused by Gram-negative or Gram-positive bacteria a more specific treatment can be chosen, which will result in higher cure rates. While for CM cases with almost equal posterior probabilities on the Gram-status a broad spectrum antibiotic treatment needs to be considered. On a farm, the availability of a probability distribution is an improvement in comparison with the current situation, where information on the causal bacteria is not provided at all.

Modeling an automated system to identify and classify stingless bees using the wing morphometry: a pattern recognition approach

Bueno, J.[1], Francoy, T.[2], Imperatriz-Fonseca, V.[2] and Saraiva, A.[1], [1]Polytechnic School of Universidade de São Paulo, Computer and Digital Systems Engineering Department, Av. Prof. Luciano Gualberto, trav. 3 n. 158, 05508-900 São Paulo SP, Brazil, [2]Medical Faculty of Universidade de São Paulo, Department of Genetics, Av. Bandeirantes, 3900, 14049-900 Ribeirão Preto SP, Brazil

This work presents a new methodology for identification and taxonomic classification of stingless bees. The purpose of this work is to model a computer system with pattern recognition approach which allows automating all identification and taxonomic classification activities of stingless bees using data from the wing morphometry. The main focus is to make use of the latest researches in computing applied to pattern recognition for identification and classification of objects with pattern database for queries. The proposed model uses a database to store the pattern of stingless bee wings to retain the historical records of successful identification and taxonomic classification. This database allows the implementation of queries over these wing patterns of species stored comparing them with the pattern obtained from the image of the wing under analysis. The pattern recognition is subject of recent researches and has experienced significant advances in iris recognition, face recognition, fingerprint recognition, and applications for military use for target recognition. What is proposed is to use the geometric morphometrics techniques to get the wing pattern of stingless bees for the species. This pattern is present in the forms of biological structures being obtained by the manual plot of biological landmarks. The number of points obtained with the landmarks is the shape of the object that must be invariant to translation, rotation and scaling. The initial stages that result in obtaining the wing pattern of stingless bees are being developed by the group of construction of portable entomological scanner for field use. The image is initially submitted to a cleaning process to remove the unwanted noise and select the area of interest for classification. With the clean image of the selected wing, the effects of binarization is applied to provide greater contrast in the detection of venation and contour. This process results in an image with precise limits, maintaining the original characteristics, such as the flaws or irregularities in the venation (synapomorphy). The binary image with venation and contour is then submitted to the landmark automatic plot. The set of points obtained with landmarks data are the shape of object that is invariant to translation, rotation and scaling. At the selection and training stage for pattern recognition, the file with landmarks data will be applied to non-linear classifiers built with artificial neural networks, and then applied to statistical classification packages. A comparison of techniques will choose the best classifier, in terms of classification efficiency and performance along with the assessment made by researchers of the Bee Laboratory. The first product of this work is the Data exchange automation interface between Geometric Morphometrics programs and statistical analysis software used for stingless bee classification. Another important result was the transfer of knowledge involved in the identification and taxonomic classification process used by biologists, designed by the group of engineering Automation Agricultural Laboratory, which allows modeling the system for automated identification of stingless bees.

Image analysis and 3D geometry modeling in investigating agri-food product properties

Weres, J., Nowakowski, K., Rogacki, P. and Jarysz, M., Poznan University of Life Sciences, Institute of Agricultural Engineering, Department of Applied Informatics, Wojska Polskiego 28, 60-637 Poznań, Poland; weres@up.poznan.pl

Knowledge of properties of agri-food products is of critical importance to understand and predict their behavior during various thermal and mechanical processes, and to design and optimize agri-food product processing technologies. Recent advances in image analysis, neural-based image recognition, shape measurement and 3D geometry modeling extend our possibilities in investigating agri-food product properties. Several original methods and computer applications were developed and applied to classify and estimate quality of various agri-food products, e.g. dried vegetables and cereal grain kernels with respect to external defects – cracks and breakage. A special procedure was developed, based on image analysis applied to consecutive sections of an object, to represent its geometry by a 3D finite element structural mesh, and then to visualize the object and its properties. Alternative approaches were also used to represent a shape by generating corresponding point clouds – a photogrammetric approach and 3D scanning approach. More accurate modeling of geometry of bio-materials resulted in increased accuracy of predicting thermal and mechanical behavior of cereal grain kernels and wood, with the use of our original finite element inverse analysis software.

A technical opportunity index based on a fuzzy footprint of the machine for site-specific management: application to viticulture

Paoli, J.N.[1], Tisseyre, B.[2], Strauss, O.[3] and Mcbratney, A.B.[4], [1]ENESAD, UP GAP, 26 Bd Dr Petitjean - BP 87999, 21079 Dijon Cedex, France, [2]Montpellier SupAgro, UMR ITAP, 2, Place Pierre Viala, 34060 Montpellier Cedex 1, France, [3]LIRMM, Department of Robotics, 161, Rue Ada, 34 392 Montpellier cedex 5, France, [4]The University of Sydney, ACPA, McMillan Building A05, NSW 2006, Australia

The problem of deciding whether or not the spatial variation of a field allows a reliable variable-rate application is of critical importance for the farmers. An other critical problem is to know if there is a particular threshold (field segmentation) which allows the minimisation of the error of the site-specific application. Therefore, there is a need in decision support tools allowing the farmers to know if it is feasible and relevant to manage the spatial variability of a particular field. Very few papers have dealt with this problem in the literature. Two original approaches were proposed. The first one is based on the definition of the Oi (Opportunity index) which uses geostatistics to derive an index allowing to rank the fields from the most opportune to the less opportune. However this last approach doesn't provide any practical information on the way the machine operates on the field and the resulting error. To answer this problem, a TOi (Technical opportunity index) has been more recently proposed. The TOi answers the question to decide whether or not the spatial variation of a field allows a reliable-rate application and to discover if a field segmentation (choice of a particular threshold) is technically feasible according to the application machinery. However the TOi still presents significant drawbacks. It requires data arranged on a regular grid which necessarily involves an interpolation method (kriging). Depending on the data resolution, this step may lead to an over or an under estimation of the opportunity to manage the variability. An other significant drawback leads to the impossibility of the TOi to take into account the inaccuracy of the data that the decision is based on and the inaccuracy of the variable rate controller (either in location due to the positioning system or in the ability of the machine to change the levels of treatment). The goal of the paper is to propose a Fuzzy Technical Opportunity index (FTOi). This FTOi considers: (1) a fuzzy footprint model of the Variable-Rate Application Controller (VRAC), which describes the area within which the VRAC can reliably operate; (2) the location inaccuracy of the data; (3) the ability (accuracy) of the VRAC to perform distinct levels of treatment. The originality of our approach is based on the use of a fuzzy estimation process to decide if a level of treatment is reliable or not on each area over which the VRAC operates. It does not need any pre-processing of the data like kriging. Tests on theoretical fields, obtained from a simulated annealing procedure and real fields, showed that the FTOi was relevant to assess how the variability was technically manageable. Tests also showed that our approach was able to consider problems of lack of data (data resolution).

Expedient spatial differentiation of technologies of precise agriculture according to productivity factors

Uskov, A.O. and Zaharian, J.G., Agrophysical research institute, Agroclimatology, 14 Grazhdansky prosp., 195220 Saint-Petersburg, Russian Federation

The methodological approach of allocation of contours is based on criterion of expediency on an organizational-economic indicator in the form of losses, or on an economic gain. Minimization of losses on each of factors of efficiency and the least expenses for realization of all cycle of technological process in systems of precise agriculture is defined by expenses for soil machining, preseeding application of fertilizers, fractional application of fertilizers during the vegetative period, chemical processing of crops on protection of plants against illnesses and wreckers. Each of listed above receptions requires own technological borders within a processed field (crops) on the basis of the information on a condition of the operated parameter or the factor. Each of factors has their 'own' contours. In many of known systems of precise agriculture the technological differentiation of contours can be estimated either according to agrochemical inspection, or on spatial heterogeneity of efficiency within a field. Depending on degree of spatial differentiation and volume of the initial information on factors which is necessary for acceptance and realization of the corresponding decisions, all variants of planned technologies in non-uniform territory can be divided into three basic classes: not differentiated strategy (NDS); in details differentiated strategy (DDS); partially differentiated strategy (PDS). Elementary of them is the NDS, focused on an average on territory value of some varying factor. Unlike the NDS strategy of DDS class assume a spatial variation of accepted agrotechnical decisions taking into account concrete conditions on each elementary platform of averaging of the measured factor. For realization of DDS it is necessary to have detailed data about spatial variability of the influencing factor. In this case the agriculturist should have the corresponding card reflecting spatial distribution. In actual practice it is always expedient to deal not with DDS, but with PDS. In each point of considered territory the decision is made individually, and it is the maximal account of heterogeneity. Other extreme case - planning for all territory without spatial variability. In the report the algorithm of computer realization of numerical experiment for choice expedient differentiation is resulted. For each planned technological operation it is necessary to appoint number of gradation on greatest resolving 'ability' of agroreception of management efficiency. Other restrictive criterion - cost of execution of the technological program at change of options of working mechanisms before each transition to the subsequent contour. Last criterion - a choice of the parameters providing planned efficiency. Set of three noted conditions will define a surface of border of expedient degree of differentiation on contours within each field.

Methodology to estimate economic levels of profitability of precision agriculture: simulation for crop systems in Haute-Normandie

Bourgain, O. and Llorens, J.-M., Esitpa, agricultural engineering school, 3 rue du Tronquet, 76134 Mont-Saint-Aignan cedex, France

Modern agriculture requires decision criteria applicable to different scales of territory in order to reconcile both productivity and the respect of the environment. Taking into account the current fluctuation of raw material costs and prices, the assessment of the economic profitability of new techniques such as precision agriculture is becoming more and more important. Different studies have shown that site specific crop management (SSCM) has no negative effects on yield levels. Most of the time, the technical results obtained were evaluated only partially from an economic point of view: limited to increased profit due to amount of fertilizer reduction. Moreover, these studies have been carried out on limited crops (mainly wheat) or specific types of input (nitrogen in general). However, very few studies on global crop systems have been undertaken due to the complexity of the work involved. The two mains aims of this project are: 1) To define a methodology and relevant technical economic indicators to try to estimate the profitability of precision agriculture for the total economic activities of the farm. 2) To apply this methodology to carry out simulations. Variable factors being soil heterogeneity and agricultural practices (the use or not of precision agriculture). Our final scale of investigation is the economic farming system. We have selected three crop systems (cereals, sugar beet-flax and potatoes). Each system is composed of the rotation of seven or eight different crops maximum. We therefore collected the information concerning the technical practices of each crop and the current prices of inputs required. We used standard economic data representative of the region. The economic factors selected were principally production costs and direct profit margin. Using Olympe (economic simulator software) we were able to work at different scales (field, crop system and farm). Thus we simulated the economic profitability of SSCM machines in the most current crop systems in Haute-Normandie. This methodology allows us to assess different levels of profitability concerning investments of the SSCM machines required in relation to soil heterogeneity levels. Furthermore, the methodology together with the specialised software had proved their efficiency to assist farmers, agricultural advisers and also regional policy decision-makers.

Evolution from 2002 to 2008 of the economic interest of precision agriculture in fertilisation management

Perrin, O. and Llorens, J.-M., ESITPA Agricultural engineering school, 3 rue du tronquet, 76134 Mont-Saint-Aignan cedex, France

Modern agriculture requires decision criteria applicable to different scales of territory in order to reconcile both productivity and the respect of the environment. Precision agriculture answers these conditions: it produces a reduction in the amount of fertilizers, while preserving the crop output level. In a context where the input prices are constantly increasing, it is important to evaluate the economic interest of this new technique. Défisol (Association of farmers from GRCETA of Evreucin) carried out experiments in precision agriculture from 2003 to 2006. These tests demonstrated a significant reduction in the use of nitrate and phospho-potassic fertilizers through the practice of precision agriculture. Some studies developed an economic argument, but this is restricted to the reduction of input use at the field level. We propose to add to this initial approach in two supplementary parts: The first aims at determining the impact of precision agriculture on the economic performance of two crops (oil seed rape and wheat). This impact is quantified in both high and low input levels and is calculated thanks to management accounting indicators applied to farms using or not using precision agriculture. The economic efficiency of the inputs is analyzed in order to determine the impact on farm autonomy with respect to the production factors. The second aims at estimating the evolution of the economic interest of precision agriculture from 2002 to 2008, a period that was marked by strong price fluctuation for agricultural inputs and outputs. For this purpose, average prices corresponding to this period are associated with the quantitative data available. The economic data extrapolated allows us to estimate the evolution of the indicators of accountancy and the economic efficiency of the farms. The economic sample data on which this analysis is based is composed of farm results obtained from the region of Haute-Normandie (France), where precision agricultural tests were carried out. The agronomic quantitative data obtained was developed by yield modelling and by geostatistic interpolation. This paper thus confirms the interest of precision agriculture in order for farmers to maximalise their economic profit, always taking into account the safeguard of the environment. Precision agriculture reduces the effects of market fluctuations and therefore has a positive influence on farm autonomy. Our study indicates that the economic interest of precision agriculture is becoming more and more important in the current context of price rises for agricultural inputs and outputs.

Modeling precision dairy farming technology investment decisions

Bewley, J.M.[1,2], Boehlje, M.D.[2], Gray, A.W.[2], Hogeveen, H.[3], Eicher, S.D.[2] and Schutz, M.M.[2], [1]University of Kentucky, 407 WP Garrigus Building, Lexington, KY 40546, USA, [2]Purdue University, 125 South Russell Street, West Lafayette, IN, 47906, USA, [3]Utrecht University, Postbus 80163, 3508 TD Utrecht, USA

A dynamic, stochastic, mechanistic simulation model of a dairy enterprise was developed to evaluate the cost and benefit streams coinciding with investments in Precision Dairy Farming technologies. The model was constructed to embody the biological and economical complexities of a dairy farm system within a partial budgeting framework. A primary objective was to establish a flexible, user-friendly, farm-specific, decision-making tool for dairy producers or their advisers and technology manufacturers. The basic deterministic model was created in Microsoft Excel (Microsoft, Seattle, WA). The @Risk add-in (Palisade Corporation, Ithaca, NY) for Excel was employed to account for the stochastic nature of key variables within a Monte Carlo simulation. Net present value was the primary metric used to assess the economic profitability of investments. The model comprised a series of modules, which synergistically provided the necessary inputs for profitability analysis. Estimates of biological relationships within the model were obtained from the literature in an attempt to represent an average or typical U.S. dairy. Technology benefits were appraised from the resulting impact on disease incidence, disease impact, and reproductive performance. The economic feasibility of investment in an automated BCS system was explored to demonstrate the utility of this model. Automated body condition scoring (BCS) through extraction of information from digital images has been demonstrated to be feasible; and commercial technologies are under development. An expert opinion survey was conducted to obtain estimates of potential improvements from adoption of this technology. Benefits were estimated through assessment of the impact of BCS on the incidences of ketosis, milk fever, and metritis; conception rate at first service; and energy efficiency. Improvements in reproductive performance had the greatest influence on revenues followed by energy efficiency and disease reduction, in order. The impact of disease reduction was less than anticipated because the ideal BCS indicated by experts resulted in a simulated increase in the proportion of cows with BCS at calving ≥ 3.50. The estimates for disease risks and conception rates, obtained from literature, however, suggested that this increase would result in increased incidence of disease. Stochastic variables that had the most influence on NPV were: variable cost increases after technology adoption; the odds ratios for ketosis and milk fever incidence and conception rates at first service associated with varying BCS ranges; uncertainty of the impact of ketosis, milk fever, and metritis on days open, unrealized milk, veterinary costs, labor, and discarded milk; and the change in the percentage of cows with BCS at calving ≤ 3.25 before and after technology adoption. The deterministic inputs impacting NPV were herd size, management level, and level of milk production. Investment in this technology may be profitable; but results were very herd-specific. A simulation modeling a deterministic 25% decrease in the percentage of cows with BCS at calving ≤ 3.25 demonstrated a positive NPV in 866 of 1000 iterations. Investment decisions for Precision Dairy Farming technologies can be analyzed with input of herd-specific values using this model.

Economic modeling of livestock diseases: a decision support tool

Bennett, R.M., Mcclement, I. and Mcfarlane, I.D., University of Reading, Agricultural and Food Economics, P.O. Box 237, RG6 6AR, United Kingdom; i.mcclement@reading.ac.uk

The UK government Farm Health Planning (FHP) project, funded by Department of Environment, Food and Rural Affairs (Defra), facilitates partnership with livestock industry working groups, encouraging livestock producers to introduce proactive measures to help manage disease. Twelve disease control models were commissioned: Bovine Viral Diarrhoea (BVD) and paratuberculosis (Johne's disease) in dairy and suckler cows; Digital Dermatitis in dairy cattle; Footrot, Ectoparasites (sheep scab and lice) and Fasciolosis (Liver Fluke) in sheep; Mycoplasma hyopneumoniae (Enzootic Pneumonia) and Porcine Reproductive and Respiratory Syndrome (PRRS) in finishing pigs; Coccidiosis in broiler chickens and Mycoplasma gallisepticum in laying hens. The novel feature of the models is that they combine the outcomes of epidemiological studies with published economic data. Each model was built using the systems simulation modeling package, STELLA 9 (www.iseesystems.com), consisting of a set of layers. The top interface layer permits users to input farm/system specific information concerning herd/flock size, market values, performance parameters, husbandry practices, costs of disease control methods and, in some cases, the efficacy of control. These values are used in the model layer of the system, in which graphical stock and flow diagrams model discrete and continuous processes over time. The assumptions made within the models are described explicitly, and are accessible to the user via linked text. The key assumptions in for example the dairy herd Johne's disease model concern the disease transmission routes, and are based on Kudahl et al., while the economic data is based on published data for UK dairy herds. Johnes disease is slow to develop and the model tracks the progress of the disease within the herd for a ten year period. For all the models the main output is in the form of tabulation of the disease costs and treatment costs. In the above example of Johne's in the dairy herd, for a herd of one hundred dairy cows with a rate of 88% calves reared and other farm specific data, the model indicates that, over ten years, a test and cull strategy reduces the total cost of the disease from £94,726 with no control to £84,421; the model suggests that an alternative strategy to optimize management is preferable, reducing the total cost of the disease to £56,903. The model outcomes are presented in detail, with predictions for associated milk losses and casualty animals. One of the most useful aspects of the models is that they can act as 'conversation tools' to discuss with farmers their disease control measures. They can also be used as demonstration tools with groups to illustrate potential costs associated with different diseases with different control measures. All of the models are available free to users via a dedicated website which can be found at www.fhpmodels.reading.ac.uk.

Integrated assessment of economic and environmental risk of agricultural production by FAPS-DB system

Nanseki, T., Sato, M., Marte, W.E. and Xu, Y., Kyushu University, Department of Agriculture and Resource Economics, Hakozaki 6-10-1,Fukuoka, 812-8581, Japan; nanseki@agr.kyushu-u.ac.jp

Agriculture has been facing big and many social and economic changes and risks recently. A structural reform of agriculture is a big policy issue not only in a country which imports farm products but also in an export country from economic view point. On the other hand, environmental risk derived from agriculture production is getting much more importance as a factor for getting a sustainable society. Therefore, we have developed a procedure for integrated assessment of economic and environmental risk of agricultural production by the FAPS-DB system. The FAPS-DB is a web-based system decision support system that is significant for overall estimation of the influence by the management strategy of the farming. The both technological and financial data are built into this system. The system is also integrated to a market price database for vegetables and fruits (Hereafter, it is called NAPASS system) through the Internet. Therefore, farming income risk caused by price fluctuation is easily evaluated. Furthermore, financial risk in terms of short term cash flow is also estimated at the same time. An environmental risk of agricultural chemicals use in agriculture is a big concerning issue in Japan. Therefore, we have added the feature for estimating this risk to the FAPS-DB system. Many types of agricultural chemicals have been used in Japan for agricultural production. There are about 4800 types of pesticides that contain about 500 active ingredients. The system is able to estimate both environmental and economic risk not only for one crop but also for a combination of crops in a whole farm management system. Several applications of the system have shown that the integrated assessment of economic and environmental risk of agricultural production is able to draw characteristics of specific crop and farming systems. This information is useful for decision making of policy maker as well as for farmers.

Harmonisation of data exchange in the IACS subsidies request procedure
Schmitz, M., Martini, D., Frisch, J. and Kunisch, M., KTBL, Bartningstraße 49, 64289 Darmstadt, Germany

Since 1993 European farmers were supported rather by direct subsidies than through guaranteed prices. This development continued with the Agenda 2000 and the reforms in 2003. The amount of aids depends on the acreage of the farmers tilled soil. Exceptions are made some crops. The European member states are obliged to implement an Integrated Administrative and Control Systems (IACS) to properly manage and control the claims for crop - and animal – subsidies. Key elements of IACS are an integrated database und the Land Parcel Identification Systems (LPIS). In Europe there are four different ways to define a referenced parcel. The method mostly used regards the 'physical block' as the reference parcel. A block is defined by geographic borders, like hedges or streets. Other definitions depend on the ownership and the cultivated crop. The choice of the reference parcel depends mainly on the historical development of the land management in the country and the farmers practices. In Germany the Bundesländer are responsible for the IACS, with the result that all four reference systems are used. The farmer has to apply for subsidies every year by filling in different forms. Basis for the request are aside from the master data the land use verification. Here the farmer is asked for geographic information about his parcel and the cultivated crop. This informations are already stored in his farm management information system (FMIS). Together with the German software companies agrocom, PC Agrar and Helm, the KTBL launched a project to automatize und simplify this procedure of applying for aids by reverting to the data kept in the FMIS. Key component of this IACS project is a server able to work on the one hand with the FMIS and on the other hand with an interface maintained by the authorities. The producers of the FMIS implement an export routine, which wrap up the needed information into an agroXML instance and send it to the project server via internet. The data exchange protocol agroXML will be enhanced by a new module and content lists, which together define all datasets required by the IACS autorities. Of course all data security standards are taken into account. The server checks the incoming data for accuracy and completeness. If some informations are unclear, a web site informs the farmer and gives him the possibility to complete the request. If required content lists are mapped to a country specific vocabulary. Then the data document is prepared for the receiving IACS interface. Subsets from the agroXML schema especially the IACS module will be used to keep the communication lean and efficient, these subsets are called profiles. In the project, two main questions have to be solved. At first, how to harmonize the data model with regard to the different requirements in the countries. Second, how to handle the possible changes in between years and how the technical solution could look like.

agroXML: a standard for data exchange in agriculture

Kunisch, M., Martini, D., Schmitz, M. and Frisch, J., KTBL, Data Management, Bartningstr. 49, 64289 Darmstadt, Germany

Farmers are subject to a multitude of obligations concerning documentation and verification of agricultural practices. Farmers' sensitivity concerning data exchange processes has increased due to publication of EU regulation 178 in 2002 and the consecutive Cross Compliance measures. The demand for appropriate technical solutions has become obvious. Using agroXML as a data exchange language, these procedures and the individual interfaces between communication partners are substituted by universally usable data exchange processes. The basic components of agroXML are the schema and the content list. The agroXML schema is based on a model of the real-world processes in agricultural production. They are represented in a tree-like hierarchy. Profiles define the required elements of the schema for a specific data exchange case. To create a profile, elements are copied from the agroXML schema to a separate file and the necessary restrictions applied to the data types in there. According to special rules data are transported in an XML instance, a file built according to the rules of the schema and the profile. Using profiles, lean, clear and easily transferable instances can be generated. Content lists are using a unified XML dictionary structure. Currently, several lists exists containing e. g. machine types, fertilizer types, pesticides and plant variety names. A Web Service is provided for checking up-to-dateness of lists. If the local copy is older than the one available on the server, the new list is provided. They ensure the uniqueness of identifiers and automatic evaluability of information. The ISOBUS specifications for machinery and farm equipment cover data exchange inside of the farm. Data exchange with external partners and between farm management information systems is conducted using agroXML. Availability agroXML is a trade-mark of KTBL. The recent version is agroXML 1.3. An extended version will be provided in January of every year. agroXML is Open Source, published under the W3C licence. Software which has implemented agroXML can be certified by KTBL and is then allowed to wear the agroXML-Logo on the cover. Currently, different companies are implementing agroXML interfaces. Several examples will be presented in the contribution. Especially for the agricultural software industry, agroXML has a high priority. Also producers of agricultural equipment, e.g. Claas and John Deere are supporting the development of agroXML. Outlook Obviously data exchange in agriculture is a topic of international relevance. Due to this several workshops had taken place at least on a European level. The aim is to develop a European Data Exchange format called agriXchange.

Enabling integration of distributed data for agricultural software applications using agroXML

Martini, D., Schmitz, M., Frisch, J. and Kunisch, M., Kuratorium für Technik und Bauwesen in der Landwirtschaft e. V. (KTBL), Bartningstraße 49, 64289 Darmstadt, Germany

Demand for information in agriculture and other sectors up in the supply chain is constantly increasing. On the one hand, food safety is of growing importance. This requires good process documentation in primary production. On the other hand, goals of agricultural policy like e. g. environmental friendliness and sustainability of production call for farsighted planning. Exchange of data and integration of information from external sources are becoming a vital factor for successful agricultural production. In the last few years, agroXML - designed and developed by KTBL in Germany and based on the eXtensible Markup Language (XML) - emerged as a standardized markup language for data in agriculture. Major parts of the agricultural domain have been modelled and cast into data structures declared in XML schema. In the meantime, applications using agroXML have been successfully implemented. Up to now, however only bilateral data exchange has been conducted. That means, that two partners set up an agroXML input and output interface, which upon demand offers or reads the necessary data. While it is possible and in some use cases necessary to conduct data exchange in this rather message oriented manner, such an architecture does not yet leverage the full potential of XML. Distributed data XML was designed by the W3C to allow for semantically rich, interconnected documents. It is the only widely spread technology having the potential to enable distributed data storage and retrieval on the farm itself as well as between the farm and external partners. Using an interplay of different internet technologies like the Hypertext Transfer Protocol and XLink for linking documents, data can be fetched on demand and integrated into local data repositories. Using only these simple mechanisms, web services and intelligent agents can be implemented in a lightweight and efficient manner using the method of Representational State Transfer (ReST) without having to use complex messaging protocols. Especially suited are these technologies for example with regard to integration of external data about agricultural supply items like pesticides or veterinary drugs or for building data structure which as a representation of the real world are better implemented in graphs instead of the XML-inherent trees. An additional benefit of this simple methodology is the device independence. As the necessary building blocks and tools are available for a wide range of hardware platforms from small handhelds to powerful servers and for almost every programming environment, ubiquitous data capture and availability is possible. In consequence, reuse and accessibility of data across different applications is achieved. Prototype During the IT FoodTrace project, it was shown, how these technologies can be used to implement an interface and link and retrieve information from different sources. The existing agroXML schema files have been extended by further ones offering elements for marking up livestock farming data. Additionally, the necessary attributes to link data together using XLink have been integrated. A prototypical infrastructure has been built showing how documents can be retrieved and utilized by applications without having to handle remote resources differently from local ones.

Populating an observatory of e-government services for rural areas: analysing the Greek services
Manouselis, N., Tzikopoulos, A., Costopoulou, C.O. and Sideridis, A., Agricultural University of Athens, Iera Odos 75 str., 118 55 Athens, Greece; nikosm@ieee.org

Governments throughout the world are facing the challenge of improving the efficiency, productivity and quality of their information and knowledge services to citizens and enterprises. Information and communication technology (ICT) can help public administration to cope with such challenges. However, this should be combined with organizational changes and new skills, in order to create more cost effective services that will improve transparency and quality of public services. It calls for a new generation of electronically supported governmental services, termed as e-government (eGov) ones. Although steadily developed, eGov has not reached yet to that degree of deployment and adoption that will lead to the radical change of the way people live and work. For instance, enterprises (mostly small and medium ones) in rural areas have difficulties perceiving the importance of eGov services and the way such services may be used to help them in their every day business activities. On the other hand, eGov services sometimes seem to be centrally developed and deployed, not taking into consideration the particular needs and problems that professionals in rural areas have. Away from the central public authorities, regional (also called rural) enterprises -mostly small and medium ones (SMEs) in such areas- do not have direct, physical access to public services that are essential in order for their business operation and flourishment. Examples include services offered by several types of governmental agencies/authorities, ranging from taxation offices, legislative authorities, local authorities, or chambers of commerce. Several times, the reason is that professionals and citizens in rural areas are not aware of electronically available public services, or do not know how to effectively use them so that they may reap benefits in their everyday business activities. In this direction, the Rural-eGov Leonardo da Vinci (LdV) project has been deployed, aiming to familiarize rural SMEs with the use of eGov services. One of the central aims of this project, is to design, develop and deploy the 'Rural-eGov Observatory', an online point of reference which SMEs can continuously access for relevant information and eGov services that are covering their regions. In this paper, we will present the population of the Rural-egov Observatory with a collection of 56 eGov services for Greek SMEs. Then, we will analyse this collection, trying to identify conclusions about the way existing eGov services can support Greek SMEs in rural areas. We assess these eGov services upon a a number of quality criteria, such as their accuracy and the reliability of their provided information, as well as, their usability and attractiveness. Moreover, their relevance and importance for rural SMEs' business operations is examined.

Information modelling in dairy production

Lindstrøm, J. and Sørensen, C.G., Faculty of Agricultural Sciences, University of Aarhus, Department of Agricultural Engineering, Schüttesvej 17, 8700 Horsens, Denmark

Agricultural production is often managed on a rather crude level. Both farm management and environmental protection could be improved significantly by a more detailed management and regulation using novel IT. A major obstacle in using more IT in agriculture is that most of the IT solutions provided have non-compatible software forcing the farmer to operate several IT units, which is time-consuming and costly. In this project, the objective is to describe how to organize system architecture for an ICT management tool for Danish dairy farms. Dairy production is very dynamic and thus it can not be properly depicted by pure tool-systems or by normal linear models but must rely on methods like object oriented analyses and design (OOAD). Various models can be created to show the static structure, dynamic behavior, and run-time deployment of these collaborating objects. A functional IT system must provide: 1) a useful model of the problem area, 2) be integrated into the users surroundings, 3) run on a fixed technical platform and 4) be a well acting whole of cooperating parts. A model of all the processes on the farm, including all stable operations, supply handling and delivering information to purchasers, consumers and authorities is outlined. This includes a precise description of the needed in order to make an ICT tool to assist in all managing aspects of dairy farm, and this description can then be used to write the program code. When programmed, the ICT system will offer a one-in-all management tool for dairy farming involving one single computer, from which all automated machinery can be controlled. All sensors on the farm will transmit their data directly to the office computer, and all un-normal events gives an alarm either as a text message to the farm manager's cellular phone, or as a pop-up on the screen. The modeling of all the processes and informations flows is essential in order to design the ICT tool. Also, the designed ICT system should only provide real-time relevant information to avoid information overload. Delimitations have to be considered carefully because they will impact the usability and function of the ICT tool. Arable farming is part of most dairy farms, but arable farming and dairy production may be viewed as two separate productions. On the other hand, dairy production is dependent on arable farming for feed supply ect.. One solution could be to use the system architecture proposed by this project to make a compatible ICT system for arable farming. Most agricultural DSS is based on a linear decision model, but studies indicate that farmers make decisions using non-linear models.

Computational architecture of OTAG prototype

Visoli, M. and Ternes, S., Embrapa Agriculture Informatics, Av. André Tosello, 209, 13083-970 Campinas, SP, Brazil; visoli@cnptia.embrapa.br

OTAG project (Operational Management and Geodecisional Prototype to Track and Trace Agricultural Production) is an Specific Support Action of the 6o. Framework Programme of the European Community. The goal of OTAG is to provide conditions to know the relative risks concerning to a bovine traceability. The project is based on the existing knowledge in Europe and Canada concerning to information systems and geo-decisional tools, as well to the interaction among experts and user groups from South Cone, Canada and Europe. The follow institutions are partners in that project: Institut the Recherche pour l'Ingénierie de l'Agriculture et de l'Envirnonnement (Cemagref, France), Centre de Coopération Internationale de Recherche Agronomique pour le développement (Cirad, France), Brazilian Agricultural Research Corporation (Embrapa, Brazil), Université de Laval (Canada) and Cooperative Program for the Technological Development of the Agro-food and Agroindustry in the Southern Cone (Procisur, Argentine). OTAG prototype conceives the use of electronic devices for measuring the animal geo-localization in the herd. This information is associated to data about farm management, such as feeding and production systems in use. The information being collected in each farm of OTAG system are grouped in a big database, that allows to get knowledge about animals movement inside the farm and between farms. These data are a valuable information source for providing an efficient traceability of animals, and as a consequence, a better sanitary control and the increase of alimentary security. Based on these aims, the computational architecture proposed is organized in three layers. The layer 1 concerns to the data collection from electronic devices in the paddock. The animals have a necklace with GPS device and the paddock has sensors for automatic collection of weight, vaccination register, and temperature measurement of each animal, among others. The information produced for the electronic devices are sent to a computer located in the farm headquarters. For this task, we intend to use the open standard XML, making possible to use electronic devices of any manufacturers. In fact, to define and to improve a standard that can be used by any manufacturers is one of the challenges for OTAG Project. The layer 2 is responsible for storing data which is sent by the electronic devices and associated with farm management data. The user interface of this layer will allow the farmer to manage data and to extract reports about his cattle production by the Internet. The data of each farm will be send for the layer 3 by using web-service technology. On the layer 3 the information of all the farms of OTAG system are joined in the same database. Thus, the information concerning to animal movement inside farm and between farms can be analysed by the use of techniques for dealing with geo-referenced information. The use of tools for receiving, treating and organizing a great amount of data are necessary too, to get an acceptable response time for the reports and queries to be asked for the final user. With that aim we intend to analyse the use of business intelligence tools and also Spatial OLAP technology. The possibility of integration with external databases is also foreseen. This can contribute for a better analysis aiming the decision making about the farm production. For instance, the integration of OTAG prototype with databases about ground, pastures and climate can increase the quality of information concerning to beef cattle production.

Robust performance: principles and potential application in livestock production systems

Van Der Veen, A.A.[1,2], Ten Napel, J.[1], Oosting, S.J.[3], Bontsema, J.[4], Van Der Zijpp, A.J.[3] and Groot Koerkamp, P.W.G.[1,2], [1]Wageningen UR, Animal Science Group, P.O. Box 65, 8200 AB Lelystad, Netherlands, [2]Wageningen University, Farm Technology Group, P.O. Box 17, 6700 AA Wageningen, Netherlands, [3]Wageningen University, Animal Production Systems Group, P.O. Box 228, 6700 AH Wageningen, Netherlands, [4]Wageningen UR, Plant Research International, P.O. Box 16, 6700 AA Wageningen, Netherlands

In livestock production systems (LPS), the predominant strategy to maintain functionality is to control variation in the production environment, by controlling internal system conditions and keeping away disturbances and perturbations. Although this has proven to be a successful approach, the drawbacks and constraints of completely relying on this approach, such as infectious animal diseases, overburdening of animals, loss of biodiversity and a lack of public support, are accumulating. We may need to reconsider the way in which LPS are designed and function from the perspective of controlling variation. The aim of the present study was to make an inventory of strategies that can be applied in LPS to maintain the system's functionality in a dynamic environment. LPS are complex systems with natural, technical and social sub-systems. By means of a literature study, we therefore explored these system fields first for terminology, concepts and strategies regarding stability and control of variation. On the basis of the strategies in each of the system fields, we developed a conceptual framework for robust strategies in the complex LPS. We found that robustness refers to the way in which systems are able to function when external systems conditions change beyond the range of conditions for which the system was designed. These changes in environment are referred to as perturbations and disturbances in system terminology. Robustness involves two aspects, resistance and flexibility. A system with a highly controlled environment has become resistant to certain perturbations, as they no longer influence the system's performance. Therefore, an optimal performance strategy can be used for this situation. This involves 1) uniformity and homogeneity, 2) efficiency and 3) enlarging of scale in the design of the system to obtain an optimal performance. However, no environment can totally be controlled. Failure to comply with protocols, technical failure and new unforeseen perturbations, such as upcoming unknown diseases or weather changes, may pose a threat to the performance of LPS in the future. Hence, a more risk-averse strategy for LPS is to reduce the consequences in the presence of the causes; it minimizes the impact of external systems conditions on the performance of the system. The latter will require the system to change its mode of operation in a flexible way and tends towards a robust performance strategy. This strategy is closely related to the flexibility of the system; the rate of 1) diversity and heterogeneity, 2) redundancy and 3) a modular design determine the ability to handle and to adapt to new circumstances. The framework that resulted from our study shows that control of the production environment and the reduction of variation in the production process is less needed for systems that are working according to a more robust performance strategy, instead of the optimal performance strategy.

Applying enterprise application architectures in integrated modelling

Knapen, M.J.R., Alterra, Wageningen University and Research Centre, CGI, Droevendaalsesteeg 3, 6708 PB Wageningen, Netherlands

Software systems that support integrated policy assessment need to work with models from different domains and provide a framework to link these models together. They need to store model results, intermediate and raw data, and have one or more user interfaces for input and presentation. Such software can be built with an inside-out focus, emphasising the model linking, but also from an outside-in perspective. Here the features and technical development is primarily driven by usability and business requirements. The inside-out view can very well satisfy the needs of the modellers and framework builders involved, but might not be enough for all other requirements of the many stakeholders in an integrated assessment, and in the project at a larger scale. Thus the outside-in view must also be considered for the software to be successful. Looking from this perspective the software system is similar in basic functionality to other data intensive enterprise applications and common architectures and design patterns could be used for its construction. Enterprise applications (e.g. used for banking or insurance) typically have an architecture that separates functionality into 5 distinct layers. This includes a persistence layer for storage of domain object state, the domain layer for the domain model and domain logic, a services layer that controls transactions and contains the business logic, an application layer for use-case workflows, syntactic validation and interaction with the services layer and finally a presentation layer for the user interfaces. Following such separation of concerns, helped by some well known design patterns like data transfer objects, service layer, command pattern and CRUD (a pattern that organizes the persistence operations into Create, Read, Update and Retrieve operations that are implemented by a persistence layer) will improve the maintainability, possibilities for re-use and systems interoperability. In this paper the software architecture of the SENSOR and SEAMLESS 6th framework EU projects will be used as case studies to illustrate the use of the layered architecture and the mentioned design patterns. Both these projects deal with environmental integrated assessments and include the construction of a decision support software system. They will illustrate the increased importance of the outside-in perspective and following standard software engineering practices to improve the interoperability and possibilities for re-use of components of the systems build.

The integration of open standards for enterprise service bus in a solution for agribusiness and enviromental applications development

Murakami, E.[1], Santana, F.S.[2], Stange, R.L.[2] and Saraiva, A.M.[2], [1]UDESC, CCT, Campus P. Avelino Marcante - Joinville - SC, 89223-100, Brazil, [2]USP - POLI, Av. P. Luciano Gualberto, T.3, n. 158 - SP, 05508-900, Brazil

Evolution of agribusiness and environmental applications has shown that solutions based on services technologies allow software reuse and faster development of related applications. A SOA-based software infrastructure to integrate services for both agribusiness and environmental purposes has already been developed. Precision agriculture data filtering and ecological niche modelling are among the applications which are already available for free usage in the Internet. The infrastructure required a software bus to support the development of service-based applications and the proposed solution, AgriBUS, was based on Enterprise Service Bus (ESB) technology. ESB represents a core part of any SOA solution because it is a technology which applies eventdriven and loosely coupled service exchanges, within a highly distributed universe of routing destinations and multiple message protocols. Applications correspond to orchestration of services abstractly decoupled from each other, connected together through ESB by interface descriptions. ESB may provide a web services hosting environment, with transport protocols based on Internet standards (e.g.: HTTP and FTP). Though AgriBUS is an open solution, it was not build purely based on open standards. This work presents a proposal to replace AgriBUS by a Java Business Integration (JBI) based solution. JBI is a specification developed on Java Community Process and describes how to connect components to SOA using ESBs. The Open ESB project was initiated in 2005 and financial sponsors are Sun Microsystems and other important companies and employees. Open ESB JBI is a solution entirely built using open standards. ESB replacement improves previous solution and makes it compliant to global standards. It will enable services migration among different ESBs by using standards, such as SOAP, WS-*, BPEL and other miscellaneous XML standards. In addition, as Open ESB is designed and maintained by a reputable software community, technical evolutions in ESB research will be easily incorporated to the existing solution, maintaining its interoperability properties.

Evolution of a SOA-based architecture for agroenvironmental purposes integrating GIS services to an ESB environment

Santana, F.S.[1], Stange, R.L.[1], Gushiken, I.Y.[1], Murakami, E.[2] and Saraiva, A.M.[1], [1]USP - POLI, Av. P. Luciano Gualberto - 3, n.158 - SP, 05508-900, Brazil, [2]UDESC - CCT, Campus P. Avelino Marcante - Joinville - SC, 89223-100, Brazil

The relationship between agricultural and environmental problems has already been demonstrated and the convergence of solutions is an unstoppable trend. This convergence allows the design of a common, reusable, integrated and interoperable architectural solution. Previous works discussed this aim, which might hardly be achieved with monolithic systems. Instead, a solution based on Service-Oriented Architecture (SOA) was developed. The solution applies open standards and free software. Ecological niche modelling, precision agriculture data filtering and cluster services are among the functionalities already available in the Internet. A challenge of SOA-based environments is the management of services distributed worldwide. In order to meet this challenge and reduce programming efforts to integrate legacy solutions, an Enterprise Service Bus (ESB) was implemented as part of the solution. ESB includes: orchestration, routing and intermediation of services and transformers, adapters and bridges among different middleware technologies. ESBs are also powerful integration infrastructures which may be the basis to implement many service-based software requirements. One important agro-environmental requirement is related to the ability to deal with maps, thus adding Geographic Information System (GIS) functionalities and the adoption of GIS standards, defined by the Open-GIS Consortium (OGC), will improve interoperability. This paper presents the integration of GIS services to the developed architecture. Solution includes OCG Web Services (OWS), Geography Markup Language (GML) and W3C standards, such as Scalable Vector Graphics (SVG), for XML image treatment. Besides the relevance of offering GIS services itself, this solution improves the overall SOA architecture in other aspects, such as network performance, since transmitting XML files is usually faster than transmitting images, and the interoperability of solution by applying standards. Other details of the global solution are also presented to discuss architectural advantages and increase opportunities for future partnerships in these research areas.

Business process modeling of the pesticide life cycle: a service-oriented approach
Wolfert, J., Matocha, D., Verloop, C.M. and Beulens, A.J.M.,

In the food economy, information sharing becomes a competitive factor. Agri-food companies increasingly participate as networked enterprises in multi-dimensional, dynamic and knowledge-based networks. They have to make new connections rapidly and employ 'up-to-the-minute' information smoothly in business operations. Standardization in information integration of processes, applications, data and physical infrastructure are important to realize this. For setting-up and changing integrations quickly, a rapid (re-)configuration approach is needed in which information integrations are set-up from standard components. This requires component-based information systems, independent components, standardized interfaces between components, a central repository of published components and standardized procedures for selection and implementation of components. The business processes in real life must be leading in designing and using information systems. A combined approach of Business Process Management (BPM) and Service-Oriented Architecture (SOA) is considered to be very suitable to meet these requirements. This paper describes how BPM and SOA can be used for information modeling of farm management using the pesticide life cycle as a case study. Decisions on pesticide use are constrained by legal restrictions, social requirements and market conditions. Briefly, the basic principles of BPM and SOA are explained. This is applied to farm management and the pesticide life cycle in a series of aggregated models ranging from strategic, tactical to operational management levels. First, a value chain model is developed that defines the relevant business partners and the main interconnecting processes and product flows. Then, a business context model depicts the whole set of business processes at a high aggregation level adopted by a company in its business context. It concentrates on the main physical flow, relation with business partners, stock points and business process categories. Finally, the business process models describe the business processes at more detailed level using the formal business process modeling notation (BPMN). For pesticide management, processes are modeled with respect to received materials, crop growth, production planning and administration. The modeling is done in the an Enterprise Integration software suite, named Cordys. This application can be used to automatically generate business process execution language (BPEL) from the BPMN models that form the basis for standard web services. In this way, a service-oriented architecture is generated from the business processes. This architecture is the basis for actual information systems that are used by the various actors in the value chain. Translation into SOA and implementing the actual information system is outside the scope of this paper. It is concluded that using BPM and SOA, it is possible to make a model of the pesticide life cycle in farm management, covering the complete range of business processes in the chain. The model illustrates the use of standardized components that can be re-used in other contexts like fertilizing, sowing, etc. In this way, the model forms a basis for setting up a complete architecture for total farm management. This should be further verified in other case studies and implementation of the model into real information systems based on web services.

Plantform: smart, agile and integrated enterprise information systems for ornamental horticulture
Verloop, C.M.[1], Verdouw, C.N.[1] and Van Der Hoeven, R.[2], [1]LEI Wageningen UR, Plant Systems Division, P.O. Box 29703, 2502 LS Den Haag, Netherlands, [2]Plantform, P.O. Box 666, 2675 ZX Honselersdijk, Netherlands

Companies in ornamental horticulture have to face developments as increasing power of the retail, growing supply chain complexity, increasing costs and more stringent societal constraints. Important strategies that firms are applying to cope with these challenges are scale enlargement, increasing pace of innovations and transition towards networked enterprises that participate in demand-driven, dynamic and knowledge-based networks. In these complex chains and turbulent environment, it is of great importance to keep business processes in control. This imposes high requirements on the supporting information systems, particularly regarding flexibility, integration and incorporation of intelligent techniques for optimization and simulation. Furthermore, ornamental horticulture deals with 'living products', which complicates the processes and causes extra sector-specific requirements on the techniques and solutions used. It was found that the used information systems in Dutch ornamental horticulture do not sufficiently meet these requirements. They are characterized by island automation, i.e. isolated software packages for specific applications that are poorly integrated. This results in insufficient management information and a lot of manual data entering with a high risk for errors. Mostly, the used systems are developed by small local suppliers that focus on sector-specific software packages. As a consequence, the functionality fits well to sector-specific characteristics, but especially integration possibilities and robustness are weak points. However, comprehensive ERP-packages demonstrated to lack sector-specific functionality and the required flexibility. The typical specifications on the business processes cause the ornamental horticulture business to be a relative small market, but with high requirements on entrepreneurship, technology and innovation. Therefore, a group of Dutch pot plant growers have initiated Plantform. This is a Dutch entrepreneurs association in which about 35 growers cooperate in realizing integrated enterprise software that meets the requirements of ornamental horticulture. Plantform now has developed the functional requirements, the component definitions and the interface specifications for this system. Next, it has executed two pilots to implement this architecture, one based on a standard available software package, the other in tailor-made software. The paper describes the architecture and implementations of the Plantform enterprise information system. It focuses on integration of knowledge-intensive modules for analysis and simulation with the backbone that handles basic transactions for sales, procurements and production management. It includes a literature review of different types of simulation and their applicability in enterprise information systems.

Designing an information and communication technology system to train private agricultural insurance brokers in Iran

Omidi Najafabadi, M., Farajollah Hosseini, J., Mirdamadi, M. and Moghadasi, R., Azad University, Agriculture, Science and Research Branch, 14538 Tehran, Iran; maryomidi@gmail.com

The Agricultural Insurance Fund (AIF) in Iran has hired private sectors named 'brokers' to implement an agricultural insurance scheme. AIF has spent considerable time, effort, and money to train brokers. Unfortunately, many of their investments are met with disappointing results and traditional education is no longer effective. To overcome this challenge, an information and communication technology (ICT)-based training system seems the best solution. This study was conducted to design an ICT system to train agricultural insurance brokers in Iran. To achieve this aim, a theoretical framework is presented, based on previous research. Our results are given using newly developed and sophisticated statistical techniques. namely, ordinal logistic regression indicated that the factors which affect on designing an ICT-based training system are technical, financial, instructors, and learner factors; while the ordinal factor analysis classified the challenges into four latent variables named organizational, social, human, and technical challenges. Finally a conceptual framework is presented for the ICT training system.

Architecture of information systems for automated arable farming

Thysen, I., University of Aarhus, Agroecology and Environment, P.O. Box 50, 8830 Tjele, Denmark

Leading manufacturers of agricultural machinery and equipment are already far in implementing pervasive computing technologies in their products. By this development, the control and regulation of the individual components of machines and equipment are handled by embedded and mutually communicating computers. On top of this, communication with the surroundings is established through wireless communication, and geo-referencing is obtained via the Global Positioning System. These technologies together provide for a beginning automation in arable farming. For example, a sprayer can be controlled by instructions sent over the Internet from the farmer's desktop without interventions from the operator, and documentation of the actual sprayings can be sent back. Products currently on the market include auto-guidance of vehicles in the field, auto-guidance of implements in row crops, boom control sprayers, and application rate control in fertilizers and sprayers. Each product is usually contained in a separate box with its own display, even products from the same manufacturer, and some furthermore require a separate GPS unit. As the number of products installed on a tractor grows, the operator will be facing a confusing view of boxes and displays. The manufacturers want plug-and-play products to minimize installation efforts, and containment in separate boxes is then an obvious choice. Communication between internal and external applications is cared for by cabling and removable disks of various kinds. From a wider perspective, however, this is hardly an optimum architecture of an information system for automated arable farming. The purpose of this paper is to develop suitable architectures for this specific problem domain taking all its important aspects into consideration, including available technologies, development and support issues, farming logistics, information resources, stakeholders' interests, and human resources. The suggested architecture of an information system for automated arable farming has three main layers: management, operation and machine layers. The management layer is serving the farm management's decision making and supervising activities. This layer contains a farm management system combined with decision support systems and information resources, including specifications for computerized machinery. The main part of data and logic are in this layer. The operation layer is serving activities related to field operations. The main tasks are 1) communication with top layer, 2) tracking the vehicle's current position, 3) auto-guiding the vehicle's route, and 4) passing geo-referenced instructions received from the management layer to the machine layer. Data and logics in this layer are mainly for real-time adjustments of the farm work. The machine layer is serving the field work being performed. The main task is to adjust the machine's behaviour to instructions received from the operation layer. Another important task is to adjust the machine to actually perform as instructed; logic in this layer is predominantly related to this task. Implementation of such architectures is only realistic if an ample environment for online exchange of resources is established.

Implementing systems usability evaluation in the design process of active farm management information system

Norros, L.[1], Pesonen, L.[2], Suomi, P.[2] and Sørensen, C.[3], [1]VTT Technical Research Centre of Finland, P.O. Box 1000, 02044 VTT, Finland, [2]MTT Agrifood Research Finland, Plant Production, Vakolantie 55, 03400 Vihti, Finland, [3]Aarhus University, Department of Agricultural Engineering, Blichers Allé, Postbox 50, 8830 Tjele, Denmark

Nordic knowledge and know-how in the area of agriculture and ICT was extracted and further developed in the joint Nordic project 'InfoXT - User-centric mobile information management in automated plant production'. The aim of the project was to draw up basic recommendations and guidelines for a novel, intelligent, integrated information and decision support framework for planning and control of mobile working units. The core of the research was in the user centred development of Farm Management Information System (FMIS) that supports mobile work in crop production. Systems usability was the key dimension in the project. 'Usability Case' approach for usability design and evaluation, developed by the Human factors group in VTT Technical Research Centre Finland, was employed in the design work. The project covered all five Nordic countries, and thus, in order to give possibility to all potential system users involve in the system design process, as the used method required, the Internet was used as communication media. The scenario video describing the system's key functional features and the evaluation questionnaire for giving feedback to designers was available on the Internet to anyone interested. Results from the Internet scenario-questionnaire-method provided valuable information for the continued evalaution, and so, collecting feedback from the audience via Internet was continued after finishing the project itself. Data will be utilised in further development of the concept. In this paper, the method used in implementing systems usability evaluation in the design process of FMIS as well as the resulting Active FMIS (AFMIS) are introduced and experiences using the Internet as communication media in evaluation process are discussed.

Migration of decision support systems to the mobile environment

Karetsos, S. and Sideridis, A., Agricultural University of Athens, Informatics Laboratory, Division of Informatics, Mathematics and Statistics, Dept. of Science, 75 Iera Odos str, 118 55, Athens, Greece

Recently, a research programme has been initiated at the informatics Laboratory of the Agricultural University of Athens aiming at design prototypes for major field crops decision support systems (DSSs). These systems have been tailored to fit market needs and farmers experience. They provide the means for the management of major field crops in order to succeed well balanced land use. Potential users of these systems are policy makers, farm advisers, agronomists, land owners and farmers who benefit form the systems. Therefore their requirements are the following: Respect directions of the Common Agricultural Policy (CAP). Provision of solutions that fit to the farmers experiences. Encouragement for the production of high quality products. Guarantee farmers income. Protection of the environment. Support during the whole cultivation period. Since the computer usage and Internet penetration presents very low rates in rural areas and especially to farmers, usage of DSSs seem to be not so tangible. The digital divide problem that remains acute for people living in rural areas forces researchers to exploit more friendly and immediate ways for providing digital services. Moreover, emerging trends make clear that in the near future there will be a strong demand for multi-channel service delivery. In this direction, the boom of the use of mobile devices, including internet-ready mobile phones, smart phones, and personal digital assistants (PDAs), are forcing researchers to develop systems that can be available through mobile phones. The penetration of mobile devices shows high rates even in population groups like farmers. Therefore, the scope of this paper is to present the migration of already developed DSSs to the mobile environment. More specifically, it presents the phases of (a) analysis, (b) design, (c) implementation and (d) evaluation. During the analysis phase the user requirements are defined. In the design phase, the system architecture and its components are carefully selected. The phases of implementation and evaluation refer to implementing, testing and evaluating the system. In addition it discusses the potential impacts of such systems on agriculture in countries like Greece.

Optimization of maize harvest by mobile ad-hoc network communication

Jensen, A.L.[1], Hammarberg, J.[2], Dobers, R.[2], Kristensen, L.M.[3] and Thysen, I.[1], [1]University of Aarhus, Dept. of Agroecology and Environment, Blichers Allé 20, DK-8830 Tjele, Denmark, [2]The Alexandra Institute Ltd., Aabogade 34, DK-8200 Aarhus N, Denmark, [3]University of Aarhus, Dept. of Computer Science, Aabogade 34, DK-8200 Aarhus N, Denmark

Optimization of most farming operations requires high quality information; precise, relevant and valid for the current time and space. This paper will present an experiment where we try to optimize the maize harvest by the use of wireless network communication between the various machines involved in the harvest operation. The maize harvest operation involves one or more harvesters and a number of tractor-driven carriers catching the cutted crop produce and transporting it to the storage place. It is important that the harvesters are kept engaged, so as soon as one carrier is filled, the next empty carrier will take over its place, and the first carrier will return to the storage place. The minimum number of carriers required for the harvest operation depends mainly on the distance between the field and the storage place. The hypothesis of the experiment is that in some cases the harvest can be performed with less carriers when information about the current position of harvesters and carriers is available to all machines and to the manager. This information can help the driver of the empty carrier to locate the field and the position in the field where the harvester will be when the current carrier is full and bound to be replaced. The experiment will test a hybrid of modern wireless communication technologies. A novel and promising technology is mobile ad-hoc networks using a publish-subscribe paradigm in contrast to the well-known client-server paradigm. Whereas the latter paradigm requires that the clients know the server in advance, the former paradigm enables automatic creation of a network between units within range. However, the transmission range of the ad-hoc network is likely to be too low with currently available hardware. To ensure robust communication the ad-hoc network communication is supplied with mobile wireless network in the experiment. The experiment will be performed in September-October and the results will be presented in the paper.

The vineyard digital dashboard

Neto, M.C.[1], Lopes, C.[2], Maia, J.[3] and Fernandes, L.M.[4], [1]Institut Superior de Estatística e Gestão de Informação, Universidade Nova de Lisboa, Campus de Campolide, 1070-312 Lisboa, Portugal, [2]Instituto Superior de Agronomia, Tapada da Ajuda, 1349-017 Lisboa, Portugal, [3]Centro Operativo e de Tecnologia de Regadio, Quinta da Saúde, 7801-904 Beja, Portugal, [4]Agri-Ciência, Rua dos Lusíadas, Nr. 72 - 1°, 1300-372 Lisboa, Portugal

The evolution that is nowadays taking place in the information and communication fields, namely in mobile computing and remote monitoring, constitutes a very interesting challenge to the agricultural sector. This reality places agronomic knowledge in centre stage as these technologies are dramatically improving data collection and storage capacities, challenging the farmers and the agricultural field experts to develop processes that efficiently transform data into information and knowledge and are able to support the everyday decision making at farm level trough the use of environmental, plant and operational real time information. In this work we will present a project under way in a vineyard in Portugal where we are exploring the potential of the most recent technological innovations available in the market to build the i-Farm, the information and knowledge society intelligent farm. i-Farm (intelligent farm) applies in a vineyard the potential offered by using in an integrated way mobile solutions, sensor networks, wireless communication and digital imagery materialized in a information system that supports farmer real time decision making in the field and in the office supported by a digital dashboard. The i-Farm project delivers to the decision maker a digital dashboard supported by a unique knowledge repository (datawarehouse) containing information from multiple sources (crop, environment, soil, operations, etc.) enabling accurate and timely decisions. For the project development a Business Intelligence approach was used.

Interchangeability of traditional trust elements by electronic trust elements: implementation and evaluation in the case of an e-business platform

Haas, R.[1], Meixner, O.[1], Ameseder, C.[1], Canavari, M.[2], Cantore, N.[2] and Fritz, M.[3], [1]University of Natural Resources and Applied Life Sciences Vienna, Institute for Marketing & Innovation, Feistmantelstr. 4, 1180 Wien, Austria, [2]University of Bologna, Viale Giuseppe Fanin 50, 40127 Bologna, Italy, [3]University of Bonn, Meckenheimer Allee 174, 53115 Bonn, Germany; m.fritz@uni-bonn.de

One of the most prevalent issues in the introduction of an e-commerce system along a supply chain is the ability to establish trust between trading partners. The most important elements for establishing trust in b2b relationships – identified in previous studies - will be reinterpreted and implemented into existing e-commerce tools in collaboration with SME business leaders in different countries. The evaluation of these trust elements in e-commerce tools will be done by executing an experimental setting using a choice based approach. Results of this study will demonstrate the compensation potential between traditional and electronic trust elements and will identify opportunities to establish trust in e-enabled supply chains.

Importance of trust building elements in B2B agri-food chains

Meixner, O.[1], Ameseder, C.[1], Haas, R.[1], Canavari, M.[2], Fritz, M.[3] and Hofstede, G.J.[4], [1]University of Natural Resources and Applied Life Sciences Vienna, Institute for Marketing & Innovation, Feistmantelstr. 4, 1180 Wien, Austria, [2]University of Bologna, Viale Giuseppe Fanin 50, 40127 Bologna, Italy, [3]University of Bonn, Meckenheimer Allee 174, 53115 Bonn, Germany, [4]Wageningen UR, Logistics, Decision and Information sciences, Dreijenplein 2, 6706 KN, Netherlands; rainer.haas@boku.ac.at

In marketing literature trust is perceived as a pivotal aspect of business relationships. However, trust is still a concept that is in need of clarification. Therefore, the purpose of this paper is to measure the importance of trust building elements important for establishing a trustful relationship between trading partners. Required data was collected by a number of qualitative expert interviews in different countries by all partners of the e-trust project focusing on an early stage of business relationships in agri food chains. For the measurement the analytic hierarchy process (AHP) was applied by using a predefined structure (hierarchy). Results of this study help to understand the importance of different trust elements in b2b relationships. For the first time the study demonstrates the influence of the cultural background on inter-organizational trust. The estimation of the importance of trust elements in inter-organizational relationships will be useful for the implementation in b2b e-business applications as it will help to identify the trust building elements that should be considered first of all when integrating b2b applications into business procedures.

A process perspective on trust building in B2B: evidence from the agri-food sector

Hofstede, G.J.[1], Oosterkamp, E.[1] and Fritz, M.[2], [1]Wageningen UR, Logistics, Decision and Information sciences, Dreijenplein 2, 6706 KN Wageningen, Netherlands, [2]Food and Resource Economics, Meckenheimer Allee 174, 53115 Bonn, Germany; Gertjan.Hofstede@wur.nl

Interviews were held with companies in four sectors in the european Agri-Food sector and beyond. These centered on the role of trust in the formation of new business relationships. This article is a meta-analysis of the data with a process perspective. The aim is to find out which steps precede which others, and how variable this order is. The order may vary with country, with sector, or along other lines, and this variation will be analyzed.

Potential application of ICT in agricultural and food sector: performing trust and new trade relationships in e-business environments

Fernandez, C., De Felipe, I. and Briz, J., E.T.S.I. Agrónomos-Universidad Politécnica Madrid, Agriculture and Economics, Avda Complutense S/N- Ciudad Universitaria, 28040 Madrid, Spain; mcristina.fernandez@ upm.es

Since some decades, agrifood products exchanges between business partners can be carried on using new Information and Communication Technology tools (ICT). Anyway, their adoption in the agrifood sector appears to be hindered, both because of consolidated dynamics in developing B2B transactions, and because of the peculiarities of the agrifood products themselves. The need for physical acquaintance and for direct relationships, the problems connected with food safety assessments and the necessity of standardized products should be matched with the diverse technologies devices implied in an electronic transaction which provide trust in an e-business environment. The analysis presented in the paper show the potential ICT tools and devices used in electronic transactions considering the trust performance in e-business environment, the attitude towards e-commerce of potential ICT users on agrifood sector and their relations with the trust performance and an overview on agrifood e-marketplaces and their evolution in the last 5 years.

AQUARIUS: a simulation program for evaluating water quality closure rules for shellfish industry
Conte, F.S. and Ahmadi, A., University of California Davis, Department of Animal Science, One Shields Ave, Davis, California 95616, USA; abahmadi@ucdavis.edu

The United States shellfish industry is regulated under the National Shellfish Sanitation Program; administered federally by the U.S Food and Drug Administration, and at the state level by departments of health or agriculture. Shellfish authorities are empowered to close shellfish harvest if water quality drops below food safety levels. Because monitoring for all human pathogens in growing areas is not feasible, fecal coliform bacteria are used as indicator organisms for the potential presence of pathogens from fecal contamination. Every decade, sanitary surveys are conducted during adverse pollution conditions to establish equations and rules for conditionally approved growing areas to predict rainfall levels when fecal coliform levels might exceed the 'NSSP 14/43' safety standard. Modifying these rules requires an extensive sampling programs and analyses. As watersheds change, there is pressure to reclassify growing sites and to modify closure rules. The AQUARIUS program is the first tool developed to directly evaluate closure rules for the shellfish industry, and to perform a series of 'what-if' scenarios for selected variables. AQUARIUS uses the actual rainfall data to simulate the open/close status of a given growing site for any length of time under two closure rules: one the current rule and the other proposed new rule. AQUARIUS then uses the actual fecal coliform data from the site to calculate the 'NSSP 14/43' standards for three situations: 'site open under current rule', 'site open under new rule', and 'site open during critical period'. AQUARIUS then runs a series of statistical procedures to compare the fecal coliform level of 'site open under current rule' versus 'site open under new rule'; and to compare the fecal coliform level of 'Site open under current rule' versus 'site open during critical period'. Based on the results of SSP 14/43' standards and T-Tests, the new closure rule is either accepted or rejected.

A multi-criteria decision support model for assessing conservation practice adoption

Vellidis, G.[1], Lowrance, R.[2], Crow, S.[3], Bosch, D.[2] and Murphy, P.[4], [1]University of Georgia, Biological & Agricultural Engineering, 2329 Rainwater Rd., Tifton GA 31793-0748, USA, [2]USDA-ARS, Southeast Watershed Research Laboratory, 2378 Rainwater Rd., Tifton GA 31793-0946, USA, [3]PlaceMatters, 1536 Wynkoop St., Suite 307, Denver, CO 80202, USA, [4]InfoHarvest, 108 South Washington St., Suite 200, Seattle WA 98165-2055, USA

The Little River Experimental Watershed (LREW) in Georgia, USA is one of 13 watersheds being evaluated by the USDA-CSREES Conservation Effects Assessment Project (CEAP). CEAP's goal is to assess the effect of three decades of conservation practice implementation on water quality in the USA. The success of conservation practices is determined by many factors including biophysical processes, landscape features, and socioeconomic factors. In our evaluation of the LREW, we are using a facilitated modeling approach to illustrate the interactions among these factors. Facilitated modeling is a multidisciplinary approach to problem solving and consensus building. The process combines systems modeling and consensus decision making to enhance collaboration and problem solving. The approach allows participants to develop a shared understanding of how a system of interest works in a relatively short time frame. The system process model will display the interactions that lead to the production of water quality benefits from application of conservation practices on a watershed scale. Our first step was to develop a Multi-Criteria Decision Support Model (MCDSM) which is being used to identify the relative importance of biological, physical, social, and economic factors affecting a farmer's decision to implement conservation practices. Ultimately, these farmer decisions determine what conservation practices are implemented, where they are implemented, and when they are implemented. By using the MCDSM we can not only determine which factors determined the adoption of existing conservation practices but we can also evaluate alternative scenarios. The water quality effect of these alternative scenarios will be evaluated with the SWAT model. This paper reports on the development of the MCDSM and its application to the Little River Experimental Watershed.

Multivariable decision-making method in landscape planning: watermills in Pápateszér, Hungary

Barabás, J.[1], Fülöp, G.Y.[2] and Jombach, S.[3], [1]Corvinus University of Budapest Faculty of Horticulture Science, Department of Mathematics and Informatics – tenured professor, Villányi út 35-43. P.O.B. 53., 1118 Budapest, Hungary, [2]Corvinus University of Budapest Faculty of Landscape Architecture, Department of Landscape Planning and Regional Development - student, Villányi út 35-43. P.O.B. 53., 1118 Budapest, Hungary, [3]Corvinus University of Budapest Faculty of Landscape Architecture, Department of Landscape Planning and Regional Development - Phd student, Villányi út 35-43. P.O.B. 53., 1118 Budapest, Hungary

Nowadays practice shows, that landscape planning and management needs methods that make decision-making more objective, faster and easier to communicate. However, we cannot give up the idea of painstaking analysis of our surroundings. Just a deep surveying and analysis can be the basis of efficient environmental planning. Therefore we worked out a decision-making method that uses multivariable analysis, TwoStep Clusteranalysis, to make assessment easier and more authentic. An analysis used at the beginning of landscape planning process has to be multivariable. Not just the observed features but even the scale of observation is various. To demonstrate how many-sided such an assessment can be, in the first adaptation of the method we used nearly seventy variables that meant nearly ninety measurements in a strict system. That is why we applied cluster-analysis that can handle the problems of a huge database. To be able to communicate the results of the assessment we used COWS decision-making matrix, so the variables construct a strict system. Even though this strictness of the incoming data, while we realized all the parts of this decision making method, we tried to keep in mind that the final result must be flexible enough to be adaptable in various fields of life. We adapted first this method analysing the state, location, capability and renewability of 25 watermills in Pápateszér. It is the settlement in Hungary where once the most watermills worked. However, nowadays the mills are not in the condition that can demonstrate their fame as agricultural relics. We chosed the topic of assessing the watermills in Pápateszér because a NGO, (REFLEX Environment Protection Association – first environment protection association in Hungary) would like to rehabilitate one of them as a monument of the local milling industry. The aim of the association is to get people closer to traditional agriculture and to the process that brings bread on our table from the fields in our surrounding. In addition to this, a mill-rehabilitation would serve aims of rural development and development on tourism as well, so our assessment touches upon also these topics. In our presentation, we will show how the method works, what its system stands for, what are the advantages and disadvantages of its use, and even what are its perspectives in other researches. We recon, that methodology never exists without its successful adaptation. So, on practical ground, we will present the result of the first adaptation of the method and our proposal in the case of the watermills in Pápateszér.

Evaluation of SWAT for a small watershed in Eastern Canada

Ahmad, H.[1], Havard, P.[1], Jamieson, R.[2], Madani, A.[1] and Zaman, Q.[1], [1]Nova Scotia Agricultural College, Department of Engineering, 39 Cox Road, B2N 5E3, Truro, NS, Canada, [2]Dalhousie University, Department of Biological Engineering, Sexton House, 1360 Barrington Street, B3H 4R2, Halifax, NS, Canada

The Annapolis Valley contains one third of the total farmland of Nova Scotia, Canada. The current landscape of the Valley has been influenced by human activities, physiography, and climate. Urbanization in some of the large towns and intensive livestock operations have created environmental problems in the area, which includes deterioration of surface water and groundwater quality. Phosphorous (P) enrichment and leaching of nitrate and other pollutants are the main water quality issues in the Valley. Continuous monitoring of water quality parameters is expensive and difficult. Several simulation models have been developed to examine the impact of different land use practices on water quantity and quality of the agricultural wetlands. One of these models, the Soil and Water Assessment Tool (SWAT), simulates the impact of land management practices on water, sediment, and agricultural chemical yields in watersheds with varying soils, land use, and management conditions over time. The purpose of this study is to evaluate SWAT for a small watershed in the Annapolis Valley. Thomas Brook watershed in the Annapolis Valley covers an area of 7.60 km2 was selected to examine the rising nutrients impact on the surface water quality. Delineation of the watershed boundary followed by further discretization into watersheds was conducted using a 10 m resolution digital elevation model (DEM). The stream layer was further modified based on field surveys especially in the south-west corner of the brook. The model was then parameterized with both the GIS-based spatial data such as DEM, soils and land uses for hydrologic response units (HRUs) within each subwatershed and temporal input variables like temperature and precipitation. Parameters like curve number, soil evaporation compensation factor, maximum snow cover, maximum and minimum snow melt factor, and baseflow alpha factor were adjusted within the permissible limits to calibrate the model. Calibration and validation of a model is a key factor in reducing uncertainty and increasing user confidence in its predictive abilities, which make the application of the model effective. The study was conducted using 29-months flow records from May 2004 to Sep 2006. Flow data from May 2004 to Dec. 2005 were used for SWAT calibration and Jan 2006 to Sept. 2006 for validation. Some adjustments were made in the flow records. SWAT generally simulated well the ET, water yield and monthly stream flow. Model calibration was assessed by coefficient of determination (R^2) and Nash-Sutcliffe efficiency (NSE). The simulated monthly streamflow matched the observed values, with the R^2 value of 0.856 and NSE value of 0.676; for calibration. For validation these values were 0.771 and 0.618 for R^2 and NSE, respectively. It was concluded that the calibration and validation of the model may be more authenticated with long term flow data.

Land use conversion and the hydrological functions effects during the dry season in Atibainha watershed, Brazil

Rosa Pereira, V.[1], Teixeira Filho, J.[1,2] and Dal Fabbro, I.M.[2], [1]State University of Campinas Geocience Institute, Geography Department, Cidade Universitária Zeferino Vaz, 13083-870 Campinas, Brazil, [2]State University of Campinas, Faculdade de Engenharia Agrícola, Cidade universitaria Zeferino Vaz, 13083-870 Campinas, Brazil

The identification of possible impacts on hydrological functions of deforestation and conversion to other land use is important to offer directions for public environment policy. This paper is a research result, which objective is to identify the relationship between the land use conversion from forest to pasture and silviculture during 20 years in the water yield during the dry season at the upstream Atibainha watershed reservoir (140 km^2). The study area is account for 20 million people water supply in the most economic and developed region in Brazil, the Sao Paulo and Campinas Metropolitan Region. So, the alterations in the water yield and hydrologic function in the study area represents a huge socio-economical impact. The land use maps were derived from 1986, 1999 and 2005 Landsat scenes classification to identify deforestation and land use conversion in km^2. The hydrological analysis were derived from 30, 45 60 and 75 consecutive dry day daily discharge and total water year precipitation during 1986 and 2005. As a methodology, linear regression analyses were used to identify the temporal changes in the discharge and precipitation trend. The results indicate that forest land cover decrease from 74% to 45% and pasture and logging increase from 20% to 29% and 5% to 25% respectively. During the same period, discharge presented a trend to decline in 30, 45, 60 and 75 consecutive dry days presenting a decrease to 61%, 56%, 55% and 60% respectively. The precipitation presents a decrease from 20% during the period. To sum up, the results indicate that exist a water yield decrease trend during the dry season at Atibainha watershed. The results of this study are coherent with another studies found in the literature, according to which forest land use conversion to another crops diminishes the soil infiltrations capacity, resulting the increase of runoff conditions and the water yield in the rainy season and consequently, the water retention diminish in the dry season. In conclusion, the results offer conditions for public environment policies to adopt the right actions to water resources conservations practices.

Decision support for grazing management: evaluation of suitability and estimation of potential on alpine pastures for sheep and goats

Blaschka, A., Guggenberger, T. and Ringdorfer, F., Agricultural Research and Education Centre Raumberg-Gumpenstein, Raumberg 38, A 8952 Irdning, Austria

In the past, the use of alpine pastures during summer was an economical resource for farmers, which became neglected due to intensification of agriculture after World War II. Nowadays they are becoming important again, mainly because of its economical impact as a resource of forage available without external dependencies, linked to the typical, regional production and nature conservation reasons. The extensive characteristics with its contribution to biodiversity and landscape variability represent land use systems that fit to the requests of society on agriculture. Small ruminants are particularly suited for low-intensity farming and for multifunctional use of marginal resources. This would suggest that it is reasonable to abandon the usual residual approach in favour of sheep or goats, i.e. giving them what is left after cattle needs have been met. The focus of the model presented is to create practical approaches for the estimation of suitability as pasture for sheep and goats that can be used in the entire Alpine region. To achieve this aim, the fields of vegetation ecology and plant production, animal nutrition and geo-information sciences were linked to their model parts. The entire working model was built into a cohesive software package called ENEALP. It deals with the analysis of flows of energetic material on alpine pastures (ENE = energy, ALP = alpine). This quantitative model is based on the assumption that an energy content of forage of more than 8 megajoule (MJ) of metabolisable energy (ME) has to be reached to achieve suitability for summer grazing. But also the practical suitability of the species of animal used (slope inclination and accessibility) and the availability of water is of great significance and considered. The aim is to provide information on additionally usable capacities in existing pasture regions. As a planning tool, for example, ENEALP calculates the number of animals that can be driven up to the pasture in addition to already existing farming. Data from the study area can be combined with expert systems (scientists, farmers, shepherds) as wished. Sheep and goats are slipping into a new role in land management in mountainous areas, which requires adaption and new solutions for a sustainable management of alpine pastures. Additionally, in connection with rising forest lines, grazing offers the only effective protection of the biodiversity of these pastures. Part of this research was done in the frame of the INTERREG IIIB Project 'Alpine network for sheep and goats promotion for a sustainable territory development' - ALPINET GHEEP.

A decision support system to predict plot infestation with soil-borne pathogens
Goldstein, E.[1], Hetzroni, A.[1], Cohen, Y.[1], Tsror, L.[1] and Zig, U.[2], [1]ARO, Agricultural Engineering, P.O. Box 6, 50250 Bet dagan, Israel, [2]Yaham Enterprise, Hevel maon, 85465 Magen, Israel; amots@volcani.agri.gov.il

Intensive growth of field crops requires implementing a meticulous crop rotation scheme to minimize the damage from polyphagous soil-borne pathogens. Plot allocation requires considering numerous factors such as infestation history, crop and variety susceptibility, source of reproduction material, soil characteristics, growing season and chemical treatments. A decision support system that provides the decision makers with all the relevant information will optimize crop allocation with respect to disease risk and reduction of pesticide use. Intelligent decision making regarding crop rotation and disease control that takes into consideration the agricultural, environmental and commercial constrains is dependent upon the availability of substantial up-to-date data and knowledge of the plot history. The challenges are in data acquisition, repository maintenance and the effective feeding of the decision making procedure with relevant information. Contemporary quality assurance techniques and international export regulations provided an opportunity to exercise the consolidation of field crop data into a comprehensive database. This approach was adopted by several major growers in the region that have committed to maintain and support data management. The design and assimilation of data repository at the early stages of the project has previously been described. The purpose of this study is to develop a decision support system for crop allocation by ranking fields according to disease prediction. At this stage, the research was focused on *Verticillium dahliae* and *Colletotrichum coccodes* on potato. Comprehensive historical database was characterized and constructed. Biotic and abiotic factors affecting the manifestation of soil-borne diseases were selected by interviews and discussions with four domain experts (including researchers and growers) and from literature. These factors were weighed and ranked by multi-criteria pairwise comparison and weighed linear combination. The experts' rankings were checked for consistency. These rankings were used to predict the susceptibility of each plot. Plot records between 1998 and 2007 were used to validate the decision support system, using at least the recent 4-5 years data. Multi criteria pairwise comparison resulted in acceptable rate of consistency with only two experts for *V. dahliae* and with one expert for *C. coccodes*. The accuracy in predicting soil pathogens was moderate.

Increasing profit of a co-digestation plant utilizing real time process data
Van Riel, J.W., Andre, G. and Timmerman, M., WUR / Animal Sciences Groups, BusinessUnit Veehouderij, Edelhertweg 15, 8219 PH Lelystad, Netherlands

There is a potential for improvement of the management of a co-digestion plant, which would make the installation more economic sustainable. Profit of a co-digestion plant can be increased by making use of temporal variation in input efficiency of the influent. Current decision support systems do not account for this variation. Suitable models for on-line estimation of the efficiency are dynamic linear models (DLM), a Bayesian approach to time series analysis. An integrated system for decision support using DLM was developed and tested for co-digestion of organic biomass with animal manure in a full-scale biogas plant. Due to market circumstances of supply and demand of organic streams the availability of biomass products for digestion will vary. In the anaerobic digestion process multiple bacteria are involved. This makes co-digestion a complex and dynamic system which is hard tot understand, interpret and optimize. To maximize profit from a co-digestion plant the challenge is to optimize continuously the daily input and it's composition. In our study we use real time process data to estimate the effect of daily input and composition on gas production of the plant. Finally the optimal settings of input allocation were calculated such that the economic returns minus input costs were maximized. The application was developed and tested on a research farm. New optimal settings for daily input were calculated and implemented weekly. Preliminary results of the application show an increase in gas production efficiency and a 'hands-on' approach of plant management.

Improvement of work processes in Gypsophila flowers

Bechar, A.[1], Lanir, T.[2], Ruhrberg, Y.[2] and Edan, Y.[2], [1]ARO - Volcani Center, Institute of Agricultural Engineering, P.O. Box 6, 50250, Bet-Dagan, Israel, [2]Ben Gurion University, Department of Industrial Engineering and Management, P.O. Box 653, 84105, Beer-Sheva, Israel

Flower production in Israel results in an annual export of 1 milliard flowers with a redemption of about € 100,000,000. Gypsophila (*Gypsophila arrosti*) comprises 7% of the total growing area (5000 ha) and requires extensive manual labor which is a major bottleneck in the growing process and impedes cost reduction. This research aims to improve efficiency and to reduce manual labor in the major work processes, harvesting and packing. Due to the rapid decrease in flower quality during harvesting the conveying time in Gypsophila is critical. Work study elements of the work processes were defined and data were collected in 12 plots along 80 hours using work measurement techniques. A simulation model was developed to compare 5 alternative work methods to the current state using four performance measures: (1) worker yield; (2) the portion of the conveying time from the total process time; (3) production to area unit; and (4) yield quality. The model was validated, using visual analysis and statistical tests (T-Test, Kleijnen Test). An optimization model was developed to determine the number of workers in each station, the productivity and working time for different optimization targets (maximum daily yield, minimum time, and minimum number of workers). About 2,400 different combinations were tested in the optimization model. Sensitivity analyses were conducted to examine the influence of small changes in the number of workers, working day duration, and, opening and closing times of a second packing line. Results indicated that up to 20% of the total work is invested in conveying the flowers between the different stages and stations. An improved working method yielded increased productivity of up to 9% and reduced conveying time in the field by 26%. Reduction of the row length can further reduce conveying time by 15% and, by reducing the packing operations from the harvesting workers and transferring the packing stations from the field to a packing house the production of these operations will increase by 25%. The optimization model can serve as a decision support tool for the farmer on a daily basis. This can yield improved allocation of workers to different tasks, increased production, and reduced time and manual labor.

The development of a sensor and crop growth model based decision tool for site specific nitrogen application

Goense, D. and De Boer, J., Wageningen University Research Centre, Animal Sciences Group, Edelhertweg 15, Lelystad, Netherlands

The decision on when and how much N-fertilizer too apply was always difficult, and this is aggravated with the possibility of precision agriculture. Remote sensing based approaches are in development, but do not account for future crop developments as determined by coming weather conditions. A model based approach has the possibility to evaluate application strategies undr different future weather scenario's. The objective is to develop in interaction with six farmers, growing winter wheat and starch potatoes, a decision tool for site specific fertilizer application which is based on sensor data to correct the model prediction of the actual situation and to optimize model parameters in retrospect. The tool is based on an information infrastructure that uses standards as much as possible. The basis of the decision tool is the combined crop growth, soil-moisture and nitrogen model WAVE. A shell is build in java, which provides the user interface for selecting the fields and locations in the field for which to optimize. This shell is able to collect required input data either from local data files or an external server by setting a switch. Important input data are the profile descriptions of the soil, which for the topsoil are based on isotope based sensor data and for the subsoil interpolated soil auguring data. Interpolation is done by the server. Ground water is monitored regularly, on some fields soil moisture by wireless sensors and remote sensing data is collected two or three times a season. Visual observations are made on crop health and weeds, and for validation also biomass is determined. Al fields are yield mapped. For data exchange objects are used from ISO11783 Part 10, but with a different structure for the treatment zone object. In model runs the actual weather was used and for the first coming 15 days the main run of a long term weather prediction. For the period thereafter, the weather of the last ten year was used. A number of fertilizer scenario's in respect of timing and quantity could be specified. The tool calculates optimal scenario based on actual prices of the product and fertilizer. The approach of Data Access Objects (DAO) proved to be very efficient in switching between different data sources. ISO 11783 Part 10 is not suited to to transfer site specific variable data as all is connected to device elements and an international standard on the decision support level is not present yet. Wirless sensor data can successfully exchanged following ISO11783, but it for yield mapping and tractor mounted remote sensing still only proprietary data files are available. following Correction of actual conditions requires too much principal changes in the WAVE model itself, so we limited ourselves by changing some model parameters during the season. Using the model over six farms showed clearly that farm or field specific parameter optimization is required. Standard soil type dependent parameters are not adequate as local soil bulk density influences hydraulic parameters and differences in quality of organic matter is not accounted for. Farmers realized that the decision tool is still in development, but appreciated the predictions made, and their comments were very valuable for the developers.

Fuzzy logic inference system (FIS) based on soil and crop growth information for optimum N rate on corn
Tremblay, N., Bouroubi, M.Y., Panneton, B., Vigneault, P. and Bélec, C., Agriculture and Agri-Food Canada, Research Center, 430 Gouin, Saint-Jean-sur-Richelieu J3B 3E6, Canada

Actual optimal N rates for corn are hard to determine since they are driven to a great extent by weather, soil and crop management factors in interaction. Still, the need for optimizing N rates is gaining in importance due to environmental and economical reasons. New tools and strategies must be proven effective and implemented for that purpose at the farm level. For instance, it is known that the crop establishment phase does not require high levels of nitrogen availability in the soil. The 'all at sowing' fertilization practice leads itself to high soil mineral N concentration at this early stage where it is particularly at risk of losses by leaching. However, the alternative practice where a complement to the quantity provided at sowing is saved for 'in season' application reduces the likelihood of N losses since most of the N dressing is provided just prior to the period of active uptake by the crop. It also allows for the plants to be used as indicators of actual local N availability through measurable changes in their chlorophyll, LAI and biomass status. The spatially variable nature of optimal N fertilization requirements within fields, has indeed led to the development of systems relying on the remote assessment of crop chlorophyll/biomass status to drive variable rate applications. Reports on the performance of these have shown room for improvements based on complementary information such as soil conditions. Since the relationships governing the interactions among the parameters of importance are not precisely known, we hypothesized that a set of a fuzzy logic inference system (FIS) that would adequately combine critical plant and soil based spatial information leading to a significant improvement of N recommendations over current grower's practices. Identification of key soil and plant parameters was based on experiments with N rates ranging from 0 to 250 kg/ha conducted over three seasons (2005-2007) on fields of contrasting edaphic and topographic features. For soils, the following parameters were selected: apparent soil electrical conductivity (EC_a), elevation (ELE) and slope (SLP). The NSI (Nitrogen Sufficiency Index; the ratio of NDVI status to the one of a fully fertilized reference plot) was used as the plant-based parameter of interest. Weather-based information was also included in the system in the form of precipitations (PPT). N rate optimization by the FIS was defined against maximal corn growth in the weeks following in season N application. This mid-season growth was assessed through remotely sensed imagery by a CASI (Compact Airborne Spectrographic Imager) at the 1 m^2 pixel resolution. Best mid-season growth was experienced in areas of low EC_a, low SLP and high ELE. Expert knowledge was built up into a set of rules involving three defined soil parameters levels: low, medium and high. For example, for low EC_a, maximal mid-season growth was obtained without in-season N. For high EC_a, minimal N rates of 87 kg/ha are needed to maximize mid-season growth. A simulation was achieved to compare fertilizer savings using the full set of information available (NSI, EC_a, ELE, SLP, PPT) or sub-ensembles of them, as compared to grower's practice. The purpose was to evaluate the usefulness of including or not plant-based, soil-based and weather-based information.

SMA, an agrometeorological system for crop monitoring

Faria, R.T.[1], Tsukahara, R.Y.[2], Silva, F.F.[1], Gomes, C.D.[1], Caramori, P.H.[1] and Silva, D.A.B.[1], [1]Instituto Agronomico do Parana, Agricultural Engineering, Rod Celso Garcia Cid, km 375, 86020-918 - Londrina, PR, Brazil, [2]Fundacao ABC, Agrometeorology, Rodovia PR-151, km 288, 84165-700 - Castro, PR, Brazil

Updated information on crop growth conditions are required by farmers to make operational decisions on performing agricultural practices, such as soil tillage, sowing, irrigation and harvesting. Other sectors, such as government, insurance companies and cooperatives are interested on production estimates to decide on measures related to food security, insurance price and crop input supply. The State of Paraná is the most important agricultural producer in Brazil, but yield is highly variable because of frequent water deficit and occasional frost. In order to provide decision makers with information on conditions for crop growth, an agrometeorological system for monitoring crops of social and economic importance was developed. Such a system is named SMA-C, a Portuguese acronym for `Sistema de Monitoramento Agroclimático de Cultivos`. It runs in a Windows platform and has a modular structure including an interface developed in Delphi 7.0, the Firebird 2.0 SQL database, and Python scripts to communicate the system with Arcgis 9.2 geographic information system. The system interface coordinates execution of the functional modules and helps system administrator to define simulation inputs, configure system parameters and visualize results. The functional modules are: a) Climate module, to import and prepare meteorological inputs for simulations, besides calculating reference evapotranspiration, b) Soil water balance module, to calculate soil water availability and water stress index using meteorological data and plant characteristics, c) Phenology module, to calculates crop development as a function of photothermal indices calculated with meteorological data as input, c) Crop yield module, to calculate crop yield decrease using a crop production function, d) Diseases module, for simulation of climatic risk to diseases infection using meteorological data as input. For a specific area (field, region or State), simulations are performed for each virtual weather station (VWS) corresponding to the node of a grid with density defined by the system administrator. The VWS in the project area are created using a SMA-C routine, which records geographic coordinates and corresponding soil type and altitude into system database. Another SMA routine is used to schedule the following procedures: a) importation of hourly or daily meteorological data from real weather stations (RWS), b) consistence of imported data; c) integration (from hour to day) or desegregation (from day to hour) of data, and d) interpolation of data from to RWS to VWS using a GIS interpolation routine of SMA. Besides meteorological data, the system requires soil water retention characteristics, plant growth and phenological characteristics, maps and geographical parameters. Different combination of planting dates and cultivar can be simulated for the project area. SMA-C is operated by the Instituto Agronômico do Paraná's (IAPAR) to obtain the results in form of tables, graphics and maps to be presented at IAPAR`s WEB site, as shown by http://www.iapar.br/modules/conteudo/conteudo.php?conteudo=667.

How long should your corn-cob pipe be: modeling yield*quality interactions in Sweetcorn

Taylor, J.A., Hedges, S. and Whelan, B.M., The University of Sydney, Australian Centre for Precision Agriculture, McMillan Bld A03, 2006 The University of Sydney, NSW, Australia

Midseason variable rate nitrogen has been successfully implemented in maize crops. Despite the similarities between maize and sweetcorn, adoption of midseason differential management in sweetcorn has not been strong. This is due in part to the absence of a crop model to drive decision support systems. In late 2007 the first sweetcorn crop model was published, an adaptation of the maize component of the CERES model. This model related various crop and management parameters to both yield and quality. To test the assumptions of the model, the applicability of such a model site-specifically, and to determine which input parameters are most relevant in Australian production systems, a study was undertaken on a sweetcorn system near Sydney, NSW Australia. The key variable in the study was plant density which is acknowledged to be a major driver of production. The methodology for the study was based on the concept of management classes. Existing a priori data (an ECa survey and elevation data) was clustered and crop sampling was stratified to these clusters. As the season progressed and new data was made available the management classes were updated. During the season, information on plant development (germination rate, tillering %, plant height, plant dry weight and plant wet weight) and information on canopy response (via two CropCircle surveys and a multi-spectral aerial image) was collected. At harvest information on yield, cob length and diameter, cob moisture %, secondary tillering %, plant density and insect damage was measured. A regression model was used to predict final yield from the ancillary data (midseason canopy reflectance and a priori ECa). The results of this model were poor. When the plant density information at harvest was incorporated into the regression equation much better fits were observed for individual fields as well as for a global model across all fields. The global model accounted for 66% of the variation in the yield. For an effective prediction of yield potential in sweetcorn it would appear that knowledge of plant density is imperative. At present a sensor or means to provide this information is not available. The NDVI provided information on photosynthetically active biomass at a location but not on plant density. Regression models were also run to predict cob quality (length) using moisture %, plant density and yield. Good fits for these models across the fields were observed indicating that yield could be manipulated to achieve a desirable cob quality. The economics of quality vs. quantity are unclear but, given a desired moisture end point and a known plant density, it may be possible to optimize the yield*quality interaction. Alternatively, the quality prediction model could be used to drive a differential harvesting strategy.

To determine the challenges in application of ICT for agricultural extension in Iran
Farajallah Hosseini, S. and Niknamee, M., Islamic Azad University, Agrciultural Extension and Education, Poonak, Hesarak, 128769494, Iran

About one third of Iranian Population lives in rural areas and they have very limited access to the information. ICT can bring about important changes to the living conditions of rural population and in this regard, agricultural extension plays an important role. However, agricultural extension faces several challenges in using the ICT in Iran. This study was conducted to identify the challenges in application of ICT by extension service in Iran. A questionnaire was developed and data collected from 187 extension specialists. The ordinal factor analysis was used and the results show the classification of the challenges into six latent variables. The variables were classified in organizational, technical, social, financial, regulatory and human factors.

CAP subsidy claims: trying to encourage French farmers to abandon paper forms and to use the Internet

Waksman, G.[1], Holl, C.[2] and Coffion, R.[1], [1]ACTA Informatique, 149 rue de Bercy, 75595 Paris Cedex 12, France, [2]EP-TE, 71 rue du Docteur Netter, 75012 Paris, France

TelePAC is the Internet service proposed by the French ministry of Agriculture and Fisheries (MAF), to enable farmers to register their claims for CAP subsidies. In 2008, around 25% of French farmers are using TelePAC most often with the help of advisors and within the framework of commercial services marketed by advisory organisations. The percentage of TelePAC users is very much different from one region to the other and depends on most important agricultural productions. The French MAF asked us in 2008 to propose training sessions and assistance to users together with information and training materials, to encourage farmers to access TelePAC autonomously. In the first part of this paper, we discuss the attitudes of farmers towards TelePAC and try to identity the reasons why so many farmers do not use this solution when the French MAF is hoping that within the three next years 80% of French farmers will use the electronic solution to register their CAP subsidy claims. In the second part, we present the actions that we performed to develop the autonomous use of TelePAC by farmers and discuss the results of these actions: edition of a printed user guide, organising information meetings, training sessions, users' assistance, edition of a DVD to promote TelePAC and to describe how it works, production of a Web TV emission. The use of 'last cry' ICT tools such as web conferencing and remote control of farmers' PCs proved to be not very successful when the use of TelePAC in a training room together with other farmers and with the support of an advisor pleased the farmers very much. In our conclusion, we go back to basic old questions: Why should farmers adopt ICT? How to encourage them to use the Internet? Which marketing for agricultural ICT services?

Business adoption of broadband internet in the South West of England

Warren, M.F., University of Plymouth, Rural Futures Unit, 8-11 Kirkby Place, PL4 8AA, United Kingdom

This paper reports on two studies of business adoption and use of broadband internet in the South West of England. The first, the Broadband4Devon (B4D) project, related to small (mainly micro-) businesses in poorer parts of Devonshire, both rural and urban (though 'urban' in this context can often mean a town of less than 10,000 inhabitants), but excluding purely agricultural businesses. The second focussed specifically on farm businesses in the neighbouring county of Cornwall, where a publicly-funded partnership,'actnow', has been very active for several years in improving access to, and encouraging use of, broadband in an essentially rural area. The B4D research was based on two iterations of an email/postal survey of 380 businesses, together with 12 company case studies based on in-depth interviewing. The actnow research relied on telephone interviews with 209 farmers, though follow-up with individual and group interviews is planned. What emerges from each study is a picture of the motivations of those adopting (and, in the case of the Cornwall farmers, of those not adopting) broadband internet; of the uses to which they put the technology; and of the value they thus derive. From this it transpires that it is the basic benefits of broadband that are valued most by adopters, and that are most tempting to potential users. The 'killer applications' are not voice-over-internet, video-conferencing or interactive gaming, but email that is 'always-on' and can carry large file attachments; a phone line that is no longer dominated by dial-up internet use; the ability to find information on the Web rapidly; and the opportunity (not taken by all) to have a simple web presence of one's own.

Italian rural ICT (computer and internet):deployment and accessibility
Namdarian, I., INEA - Istituto Nazionale di Economia Agraria, Unit5, Via Barberini, 36, 00187 Rome, Italy

The European Strategic Guidelines for Rural Development 2007-2013 increase the focus on ICT take-up in rural areas in line with EU priorities in the field of the Information Society. In particular, 'there is a need to accompany changes in rural areas by helping them to diversify farming activities towards non-agricultural activities and develop non-agricultural sectors, promote employment, improve basic services, including local access to Information and Communication Technologies (ICTs)'. Rural Italy, like many other non-urban parts of Europe, is very limited in terms of providing broadband Internet access to business and home users. The connectivity is very limited outside major cities, and people and business managers in rural areas still have to rely mostly on slow dial up access. RAIANET, 'Rural Areas and Internet Access in Italy,' is a survey (granted by EUROSTAT-TAPAS -Technical Action Plans for Agricultural Statistics -2007 action - Commission Decision N°2007/84/CE) that provides a portrait of rural Italy's ICT and Internet users. The study is based on a survey that considers basically the deployment and availability of ICT and Internet, users and their attitude towards ICT and Internet in 5 pilot Italian rural areas. Within this frame the Community interest is emphasised by the fact that in addition to the importance of various application of these new data obtained by the survey and the improvement of the existing ones, the methodology can be applied to other European rural areas because it is not strictly linked to the Italian rural situation but to any other geographical rural area.

Stimulating interest in and adoption of precision agriculture methods on small farm operations

Yohn, C.W.[1], Basden, T.J.[2], Rayburn, E.B.[2], Pena - Yewtukhiw, E.M.[3] and Fullen, J.T.[4], [1]West Virginia University Extension Service, Agriculture and Natural Resources, 1948 Wiltshire Road, Suite 3, Kearneysville, WV 25430, USA, [2]West Virginia University Extension Service, Agriculture and Natural Resources, P. O. Box 6108, Morgantown, WV 26505 - 6108, USA, [3]West Virginia University, Division of Plant and Soil Sciences, P. O. Box 6108, Morgantown, WV 26506-6108, USA, [4]Fullen Fertilizer, N/A, 46 RR 1, Union, WV 24983, USA

In 2005, West Virginia University Extension provided seed money to demonstrate precision soil sampling and precision nutrient management. West Virginia University Extension Service partnered with a crop consultant and 'custom applicator' (Fullen Fertilizer) to perform this task. In 2006 11 land owners consigned 271.5 hectares to an expanded demonstration. These farms were soil sampled using conventional (0.56 hectares per sample) and precision (0.76 hectares per sample) method. WVU Extension paid for the conventionally taken samples. The individual soil data points were used to determine the number of acres represented by the sample, the soil type and yield potential for alfalfa, mixed grass and corn from soil survey. Results showed that precision sampling and application required more lime than recommended by conventional sampling. Precision management required more phosphorous and less potash on fewer acres than recommended by conventional management. Extension education through presentations, a field day and farmer to farmer talks resulted in 625 additional hectares being sampled. Over 647.5 hectares received precision application of lime, phosphorous and potash. This occurred on five of the twelve cooperators and eight new land owners. Producers have used this information to change management practices including using mineral feeders and hay placement to guide cattle to specific field areas, delineate areas for variable nutrient application using conventional equipment. Acquisition of precision tools has increased in the area. Private consultants are partnering to help farmers interpret collected data.

M-learning in agriculture: potential and barriers

Hansen, J.P. and Hejndorf, P., Danish Agricultural Advisory Service, Web & IT, Udkaersvej 15, DK-8200, Denmark; jph@landscentret.dk

Working in agriculture means working out of reach of a traditional computer. This and the fact that farmers typically have an activity based learning style indicate, that there is a potential in introducing mobile learning. That was the background for a project running 2007-2008 and supported by The Ministry of Science, Technology and Innovation, in which we investigated the potential of using m-learning in agriculture. The term 'm-learning' is used here to refer specifically to learning that is facilitated and enhanced by the use of digital mobile devices, that can be carried and used anywhere and anytime such as mobile phones, PDAs and MP3 players. In our project, focus was on utilising the mobile phone, as 98% of Danish farmers use a mobile phone. We experimented with a number of methods and technologies such as mobile audio (podcast), mobile video, documents and images, simple games, knowledge assessments (interactive test) and provoking SMS quizzes. Two alternative authoring systems was used and evaluated. Besides using the mobile phone as delivery platform, we experimented with re-using podcast in an internet based radio station established for this purpose, such that e.g. a farmer while milking can listen to agricultural information instead of music (most Danish farmers has a radio in their milking parlour). Also Bluetooth streaming of the radio station to Bluetooth enabled ear muffs were tested. The presentation will show examples of developed mobile learning modules, and we will discuss the different barriers encountered trying to get information from the head of the adviser/specialist into the ears (and eyes) of the farmer.

E-learning for agriculture and food sector in Poland
Wozniakowski, M., Orlowski, A. and Wozniakowski, T., Warsaw University of Life Scinces, Department of Informatics, Nowoursynowska 166, 02-787 Warsaw, Poland; miroslaw_wozniakowski@sggw.pl

The paper concerns technological change and implications of e-learning technologies on agriculture and food sector in Poland. Particular attention is given to technological readiness of agri-food companies in area of e-learning use. The research has been made to fill the gap in knowledge on this specific sector and its needs and capabilities in modern educational technology utilization. E-learning is seen as a means of increasing access to educational opportunities and keeping pace with rapid changes within the sector. This study examines elearning adoption trends within the agricultural and food market in Central Poland, where e-learning has reached late adoption status within other sectors, such as in the IT sector. Specific factors of the research were: (1) general use of computers (2) use of internet (3) use of it systems (4) use of e-learning (5) plans for e-learning use (6)knowledge on advantages and disadvantages of e-learning use. Major results of the research proves the hypothesis that Agri-Food sector has not reached early status yet. It is shown that increasing returns to adoption can arise if related technologies do not substitute each other in their functionalities. Hence, the probability to adopt the technology is an increasing function of previously adopted, related technologies. Numerous circumstances for e-learning in Agri-Food sector were noted, including: rising cost of traditional, classroom-based training, long distances from rural regions to centres of education, increasing internet and computer access and computer skills etc.

Thinking styles of agriculturalists

Singh, S. and Lubbe, S., UNISA, School of Computing, P.O. Box 392, Unisarand 0003, South Africa; lubbesi@unisa.ac.za

The uses of Information Technology (IT) and IT products are pervading facets of the South African farming community. IT is also actively used by agriculturalists although not as effectively in certain agricultural disadvantaged communities. There, however, exist gaps in communication channels between some agriculturalists and IT specialists when demonstrating the benefits of IT applications. One aspect to bridge the communication gaps is to understand the way agriculturalists think. The research surveyed present and prospective agriculturalists. It discusses the different thinking styles that presently exist in a disadvantaged farming community in South Africa and explain how this could be used to the benefit of the disadvantaged community. The recommendation is that one should determine what thinking style they use to ensure successful approaches in teaching the thorough application of IT in the farming sector of any individual. It also demonstrates how these people might process information – an important factor in successful farming.

Optimization model for variable rate application in extensive crops in Chile: the effects of fertilizer distribution within the field

Ortega, R.A., Muñoz, R.E. and Acosta, L.E., Universidad Técnica Federico Santa María, Departamento de Industrias, Avenida Santa María 6400, 7660251 Vitacura, Santiago, Chile

With the current available machinery, variable rate N-P-K fertilization can be done in Chile only through a physical blend applied at different rates. An optimization model was built that allows estimating: (1) the 'best' N-P-K blend, given the nutrient requirements of each place within the field, the available fertilizers for blending, and their cost; and (2) the blend rates for each place within the field. Input data needed are: (1) soil grid sampling data; (2) yield data which can be either variable (yield map) or constant (yield goal) for the whole field. Output data corresponds to the best N-P-K blend for the field and a prescription map of variable rates for the best blend. The model was evaluated on data coming from different crops and fields; model performance was measured by comparing nutrient requirements at each point of the field and the nutrients theoretically applied with the variably-applied blend; the assumption behind this evaluation was that each product in the fertilizer blend is homogeneously applied, which is uncertain; in fact a large dispersion in terms of nutrient content is observed when a physical blend is used in comparison with a chemical blend, where the variance is zero. Thus there are some cases where it would be better to use a fixed N-P-K chemical blend than applying 'the best' N-P-K physical one. Technical and economical comparisons are made for 'the best' physical blend and available chemical ones.

Sensitivity analysis of site-specific fertilizer application

Gebbers, R.[1], Herbst, R.[2] and Wenkel, K.-O.[3], [1]Leibniz-Institut für AgrartechnikPotsdam-Bornim e.V. (ATB, Technik im Pflanzenbau, Max-Eyth-Allee 100, D-14469 Potsdam, Germany, [2]Humboldt-Universität zu Berlin, Fachgebiet Precision Agriculture, Invalidenstrasse 42, D-10115 Berlin, Germany, [3]Leibniz-Zentrum für Agrarlandschaftsforschung e.V. (ZALF), Institut für Landschaftssystemanalyse, Eberswalder Strasse 84, D-15374 Müncheberg, Germany

The accuracy of site-specific application of macronutrients (N, P, K, Mg) and lime can be affected by a number of factors. However, only limited efforts were made to rank these factors with respect to their impact on application accuracy. In this study, sensitivity analysis was carried out to analyse the influence of the variability of input factors on the error of fertilizer application. The outcome provides a basis to optimize site-specific fertilizer application. Inputs regarded in the analysis were: (1) spatial variability of the fertilizer demand expressed by the variogram range parameter; (2) size of the management units; (3) positioning error (deviation from optimum track, wrong coordinates on the prescriptive map); (4) systematic spread pattern error (constant deviation from optimum shape); (5) random spread pattern error (variability within the spread pattern) Randomness of input data was generated by Monte-Carlo methods (computer generated random numbers). The statistical distributions were derived from real data. Geostatistical simulation was used to produce maps of varying fertilizer demand on a virtual 36 ha field of 1 m² resolution. Fertilizer demand was spatially averaged over grids of different sizes to obtain management units that form the prescriptive map. Positioning errors were modelled by fitting temporal variograms to observations from a dynamic GPS test. 2D spread patterns of centrifugal disc spreaders were derived by deconvolution from lateral spread patterns measured during a European spreader test at DIAS, Bygholm, DK. Spreading was simulated by 'driving' a virtual spreader over the field using different combinations of variability (errors) in the input factors. The spreading results (as-applied maps) were compared with the respective map of fertilizer demand. These differences were used in the sensitivity analyses as output variability to be analyzed. Sensitivity analysis was carried out by Fourier Amplitude Sensitivity Testing (FAST). FAST is a highly efficient Monte-Carlo based method to obtain global sensitivity indices which are independent from model properties (e.g. linearity, monotonicity, additivity). First and total order effects, including interaction of factors, can be estimated. According to the total effect sensitivity index spatial variability has the largest influence on application accuracy. Influence of the size of management units is slightly lower. A strong interaction between these dominating factors exists. Sensitivity of the random spread pattern error is considerably lower and can be rated as less important. Compared to the other factors, systematic spread pattern errors and positioning errors are unimportant. From a technical perspective, the most promising strategy to improve site-specific fertilizer application will be the accurate capturing of spatial variation of fertilizer demand and the use of high resolution management zones. In contrast, investment in a more accurate GPS or fertilizer spreader will be less effective.

Precision manure management across site-specific management zones in the Western Great Plains of the USA

Moshia, M.[1], Khosla, R.[1], Westfall, D.[1], Davis, J.[1] and Reich, R.[2], [1]Colorado State University, Soil & Crop Sciences, C013 Plant Sciences Bldg., Campus Delivery 1170, Fort Collins, CO 80523-1170, USA, [2]Colorado State University, Forest Range and Watershed Stewardship, Forestry Building, Fort Collins, CO 80523, USA

In the western Great Plains of the USA, animal agriculture is an important contributor to agricultural economy. Most livestock farms are close to water bodies and pose a high risk in contaminating the environment. Precision manure management is a relatively a new concept that converges the best manure management practices along with precision agricultural techniques, such as management zones (MZs) for crop production. The objectives of the study were (1) to evaluate variable rate application of manure in improving topsoil quality and enhancing corn (*Zea mays* L.) grain yield and (2) to evaluate and compare N mineralization rates of various levels of manure applied on soils collected from three MZs in a controlled environment. The study was conducted over four site years in northeastern Colorado under continuous maize fields under rainfed and irrigated conditions. Experimental strips were 4.5 m wide and 540 m long spanned across all MZ with treatments nested within MZs in the field. Variable rate treatment for the study ranged from 0 to 134 Mg ha^{-1} of animal manure, and farmers' recommended N fertilizer rate for grain yield comparisons. For objective 2, a 120 day laboratory incubation study was conducted at Colorado State University's Natural Resources Ecology Laboratory. Our results indicate that variable rate application of animal manure mineralize differently ($P \leq 0.05$) within MZs. Grain yield was significantly different across MZs with low MZ showing a significant increase in grain yield than that of synthetic N fertilizer treatment in the second year of study. The enhanced grain yield in low MZ can be attributed to improved bulk density, supplementary organic matter and mineralized N from applied manure in the second year of on-going study. A significant increase in topsoil organic matter was observed across MZs with a one-time application of 134 Mg ha^{-1} of animal manure increasing topsoil organic matter with 36% on high MZ. The study suggests that variable rate application of manure has a potential to improve soil quality and can be used as an alternative to or in conjunction with synthetic N fertilizer for improving or maintaining maize grain yield. The key is to find a balance between agronomically and environmentally sound manure application rates across spatially variable soils.

Sensitivity analysis of soil nutrient mapping

Gebbers, R.[1], Herbst, R.[2] and Wenkel, K.-O.[3], [1]Leibniz-Institut für AgrartechnikPotsdam-Bornim e.V. (ATB), Technik im Pflanzenbau, Max-Eyth-Allee 100, D-14469 Potsdam, Germany, [2]Humboldt-Universität zu Berlin, Fachgebiet Precision Agriculture, Invalidenstrasse 42, D-10115 Berlin, Germany, [3]Leibniz-Zentrum für Agrarlandschaftsforschung e.V. (ZALF), Institut für Landschaftssystemanalyse, Eberswalder Strasse 84, D-15374 Müncheberg, Germany

The accuracy of soil nutrient maps for site-specific application of macronutrients (N, P, K, Mg) and lime can be affected by a number of factors. While individual factors like sampling density and design or interpolation method have been evaluated in number of papers only a few attempts were made to compare these factors with respect to the accuracy of soil nutrient maps. However, to optimize soil mapping it is indispensable to identify the most important sources of error. To accomplish this, a sensitivity analysis was carried out, which analysis the influence of the variability of input factors on the error of the soil nutrient maps. Inputs regarded in the analysis were: (1) spatial variability of soil nutrients and pH expressed by the range of the variogram; (2) sampling design (bulk sampling (linear and circular sampling line), point sampling); (3) sampling interval; (4) positioning error by GPS; (5) regionalization (interpolation, attributing to sampling zones); (6) error in chemical analysis Geostatistical simulation was used to produce 'true' soil nutrient maps of a virtual 36 ha field. Soil sampling was modelled by point queries on the true maps. Two main sampling strategies were modelled: a) classical sampling by bulking 20 samples from sample lines arranged on a regular grid and b) sensor based online data collection by point sampling and instant chemical analysis. Static dGPS positioning errors were obtained via Monte-Carlo simulation using a zero-centered normal distribution with standard deviation (SD) of 2 m. To create field extend maps from the samples two regionalization methods were applied: a) attributing to predefined management zones and b) interpolation. Different levels of errors in the chemical analysis of the samples were simulated to account for the differences in the accuracy of laboratory analysis and on-line measurements. The mapping results were compared with the respective map of true nutrient distribution. These differences were used in the sensitivity analyses as output variability to be analyzed. Sensitivity analysis was carried out by Fourier Amplitude Sensitivity Testing (FAST). FAST is a highly efficient method to obtain global sensitivity indices which are independent from model properties (e.g. linearity, monotonicity, additivity). First and total order effects, including interaction of factors, can be estimated. According to the total effect sensitivity index sampling interval has the largest influence on mapping accuracy. Sampling interval interacts with spatial variability. However, spatial variability is less important compared to sampling interval. All other factors are rated as relatively unimportant. In particular, positioning errors of 2 m SD seem to be completely negligible. In conclusion, the key to improve soil nutrient maps is an appropriate sampling density. More reliable maps can be obtained by increasing sampling density even on the expense of accuracy of chemical analysis. This advocates soil nutrient data collection by online sensors like the Veris MSP ph sensor which may produce higher errors in the chemical analysis but permits very small sampling intervals.

Irrigation management zones for precision viticulture according to intra-field variability
Vallés-Bigordà, D., Martínez-Casasnovas, J.A. and Ramos, M.C., University of Lleida, Department of Environment and Soil Science, Rovira Roure 191, E25198, Spain

Precision viticulture is at present a boiling field of research since, in contrast to extensive herbaceous crops, winegrapes are considered as a high value crop and the benefits can compensate the investment in the required technology. In this respect, definition of management zones for selective harvesting from yield or vegetation indexes detailed maps is one of the most common applications of spatial information technology in precision viticulture. Different from this common application, the present research shows a case study in precision viticulture as a tool for improved irrigation. The research was carried out in a vineyard field located in Raimat (Costers del Segre Denomination of Origin, Lleida, NE Spain). This is a semi-arid area with continental Mediterranean climate and a total annual precipitation between 300-400 mm. The field, with an extension of 4.5 ha, is planted with Syrah vines in a 3x2 m pattern. The vines are irrigated by means of drips under a partial root drying schedule. At present, the irrigation sectors are have a quadrangular distribution, irrigating a homogeneous area of about 1 ha each. The sectors were designed previous to the vine's plantation in 2002 without having into account the possible spatial variability of soil characteristics. At present, yield in this field presents a coefficient of variation of 32.2%, with an average yield of 6.9 t ha^{-1}. The method to re-design the irrigation sectors was based on multi-variant statistics analysis of soil properties (pH, electric conductivity, organic matter content, calcium carbonate content, water retention availability for plants, texture and multi-temporal profile water volumetric content). The sampling density for these properties was of 8.9 samples ha^{-1}. Other data used in the analysis were the normalized difference vegetation index (NDVI) from Quickbird-2 satellite images of the years 2004 to 2007, and yield data acquired from a Canlink 3000 Farmscan monitor for the years 2004 to 2006. The multiple regression analysis between yield or NDVI and soil properties, show that there is a scarce influence of top soil properties on the spatial variability of plant vigour or yield along the years. The best spatial prediction of yield is mainly explained by the NDVI of images acquired during the berry ripening and the sand content. It explains 85.6% of variance. At the same time, NDVI variability is mainly explained by the volumetric water content of the soil profile, explaining 72.1% of its variance. These properties: average yield, average NDVI, average volumetric water content and sand content, were used in a cluster analysis performed by means of the ISODATA algorithm implemented in Image Analyst for ArcGIS 9.1 to distinguish two management zones within the field. Those zones were finally used to re-design the irrigation sectors to adapt them to the variability expressed in the two zones.

Impact of fertigation by sprinkler irrigation on variability of crop performance
Nouri, H.[1], Razavi Najafabadi, J.[2], Amin, M.[1] and Aimrun, W.[3], [1]Biological and Agriculture Department, Engineering Faculty, UPM, 43300, Malaysia, [2]Department of Farm Machinery, College of Agriculture, IUT, Isfahan, 84155, Iran, [3]ITMA (Institute of Advanced Technology), Universiti Putra, 43300m, Malaysia; jrazavi@cc.iut.ac.ir

Environmental preservation is a growing concern in industrial and agricultural sectors. As the most important consideration in environmental pollution, chemicals such as fertilizers should be applied in the optimal amount to the field. Consequently, there is an increasing need for soil fertility management. Precision farming is an approach to manage the crop needs in required quantity to meet optimal quality by applying the right amount of inputs at the right time on the right area, conserving both water and agricultural chemicals while improving production output. Hence, suitable temporal and spatial distribution of nutrient application would be a helpful step to apply precisely and uniformly the needed amount. The study was conducted in Isfahan Province, Iran. The chosen crop was sugar beet watered by sprinkler irrigation system (solid set with removable sprinklers). Due to huge investment on the implementation and maintenance of pressurized irrigation systems in most provinces of Iran, sprinkler fertigation could be a suitable fertilization technique. This paper presents the effects of spatial distribution of nitrogen fertigation on the crop performance. The effect of fertigation by sprinkler irrigation on variability of soil and leaves N content and sugar beet yield was studied. Geostatistical sampling method was selected for an accurate interpolation by kriging to produce spatial variability maps. A total of 162 soil samples were collected and analyzed for before and after fertigation. Leaves and tuber samples were also analyzed. The results show that fertigation of N by sprinkler increased nitrogen leaves uptake, tuber weight and tuber moisture content but no significant correlation was found between fertigation and yield.

Spatial variability of cover crop performance in row crops

Munoz, J.D.[1], Kravchenko, A.[1], Snapp, S.[2] and Gehl, R.[3], [1]Michigan State University, Crop and Soil Sciences, PSS Bldg., East Lansing, Michigan, 48824-1325, USA, [2]Michigan State University, Kellogg Biological station and Dep. of Crop & Soil Sciences, PSS Bldg., East Lansing, Michigan, 48824-1325, USA, [3]North Carolina State University, Soil Science, Fletcher, North Carolina, USA

Cover crops application in conventional agriculture can generate important benefits to agro-ecosystems. However, great spatial and temporal variability in cover crop performance and in the resulting supply of nutrients to the main crop is one of the limiting factors for their wide adoption by producers. Better understanding of how variations in topography and soil properties affect cover crops will contribute to increased cover crop adoption, allowing producers to tailor cover crop management to site-specific features of their fields. The objectives of this study are to quantify effects of topography and soil properties on cover crop growth and performance; to estimate spatial and temporal components of cover crop variability; and to generate empirical predictive models of the potential nutrient supply that can be provided by the cover crops. We have collected data in southwest Michigan in summer and fall of 2007 and spring of 2008 from three agricultural fields; and we will further collect data in summer and fall of 2008 and in spring of 2009 from four additional fields. All fields are in corn-soybean-wheat rotation and red clover is used as a cover crop following wheat. On each sampling date at each field we collect red clover biomass from 15-20 sites accompanied by NDVI measurements. NDVI measurements are conducted at the biomass sampling sites and at the entire fields by GreenSeeker™ optical sensor with density of approximately 570 data points per 1 ha. Elevation data are available to derive terrain attributes. Measurements of total soil carbon and nitrogen, soil texture and bulk density are collected at the studied sites. Preliminary results from 2007-2008 sampling indicate that NDVI data from GreenSeeker™ is a promising tool for predicting red clover biomass on a field scale, especially at early growth stages. Red clover biomass tends to be closer related to soil properties than to topography. Biomass was positively correlated with total carbon (r=0.61), total nitrogen (r=0.68), and silt content (r=0.45) and was negatively correlated with sand content (r=-0.46). Clover performed better at sites with lower elevation (r=-0.34) and slope(r=-0.33). Relationships between topographical characteristics and biomass, as assessed based on NDVI, were sensitive to the studied map cell sizes. Topography relationships with red clover biomass at small scales (< 5m) appeared to be relatively weak, however they became much stronger as the map cell increased to 10- 20 m. The result indicates the importance of studying factors affecting cover crop growth and productivity at scales that best reflect the magnitude of their spatial influences. Combined results from all 7 fields of the two studied years will be reported and discussed.

Precision agriculture in an olive trees plantation in Southern Greece

Fountas, S.[1], Bouloulis, C.[2], Paraskevopoulos, A.[3], Nanos, G.[1], Gemtos, T.[1], Wulfsohn, D.[4] and Galanis, M.[5],
[1]University of Thessaly, Fytoko str, 38446 Volos, Greece, [2]Dianthus ltd, Gargalianoi, 24400 Gargalianoi
Messinias, Greece, [3]Ministry of Rural Development, Prefecture of Messinia, Division of Plant Protection,
Kyparissia, 24200 Kyparissia Messinias, Greece, [4]Copenhagen University, Hojbakkegaard Alle 30, 2630
Taastrup, Denmark, [5]Monsanto Hellas ltd, 29 Michalakopoulou Street, 11528 Athens, Greece

In Greece, and Southern Europe in general, the application of site-specific management has been delayed because of small farm size, low adoption of new technology, crop subsidies from the EU and lack of relevant technology and applications for fruits and vegetables. Particularly, precision horticulture, which includes high value crops, requires highly efficient farm management under Greek conditions in order to increase competitiveness in the global market. Site-specific management may assist in this direction mainly for site-specific application of fertilizers and soil cultivation. In olive trees, there is a very small amount of references for the application of precision agriculture practices. The objectives of this study were: (1) to study the in-field variability in yield and soil parameters, (2) to estimate the effect of post emergence herbicide application to soil fertility using precision agriculture and (3) to design and apply variable rate fertilization for P&K. The study was carried out in a 9.1 ha commercial olive tree plantation for olive oil production in 2008. The orchard is located in Southern Peloponnesus, where the main crop is olive trees and produces extra virgin olive oil. The field has been planted in rows and the total number of trees is about 1,700. For yield mapping the olives were collected manually with sticks shaking the tree shoots. The olives fell on the plastic net covered ground, and placed in sacks averaging 58kg each. The farmer was recording the position of each sack or group or close placed sacks using a commercial GPS, and a total of 433 measurements were recorded. Additionally, a regular grid was created and 91 soil samples were taken (one sample per 0.10 ha). The soil parameters measured were soil physical properties, organic matter, K, pH, P, NO3-N, Mg, Zn, Mn, Fe, B, Ca. Weed control was practiced for the last 3 years with post emergence herbicides, as no-tillage practice, in the 2/3 of the field and with a rotary cultivator in the rest. The differences in soil fertility between the two practices were also determined using precision agriculture. Finally, site-specific maps were created for P&K and applied in the field. For the collected data both descriptive statistics and geostatistics were applied. The results indicated very high variability in yield and soil parameters. The farmer found the process of recording the yield easy but time consuming and other methods may have to be explored. About 22% more organic matter found to be in areas with no-tillage practices, while the olive trees were affected more with the 'iska' virus in areas with mechanical cultivation that is a very serious threat in the region under study. Finally, the farmer variably applied P & K fertilizer to the fields according to the produced application maps. The study will be carried out in the growing season of 2009 to verify the initial findings and also evaluate the effect of site-specific fertilization.

Influence of site specific herbicide application on the economic threshold of the chemical weed control
Gutjahr, C., Weis, M., Sökefeld, M. and Gerhards, R., Institut für Phytomedizin, Herbologie, Otto Sander Straße 5, 70593 Stuttgart Hohenheim, Germany

The heterogeneous distribution of weeds in agricultural fields offers the possibility for site specific herbicide application. This procedure leads to enormous savings in herbicides and has economical and ecological benefits. In the herbology department at the University of Hohenheim, A sensor system to detect weeds in cultural fields was developed. A bispectral camera is used in combination with an image analysing software to identify weeds. The aim is to realize an online herbicide application by patch spraying. This technique can only be economically useful, if accurate decision-making systems are developed. Based on field trials in wheat, barley and maize has been analysed the impact of a site specific herbicide application on to the economic threshold for chemical weed control. A linear mixed model with anisotrophic spatial correlation structure has been used to explain the yield variability within a field. It was possible to identify the yield effect of a single weed plant as well as the yield effect of the herbicide application itself. The results show that in areas with no weeds or with a low weed density the herbicide application tends to have a negative yield effect. Using site specific herbicide application technique, it is possible to avoid this negative yield effect in weed-free areas. The herbicide savings on the one hand and the inhibited negative yield effects on the other hand are benefits of the site specific herbicide application. To identify an economic threshold for site specific weed management not only the yield effect of the weeds but also the yield effect of the herbicide application must be attended.

Rural living labs: used based innovations for rural areas
Zurita, L., Atos Origin, Albarracin 25, 28037 Madrid, Spain

Rural living labs (RLL) constitute a new and not yet validated approach of enabling user driven ICT-based innovation initiatives aimed to the development in rural areas. Living labs provide a context for open innovation based on partnerships between all stakeholders. This paper presents different examples of how RLL have been set up, and presents initial results from the C@R Integrated Project. Three living labs cases are presented and compared: Homokháti in Hungary, Sekhukhune in South-Africa, and Cudillero in Spain. The process of establishing the living lab, the involvement of users, the experimentation and innovation processes, and the technical and business innovations and their impacts on the rural environment are being discussed in order to reach some conclusions about the suitability of the RLL as tools for development in the rural areas. Initial results indicate that in order to be successful, the RLL have to stick to the user centric approach and adapt to the local situation.

KodA: from knowledge to practice for Dutch arable farming
Paree, P.G.A.[1], Van Gurp, H.[1] and Wolfert, J.[2], [1]ZLTO, Projecten, Postbus 91, 5000 MA Tilburg, Netherlands, [2]WUR-LEI, Plant Systems, Postubs 29703, 2502 LS Den Haag, Netherlands

The research and technology development program 'Knowledge in the field of arable farming', KodA, aims for sustainable farm practices in the field of arable farming by putting knowledge into practice in an applicable way. In KodA, several hundreds of arable farmers, their suppliers and processors (about 12 large companies), work together to improve quality and efficiency of arable crop production. The program has a total budget of 8 MEuro, in a private-public partnership with the Ministry of Agriculture. The living lab approach is the basic method of working in KodA. The business end users, the farmers and their supply chain partners, are in the lead. Researchers, teachers and (ICT) service providers are involved in the process of formulating wishes, in which they can point at existing knowledge and possibilities and in the implementation of innovations. This approach is supported by the so-called 'KodA miles system', which is copied from the well-known air miles. Business partners in the KodA program earn KodA miles by doing own investments that fits in the overall KodA objective. They can cash KodA miles by submitting a proposal for research or development that is financed for an equal amount of money by the government. To monitor and evaluate the program, a special tool was developed based on definition of the innovative tasks and related critical situations. Innovative tasks were defined at the level of perspectives, conventions and routines. The critical situations were identified at the level of competences, designs and insights. Solving the critical situations results logically in an agenda for research and development. ICT is seen as a key enabler to achieve the program's objective. By means of ICT, the farmer is able to use and deploy knowledge, information and data in an efficient way. Development of integrated management support systems in which actual, state-of-the-art knowledge and farm-specific data are combined, is a key prerequisite for further development. Several factors hamper integration of these systems. Standardization was identified as one of the major problem areas. Several solutions were made, based on the approach of Business Process Management (BPM) and Service-Oriented Architecture (SOA). Precision agriculture is also a dominant theme within the KodA program. Several projects were conducted such as: site-specific nematode management, -N-fertilizing, -sowing sugar beets, soil and crop monitoring and yield mapping. This paper will briefly highlight the activities in KodA with special attention to the innovative Steering Procedures and to ICT and precision agriculture aspects. Finally general conclusions and 'lessons learned' will be presented and discussed.

Farmers as co-developers of innovative precision farming systems
Eastwood, C.R.[1], Chapman, D.F.[1] and Paine, M.S.[2], [1]The University of Melbourne, Faculty of Land and Food Resources, Parkville, Victoria 3010, Australia, [2]DairyNZ, Private Bag 3221, Hamilton, New Zealand

Understanding the processes involved in farmer learning and adaptation around precision farming practice is important in creating conditions for innovative precision farming systems. Six case studies of new Australian precision dairy farmers were analysed over an 18 month period with the aim of developing a greater understanding of learning and adaptation. Qualitative research methods were used, farmers were interviewed throughout the study and these interviews were recorded, transcribed, and analysed using thematic coding. The study findings highlighted that development of innovative precision dairy systems relies on recognition of farmers as co-developers of precision farming knowledge, and that successful precision farming requires a systems approach rather than a focus on technological devices. Further to this, innovation is driven not just by farmers but by the network of people who operate around them in a 'community of practice'. Strengthening the linkages between community members, and facilitating knowledge transfer around the community (for example from farmer to retailer to developer) is the key to innovation in precision dairy systems. In this study a continual on-farm development of precision dairy systems was observed from which lessons could be harnessed to benefit the wider dairy industry in Australia. While on-farm learning can filter back to national and global retailers, farmers need to be explicitly acknowledged as co-developers of innovative dairy systems. The knowledge feedback mechanisms from farmers to other members of the precision farming community require strengthening to facilitate transfer of farmer-based knowledge of precision farming techniques and adaptation processes.

Living lab 'information management in agri-food supply chain networks'
Verloop, C.M.[1], Wolfert, J.[1] and Beulens, A.J.M.[2], [1]LEI Wageningen UR, Plant Systems Division, P.O. Box 29703, 2502 LS The Hague, Netherlands, [2]Wageningen University, Information Technology Group, P.O. Box 8130, 6700 EW Wageningen, Netherlands

In the food economy information becomes more and more a competitive factor. Mostly information is owned by several parties in the Agri-Food Supply Chain Network (AFSCN) and is stored in different information systems. Seamless communication between that systems is mostly not the case. Because of the competitive factor not only technical aspects but also organizational aspects, as authority and confidence, play an important role. Involvement and commitment of all stakeholders are crucial. Because of that, in the domain of information integration a lot of actual and interesting challenges are present. LEI Wageningen UR initiated the Living Lab 'Information Management in the AFSCN' to create and maintain an environment to cope with these challenges. The most important characteristic of a Living Lab is that innovation from start to end is embedded in the context of the users. Continuous interaction takes place between end users and developers. The purpose of the Living Lab is to facilitate a structural and independent place were companies and other AFSCN parties meet and cooperate, with each other ánd with education and research, in the cadre of information integration topics. The Living Lab is facilitated by an ICT environment (hardware and software) based on currently available information systems and applications, to be used for simulations and experiments. The knowledge and expertise obtained, cause the Living Lab to become a Centre of Excellence in the domain of information management in AFSCN. There are several effects. The Living Lab is a node in the network of students, teachers, business people and scientists. Educational institutions use the Living Lab in their education program to acquire theoretical knowledge and practical experience, and to stimulate to make use of it subsequently. Business sees research and education as interesting partners in solving information integration topics. Research and education are forced to think from a practical point of view and to deliver practical solutions. Students get a realistic view on their future job and labor environment. Research institutions become more involved to education. Several parties contribute to the Living Lab. Agri-Food business partners deliver the business cases: information integration topics in real-life business. LEI Wageningen UR brings in her knowledge and overview on the Agri-Food domain and the important topics in it. ICT business partners contribute delivering licenses of their software applications, for example their BPM tool. Research projects and other projects are executed in cooperation with all project partners. Students, teachers and researchers together work on knowledge development, preferably in cooperation with the Agri-Food and ICT business partners. Knowledge and other project results will be presented in education material, (scientific) publications and other documentation. Because of the unique cooperation between business partners, research institutions, universities and other education institutions the knowledge and experience developed will be shared with all parties. Which creates a win-win situation for all stakeholders! In this paper the setup, the startup and the first results of the Living Lab will be presented.

Traceability of food of animal origin: major findings of the interdisciplinary project IT FoodTrace

Doluschitz, R.[1] and Engler, B.[2], [1]University of Hohenheim, Institute of Computer Applications and Business Management in Agriculture, Schloss, Osthof-Sued, 70593 Stuttgart, Germany, [2]University of Hohenheim, Life Science Center, Fruwirthstr. 12, 70593 Stuttgart, Germany

Increasing pressure by consumers, provoked by numerous food scandals, forced the European Union and national authorities to strengthen the regulations on food safety and traceability along food supply chains. The regulation (EC) 178/2002 can be regarded as recent policy shift towards consumer protection. Commercial members of food chains are only partially prepared to implement such regulations and to fulfil the respective requirements. Within the interdisciplinary research project IT FoodTrace a comprehensive IT-solution is going to be developed. It aims to achieve traceability and quality assurance along the food chain of 'meat and meat products'. Subprojects of the interdisciplinary research consortium cover the entire supply chain and also address cross cutting issues. Substantial added value could be gained from intensive interdisciplinary co-operation. Selected findings of at least four sub-projects will be presented: a) Information and Data Collection in Livestock Systems (Farming Cell): The aim of this sup-project is the development of an ISOagriNET compliant IT system for the acquisition and consolidation of appearing data and information in livestock systems. The way, an additional benefit to humans, animals and the environment, against the background of traceability and process control, can be generated. b) Linking animal-health-related information to an integrated animal-health system: The aim of this sub-project is to analyse data acquisition, data flows and data retrieval between selected process participants (e.g. livestock farmers, veterinarians, public authorities). The results of this research will be used for the conception and development of an IT-model, which integrates decentralized, fragmented and redundant animal-health related information (Fick and Doluschitz, 2007). c) Agro Technical Solution Model: The Agro Technical Solution Model (ATSM) will be developed as an integrated IT-System. With this system it should be possible to trace food along the supply chain. Already existing systems can be integrated. Partly, ATSM works as a database, which holds master data of the participating companies. Data exchange between members of the food chain will be carried out by interfaces. The development of the ATSM is in constant progress. The final model will be presented at the JIAC 2009 in Wageningen. d) Costs and Benefits: The main objective of this sub-project is a cost-benefit-analysis of the quality assurance and traceability system of the interdisciplinary project IT FoodTrace, by means of a profitability and adoption analysis. Statements concerning the predicted profitability and acceptability of IT based quality- and traceability systems are essential for their success in the market. Results from a Delphi-based empirical analysis will be presented at the JIAC 2009. Contents to be presented at the JIAC: The aim of a full paper will be a presentation of a) the complexity of underlying problems, b) the project structure, c) available results from numerous sub-projects and d) the added values from interdisciplinary co-operation.

A traceability system for the food service industry

Rogge, C.B.E. and Becker, T.C., University of Hohenheim, Institute for Agricultural Policy and Agricultural Markets, Universitaet Hohenheim 420b, 70593 Stuttgart, Germany

The joint research project 'IT FoodTrace' (www.itfoodtrace.de) aims to develop a traceability system for the entire food supply chain. While traceability is already a major topic in retailing, it is still in the fledgling stages in most companies of the food service industry. With its rather long supply chain, meat has been chosen as an example for the design of a traceability concept. Therefore it was necessary to have a closer look at the structure of the meat supply chain and analyse the product flow with regard to the food service industry. Since the sector is extremely heterogeneous, the next step was the development of a structure of the food service industry.Subsequently a questionnaire has been elaborated for semi-structured expert interviews with representatives of procurement, quality management, business communication, and executives from companies of the systematised gastronomy (e.g. fast-food-chains, restaurants in shops and stores), catering companies and communal feeding (staff canteens, school feeding, care-catering). Between May and August 2008, 10 interviews have been conducted throughout Germany. The topics covered were: information flow into and out of the companies, as well as the internal information flow, data management, technical solutions for the food service industry and the use of IT, traceability and traceability systems, business strategy with respect to food safety/quality and consumer trust, crisis management with respect to supply and production, expectations towards an IT traceability system. The main gate keeper for the information flow into the food service industry is whole sale. There have been a couple of approaches, induced by producers and distributive trade, to standardise master data (e.g. SINFOS). However some parts of information, specifically important to the food service industry, are still in need for standardisation (e.g. numbering of allergens). While for incoming goods the unambiguous identification is generally unproblematic, the opposite is often the case during and after processing. One important precondition is a strict batch splitting, not only for the main ingredients, but also for spices. Even among larger companies only few assign their own identification numbers to final products. While among industrial producers automatic data acquisition is widely spread, it is still uncommon in the most facilities of the food service industry. Information in IT-systems is often limited to ERP (and therefore external traceability). The understanding of traceability by the representatives often goes beyond the legal definition (one step up, one step down). However it does seldom emphasise internal traceability. Where data regarding internal traceability are collected, this is usually done manually in paper sheets. Although food safety is perceived to be highly important, the majority of companies are not willing to pay more for 'traceable' ingredients, because they do not assume their customers to be willing to pay a price premium for traceability. From the expectations towards an IT traceability system, recommendations can be given for the design of such a system. Most interviewees agree that the operator of a traceability system for the entire food supply chain should be an independent institution. There are different arguments for a voluntary or a mandatory participation.

Improving information management in organic pork production chains
Hoffmann, C. and Doluschitz, R., University of Hohenheim, Department of Farm Management (410C), Schloss Hohenheim, 70593 Stuttgart, Germany

Besides conventional pork production, organic production reported enormous, annual fluctuations in recent years (e.g. Greece: +353%; Great Britain: -46%, 04-05. EU regulation 834/2007 harmonises the reglementations for the organic production within the EU. Nevertheless, due to national distinctions, the importance of organic produced products (including pork) and customer requirements to products are different. A quite diverse spectrum in the characteristics of organic meat programs are consequences. As a result of increasing requirements to quality assurance of products (such as EU regulation 178/2002), traceability becomes more important. This comes along with the need for documentation and inter-organisational information exchange among the chain actors. Hence, the relevant information has to be selected and data has to be evaluated. These challenges also have to be managed by organic pig producers. Technical supports are inter-organisational information systems, connecting several actors (e.g. via online databases). They not only store data, but also perform a wide range of additional functional features (e.g. reports). However, costs and benefits of such systems have to be balanced. Objective The main objective of the study is to improve the competitiveness of German organic pork production and to contribute to higher standards in food safety by evaluating the status quo of information management in several European countries and to create a concept for improving efficiency of inter-organisational information management, according to the special terms of organic pig production. Methodology The study takes place in several main organic pork production countries in Europe. It is divided into three steps: In step one about 30 expert interviews will provide an overview of the status quo of organic pork production chains and inter-organisational information management in Europe. In a further step the interviews will be evaluated and outstanding results will be observed in detailed case studies. Using cost and benefit approaches the inter-organisational information systems in organic pork production chains will be analysed extensively. Finally, the case studies will be evaluated and a concept for improving inter-organisational information management in organic pork production chains will be designed. Outlook At the Joint International Agriculture Conference 2009 step one (status quo analysis) of the research will be completed. Respective results from the expert interviews will be presented. Analysing the inter-organisational information management in organic pork production chains deficits in the quality and information management will be expected. Furthermore, first results from step two (case studies) will be available and presented.

Logistic study on the recall of non-conform perishable produce through the supply chain by means of discrete event simulation model
Busato, P. and Berruto, R., University of Turin, DEIAFA, Via L. Da Vinci 44, 10095 Grugliasco (TO), Italy;
patrizia.busato@unito.it

The traceability allows, for each product, to sketch the manufacturing process through a system documentary, enabling to identify the operational structures involved, the products and the lots, to define the flows of production, packaging and distribution. An important point of the procedures of traceability is to establish the size of the lots as a function of the risk associated with the non-conformity of a product belonging to each lot. The supply-chain of fresh produce is constituted of many links: producer/grower, warehouse, packing centre, distribution centre, retailers and finally the consumer. Each of these is a system itself that interacts with the other components of the supply-chain. The non-conformity could occurs in each of these links. Because of processing plant requirement, storage requirements, and because of savings in the traceability process, often small size lots are merged together to form a large size lot at some points in the supply-chain. Larger lot size could imply higher risk for the consumers in case of recall of the produce and much higher recall time and cost for the supply-chain. When a non-conformity occurs, the time to recall the produce depends on many factors: lot size, lead time for information spreading from link to link, product transit time among links, product storage procedures and times, and the point in the supply chain where the problem occurred. To study with a system approach different scenarios for the recall procedure the authors realized a discrete event dynamic simulation model using Extendsim®. In the paper the model architecture and two example of application will be described, concerning different levels of information sharing and technology, with estimation of time requested, cost of the produce recall and potential number of customers involved.

How are public authorities and food enterprises interwoven: state of the art of information transmission in the food sector

Otto, J.[1], Frost, M.[2] and Doluschitz, R.[1], [1]Hohenheim University, Institute for Farm Management, Institut 410C, 70593 Stuttgart, Germany, [2]Federal Office for Consumer Protection and Food Safety, Unit 107, Mauerstr. 39 – 42, 10117 Berlin, Germany

Since the last decades the borders of both private and public organisations which have been agreed upon in general are being blurred. Within the food industry coordination takes place in vertical and horizontal directions leading to a quite higher degree of degradation of organisational boundaries on the one hand. On the other hand a demand for intensified interaction/communication between food enterprises and public authorities has emerged since the General European Food Law and General Hygiene Package regards food enterprises responsible for the safety of foods, but also prescribes the cooperation of both public surveillance authorities and food enterprises. This paper analyses information related to food safety that is being transmitted between enterprises and public authorities. Firstly, it reviews the literature that is at present and analyses thereby the vertical and horizontal coordination of organisations in the agro-food chain. Secondly, information transmission from food enterprises to public authorities is analysed. Thirdly, materials and the applied methods are described. Fourthly, a framework is presented that reveals the interconnectivity between the actors in the food chain and some of its stakeholders. This framework includes findings about data formats, classifications and codifications being used for information transmission and reveals problems that are to be overcome. Lastly, conclusions are drawn that also demand further research. Qualitative data analysis is applied; after the analysis of literature regarding coordination in the food sector and organisational change, expert interviews are used as a method of operation. These have been conducted as case studies in one abattoir, the Federal Office of Consumer Protection and Food Safety, the Federal Research Institute for Animal Health and a parastatal organisation that is responsible for compliance regarding the German breed law. Strengths and weaknesses that arise when organisational change is strived for are discussed and reveal possible chances of success for improved information management. This is related to the advantages and disadvantages of blurred boundaries of the actors in the agro-food chain and some of its stakeholders. Throughout the literature it has been assumed that boundaries of organisations in the food sector are being blurred. This is not only limited to actors in the food chain but also to others like public authorities. On the level of information exchange and transmission it is shown how the aforementioned actors are connected and that there are opportunities which still remain unused and impediments which should be overcome to solve the analysed problems.

Information technologies and transparency in agri-food supply chains
Frentrup, M. and Theuvsen, L., University of Goettingen, Department of Agricultural Economics and Rural Development, Platz der Goettinger Sieben 5, 37073 Goettingen, Germany

Transparency in agri-food supply chains and the required information infrastructures have gained increasing importance in recent discussions. This paper aims at better understanding the contributions of information technologies (IT) to information infrastructures in food supply chains. Furthermore, it analyzes the effects of IT on transparency in agri-food supply chains in general and in particular in dairy and pig production. Transparency in supply chains is defined as the extent to which all stakeholders have a shared understanding of and access to the information they request. Next to structural and behavioural determinants, IT as a precondition for the transfer of large amounts of information plays a pivotal role with regard to transparency. In this paper we include all relevant determinants of transparency, including IT, into a measurement concept of transparency in agri-food supply chains. In a large-scale empirical study (N=204) in the German dairy and pig production the structural and behavioural determinants of transparency are evaluated from the farmers' point of view. The study analyses in particular the current use of IT in business relationships and the relevance of IT for improving transparency in agri-food supply chains. The results point out in which way information which are relevant for producing, processing and commercialising agri-food products are exchanged between farmers and processors. They describe the level and the character of IT which are used for information transfers and show differences between the dairy and the pork sectors. The study highlights the existence of deficits in information infrastructures in both value chains. Information are still mainly exchanged in written or oral form; IT are used on a remarkable level only in information transfer processes from dairy companies or slaughterhouses to farmers (dairy: 19.0%; pig: 15.9%). In the opposite direction (from farmers to processors), IT is only used rarely (dairy: 5.2%; pig: 7.2%). Interestingly, in those business relationships between farmers and processors which are characterised by a high degree of (perceived) transparency IT is used on a comparatively high level. As a result, the hypothesis is confirmed that the use of IT positively impacts transparency in agri-food supply chains. Furthermore, the study shows that besides IT additional factors influence the information exchange between farmers and processors. All in all, deficits with regard to the communication processes between farmers and processors can be identified in both sectors. Nevertheless, the situation is more promising in the dairy sector than in pork production. The study also shows remarkable processor-specific differences with regard to the quality of communication processes. The empirical results parallel former studies that show that hardly any IT infrastructures have been implemented that meet current requirements with regard to transparency. Furthermore, the empirical database delivers starting points for an improvement of information exchange between producing and processing stages in agri-food supply chains by using modern IT. In this respect the empirical results support current efforts of leading companies in the dairy ndustry and the pig sector to set up advanced IT solutions and to improve transparency.

Interaction models in the fresh fruit and vegetable supply chain using new technologies for sustainability and quality preservation

Reiche, R., Fritz, M. and Schiefer, G., Department of Food and Resource Economics, Chair for Business Management, Organization and Information Management, Meckenheimer Allee 174, 53115 Bonn, Germany; robert.reiche@uni-bonn.de

Actors in the food supply chain have to deal with particularities, like sustainability, global sourcing, quality preservation and therefore the optimization of efficiency in their logistic strategies. The consumer demand for fresh fruits and vegetables all over the year is a special challenge of this chain, which strongly impacts all of the particularities. Hence, the adoption of new technologies can have important benefits to meet these conditions. One important benefit is the collection and sharing of informations, which can have an impact on the strategies of enterprises. In this paper, different scenarios using new technologies, like RFID, GPS and wireless sensor networks, for optimizing the logistics strategies in the fresh fruit supply chain are introduced. The modelling and evaluation of these scenarios is accomplished by simulation experiments. The results are presented at the EFITA conference.

Data warehouse for soybeans and corn market on Brazil

Elias Correa, F.[1], Pizzigatti Corrêa, P.L.[1], Aparecido Alves, L.R.[2] and Saraiva, A.M.[1], [1]Agricultural Automation Laboratory, Department of Computer Engineering - Polytechnic School – USP, Prof. Luciano Gualberto, travessa 3, n° 158, Edificio de Engenharia Elétrica, 05508-900, SP, Brazil, [2]Center for Advanced Studies on Applied Economics, School of Agriculture Luiz de Queiroz, Av. Padua dias, 11, 13400-970, Piracicaba, SP, Brazil

Brazilian agribusiness generates a large amount of trade information considering all production chains. The volume of money involved in it worldwide is huge. However there are few solutions for management information systems that support decision making based on the operational data generated by each chain, it means, systems that provide consistent information toward agribusiness. This is important not only to the farmers, but also to the government and companies that need parameters about agribusiness. So, this study aimed are modeling and create an Data warehouse based on grain market in Brazil, using information about prices and regions of soybeans and corn. This study is based on history database on agricultural market for up to ten years. The proposed paper uses methods based on dimensional modeling to model the agricultural market, bus matrix and open source data warehouse tools. This Data warehouse allows researchers to navigate using a web browser through historical data and decision-making information in a friendly way. Also, the model can be used in future works of Data warehouse to increase amount and quality of information for agriculture.

Spatial optimisation as a tool to maximise profit on mixed-enterprise farms within environmental constraints

Cristia, V.[1], Houin, R.[1] and Betteridge, K.[2], [1]AgroParisTech, Department of Life Sciences and Health, 16 rue Claude Bernard, 75231 Paris, Cedex 5, France, [2]AgResearch Grasslands, Climate Land and Environment, Private Bag 11008, Palmerston North 4442, New Zealand

Environmental regulations on farms are being increasingly used world-wide to limit nutrient emissions to the environment. In the Lake Taupo catchment, New Zealand, a nitrogen discharge allowance (NDA) has been imposed on each farm according to the amount of N leached during one farmer-nominated year between 2001-2005. This effectively prevents the stocking rate of the existing system from being increased to cover ever increasing costs of production. Our spatial optimisation model was developed to optimise the selection of potential farm enterprises suited to a given farm, such that farm Gross Margin (GM) is maximised while the average NDA for the farm is not exceeded. The model considers the areas of spatially variable resources on the farm including: soil type and fertility status, topography, elevation, slope, existing forest-planted land, infrastructure (tracks, fencing, buildings), rainfall and temperature. It also allows for farmer-imposed restrictions to be imposed on how selected resources may be used. For instance, 'existing exotic and indigenous forest blocks must be preserved', or 'no cattle can be grazed within 20 m of a stream'. Resources are mapped within a GIS system to develop a series of land management units (LMU). These LMUs are then used to determine (a) the feed requirement of various grazing livestock enterprises at a range of stocking densities and (b) the amount of N leached from each of these LMUs. In a 2,500 ha Case Study farm, we show that the GM of the existing sheep, beef and deer farming enterprise could potentially be tripled if 644 ha was converted to dairy farming and only deer and plantation Pinus radiata were used on remaining land to offset the increased leaching arising from intensive dairying. This modelling tool is envisaged for use by farm consultants working with the client to select, evaluate and rank optimum enterprise mixes based on GM. Several iterations would be run as constraints are modified until one or two top ranking enterprise mixes are identified. It would then remain for the consultant and farmer to do a detailed analysis to test the feasibility of the selected enterprise mix and to ensure that it meets with the farmer's goals, risk profile, financial position, labour skills and labour availability.

Spatial variability of spikelet sterility in temperate rice in Chile

Ortega, R.A.[1], Del Solar, D.E.[1] and Acevedo, E.[2], [1]Universidad Técnica Federico Santa Maria, Departamento de Industrias, Avenida Santa Maria 6400, 7660251 Vitacura, Santiago, Chile, [2]Universidad de Chile, Facultad de Ciencias Agronomicas, Av.Santa Rosa 11315, 8820808 La Pintana, Santiago, Chile

Spikelet sterility (blanking) causes large economic losses to rice farmers in Chile. Available varieties are susceptible to low air and water temperatures during pollen formation, making it non viable, which is the main responsible for the large year to year variation observed in terms of grain yield. The present work had for objective to study the spatial variability of spikelet sterility within a rice field and relate it to water and air temperature as well as soil properties within the field. An approximately 20-ha field was studied. A network of 24 temperature sensors, with their corresponding loggers, were deployed within the field using a sistematic non-aligned design, with the help of a GPS and GIS program. Air, at canopy level, and water temperature were measured at 1-hr intervals during the growing season. At the same positions, soil chemical properties were measured before planting. Tissue analysis for total N on the flag leaf was performed at panicle excersion. Spad readings were also collected at the same stage. At harvest, grain yield and spikelet sterility were determined. Resulted showed a large spatial variability in terms of water temperature and spikelet sterility within the field. Soil properties as well as N status in the rice plants also showed a large variability. Spatial relationships between the different factors measured and spikelet sterility are presented.

Investigations on the management of small scale variability of soil nutrients

Herbst, R.[1], Schneider, M.[2] and Wagner, P.[2], [1]Humbolt University Berlin, Faculty of Agriculture and Horticulture, Precision Agriculture, Invalidenstarße 42, 10115 Berlin, Germany, [2]university of Halle, Farm Management Group, Luisenstr. 12, 06099 Halle, Germany

The basic question for this research was to quantify multiscale observations and manage small scale variation of soil nutrient samples on a trial field in the mid of Germany. Therefore, 1258 probes were taken in 2006 by a nested sampling strategy on a hexagonal grid with a minimum distance of 12.5 m on a 65 ha field. The soil samples were analysed for potassium, phosphorous, pH, magnesia, total carbon and total nitrogen and analysed using Geostatistics. By a smooth data reduction, the minimum distance of the points was increased from 12.5 to 25 m, 50 m and 100 m. As an example, we compared the mean phosphorous content and the standard deviation as well as the coefficient of variation in % for the different scales of the lag distances 60, 120, 180, 240, 300 m. With increasing distance, the mean phosphorous content was almost stable but the semivariance was decreasing. The coefficient of variation was reduced from 25% to 20%. In the following step, we tried to quantify the differences of multiscale observations. For this, the kriged map of 15 m point distance was taken as the reference and compared with the kriged maps of 25 m and 50 m. We used a technique which combined information on the quantity and the location on different scale levels of categorical maps. For a multi based resolution of 10 by 10, the standard Kappa Index of Agreement (KIA) is 0.7. Looking on the comparison of the reference image with the 50 m and 10 by 10 multi base resolution, the KIA is only 0.4. In order to manage the observed small scale variability, we applied 'on farm experiments' on the research field. Three different base fertilization strategies were carried out: 1) uniform field treatment, according to the nutrition status, received from one soil sample of the experimental field (mixed, virtual sample of 15 repetitions over the whole field); 2) site specific base fertilization, according to the high resolute soil survey mentioned above; and 3) no base fertilization. Therefore, the field was subdivided into 15 strips. Each strip is 72 meters wide (two tramlines). Each strategy is repeated five times. By carrying out virtual soil samplings on a 5 ha, 3 ha and 1 ha grid, different scales of site specific sampling approaches were considered in the analysis whereas the 25 by 25 meter grid was taken for the validation. The results (nutrient allocation maps) for each investigated grid class were compared with the uniform sampling approach. It is elaborated in this research that the 1 ha and 3 ha grid describes much better the soil nutrient classes than the uniform field soil sampling approach. The results of the 5 ha grid is unsatisfactory with one exception: for the phosphorous content, the whole field soil sampling approach seems to be more efficient. Anyway, the analysed downscaling approach of the 'on farm experiment' will be compared in the next time with an upscaling approach in order to get a better understanding on multi scale investigations.

Selective harvesting zones from remote sensing images and yield data and relation to grape quality parameters for precision viticulture

Agelet-Fernández, J.[1], Martínez-Casasnovas, J.A.[1] and Arnó, J.[2], [1]University of Lleida, Department of Environment and Soil Science, Rovira Roure 191, 25198, Spain, [2]University of Lleida, Department of Agro-Forestry Engineering, Rovira Roure 191, 25198, Spain

Selective harvesting in viticulture refers to the separation of grapes at harvest according to differences in plant vigour, yield, soil data and/or grape quality criteria. Harvesting zones within a field are established based on, as detailed as possible, spatial information. The present research shows a case study in precision viticulture as a tool for selective harvesting. The case study was conducted in three vineyard fields located in Raimat (Costers del Segre Denomination of Origin, Lleida, NE Spain). This is a semi-arid area with continental Mediterranean climate and a total annual precipitation between 300-400 mm. The fields are planted in a 3x2 m pattern with Cabernet Sauvignon (5 ha, T system formation, sprinkle irrigation), Tempranillo (13.6 ha, Smart-Dyson formation, sprinkle irrigation) and Syrah (4.5 ha, Vertical Shoot Position formation, drip partial root drying irrigation). The fields present a yield coefficient of variation between 22.4% and 30.3%, which indicate a potential for precision viticulture applications. The research objective was to analyze the potential of high resolution satellite images and yield maps acquired by means of yield monitors to establish zones with different grape quality parameters. The analysis was made with data of the 2005 campaign. For that, a Quickbird-2 multi-spectral image was acquired at the moment of ripening (July 2005). From this image, the normalized difference vegetation index (NDVI) was computed. Harvesting data was acquired by means of a Canlink 3000 Farmscan monitor during September 2005. Two days prior to the harvesting, grape samples in a 10 rows x 5 vines pattern, with sample densities between 25-32 samples ha[-1]. The analyzed parameters were probable alcoholic degree of the juice, pH of the juice, total acidity, total polyphenols and colour. NDVI and yield maps were clustered (two and 3 clusters considered) by using the ISODATA algorithm implemented in Image Analyst for ArcGIS 9.1. The grape samples were classified in the resulting clusters according to their spatial location. Multiple rang analysis were performed to the classified samples in clusters to analyze the mean separations. The results show that the zones (clusters) derived from the NDVI map are more effective to differentiate grape quality parameters than the zones derived from the yield map. In this respect the probable alcoholic degree shows significant differences only in the case of the Syrah vineyard, either with NDVI or yield zoning. The other parameters (pH, total acidity, total polyphenols and colour) present better differentiation with the NDVI zoning. Finally, two management zones are recommended in front of three, since the results of the multiple rang analysis show that the medium NDVI cluster is not different from the high or low vigour clusters.

Heterogeneity analysis on multiple scales: a new insight in precision agriculture
Karydas, C.G., Zalidis, G.C., Tsatsarelis, K.A., Misopolinos, N.L. and Silleos, N.G., Aristotle University of Thessaloniki, Lab of Remote Sensing and GIS, Farm of the University, Building 68g, 57001 Thermi, Greece; xkarydas@agro.auth.gr

In general, heterogeneity patterns depend on the scale of observation. In site-specific management heterogeneity is conventionally recorded and mapped at the individual field scale, because variable rate inputs are applied at this scale as well. However, neither financial nor environmental benefit expected from these inputs can be assessed at the individual field scale. Instead, financial benefit has to be assessed at the farm scale, while environmental benefit has to be assessed at the catchments scale. Moreover, applications of Precision Agriculture depend on the agro-environmental policy framework usually realised at the local community's level, thus adding an extra necessary scale of observation. This study attempts heterogeneity analysis on multiple scales (i.e. field, catchments, farm, area) in order to promote the possibility for an integrated financial and environmental assessment and management, thus giving an new insight in Precision Agriculture. The test site is Archanes municipality, a Mediterranean landscape located in central Crete. The dataset comprises fine satellite imagery (pixel of 1 m) and an elevation grid, while a series of topographic factors and landscape metrics, such as slope, aspect, patch density, fractal dimension, Shannon's diversity index, crop proportion, or stream density, are used. Implementation of the hierarchy theory on thematic data in a GIS environment resulted in a series of sample classifications according to the selected factors, allocation of sites meeting certain conditions throughout the whole area, catchments comparison in agricultural landscape terms, estimation of zone spatial autocorrelation, and sub-areas characterisation in terms of similarity. The methodology is open in heterogeneity parameter selection, is repeatable when the same type of dataset is used, and gives an overview of the situation in the area while keeping the field scale in the centre of interest.

Geographic management and monitoring of livestock disease events
Janssen, H., Staritsky, I. and Vanmeulebrouk, B., Alterra, P.O. Box 47, 6700AA Wageningen, Netherlands; henk.janssen@wur.nl

In case of an outbreak of a contagious livestock disease in the Netherlands, the Food and Consumer Product Safety Authority (VWA) puts procedures into action in an attempt to prevent the disease from spreading further. These procedures entail the demarcation of an area around the outbreak location and the selection of farms within this area. These farms will be subjected to certain measures, for instance additional checks, a standstill order, vaccinations or even putting down of livestock. The current procedures for demarcating the area and selecting the farms inside the area involve some time consuming manual operations. Consequently, valuable time is lost in the first hours after a possible contamination, when chances of preventing the disease from spreading are highest Therefore, the following research question was investigated: can the process to demarcate the area and select the farms within the area be sped up using GIS? To answer the research question, a geographical information system (GIS) consisting of a number of components was designed to facilitate the demarcation of the area and the selection of farms. This GIS consists of the following components: (1) an algorithm to demarcate the area and to generate a description of the area; (2) a user-friendly internet GIS application with a number of functions: (a) managing data with regard to suspicions of disease outbreaks, (b) adapting the area and description of the area generated using the aforementioned algorithm, (c) presenting data with regard to the farms inside the area, (d) exporting data for use in other systems; (3) a GPS-enabled mobile application for fieldworkers to check if farms fall inside or outside the area. Some parts of the system were implemented in a prototype to see if the devised methodology is feasible. Special attention was given to the algorithm to demarcate and describe the area and to the selection of farms inside the area. The implementation of parts of the system proposed in prototypes allowed us to assess whether the data and the tools necessary to implement the functionality required are available The results show that both the data and the tools needed for such a system are available. The algorithm to demarcate the area and to generate a description of the area is not yet perfect; there are some cases which cannot be envisaged. Therefore, manual editing of results is needed. The ultimate gain in time depends on the amount of manual modifications needed. The automated selection of farms located inside the area caters for time-savings while results are as reliable as those of the manual procedure since the same data are used.

Creating space for biodiversity by planning swath patterns and field margins using accurate geometry

De Bruin, S.[1], Heijting, S.[1], Klompe, A.[2], Lerink, P.[3], Vonk, M.B.[4] and Van Der Wal, T.[5], [1]Wageningen University, Centre for Geo-Information, P.O. Box 47, 6700 AA Wageningen, Netherlands, [2]H-WodKa, Langeweg 55, 3261 LJ Oud Beijerland, Netherlands, [3]IB-Lerink, Laan van Moerkerken 85, 3271 AJ Mijnsheerenland, Netherlands, [4]Commissie Hoeksche Waard, Rijksstraatweg 3b, 3286 LS Klaaswaal, Netherlands, [5]Portolis b.v., Spoorbaanweg 23, 3911 CR Rhenen, Netherlands

Potential benefits of field margins or boundary strips include promotion of biodiversity and farm wildlife, maintaining landscape diversity, exploiting pest predators and parasites and enhancing crop pollinator populations. In this paper we propose and demonstrate a method to relocate areas of inefficient machine manoeuvring to boundary strips so as to optimise the use of available space. Accordingly, the boundary strips will have variable rather than fixed widths. The method is being tested in co-operation with seven farmers in the Hoeksche Waard within the province of Zuid Holland, The Netherlands. In a preliminary stage of the project, tests were performed to determine the required accuracy of field geometry. The results confirmed that additional data acquisition using accurate measuring devices is required. In response, a local contracting firm equipped a small all-terrain vehicle (quad) with an RTK-GPS receiver and set up a service for field measurement. Protocols were developed for requesting a field measurement and for the measurement procedure itself. Co-ordinate transformation to a metric system and brute force optimization of swath patterns are achieved using an open source geospatial library (osgeo.ogr) and Python scripting. The optimizer basically tests all orientations and relevant intermediate angles of input field boundaries and tries incremental positional shifts until the most efficient swath pattern is found. Inefficient swaths intersecting boundary areas are deleted to create space for field margins. The optimised pattern can be forwarded to an agricultural navigation system. At the time of the conference, the approach will have been tested on several farm fields. Anticipated additional benefits of the approach are potential integration with tracking and tracing systems and reduced soil compaction if field traffic is restricted to predetermined paths.

Agricultural-GIS-Sphere: an innovative expert system for national renewable energy and food planning

Guggenberger, T.[1], Bartelme, N.[2] and Leithold, A.[1], [1]AREC Raumberg-Gumpenstein, Departement for economics and resource management, Raumberg 38, 8952 Irdning, Austria, [2]University of Technology Graz, Institute of Geoinformation, Steyrergasse 30, 8010 Graz, Austria

Listening to national and European authorities, the lack of energy supply can be avoided in future using renewable energy. Changing from 'Energy for field', which describes the period of oil, to 'Energy from field' we have developed a GI-System to balance, estimate and control the production of food and energy based on farm level. The importance of this kind of management tool is visible in the latest discussion about rising food prices. Unregulated energy production with crops couples the price of food to the price of oil! For a sustainable production of renewable fuel or biogas, the two following topics have to be considered: The farmland suitable for renewable energy production and the effect of loosing this area for food production. Additionally the energy flux of intermediate inputs of farms must be included. First we developed a hierarchic design pattern on farm level to implement all typical functions of Austrian farms into an object-oriented data model named Agricultural Sphere Model (AGS). This model represents an expert system which connects the cycles of carbon, nitrogen, energy and produced greenhouse gases and commodities with the local farmland represented by spatial data objects. They are embedded in standard GI-Software (Arc Map, ESRI). As data source we used the anonymous IACS (Integrated administration and control system) data of 144.000 farms and over 3,000,000 plots of farmland with a total area of 21,800 km². As second step the local results are generalized in spatial homogeneous units from the European Reference Grid with a size of 500 to 500 meter. The additional supply with renewable energy from forestry is calculated in spatial cells with a size of 10 to 10 km. These units are used in the last challenge - the navigation system for food and energy. On the Austrian road and train network, we simulate first the flow of milk, meat, grain and energy to the local factories and traders. While milk, meat and renewable energy are delivered to the consumers, a lot of grain flows back into livestock farming. As a result of this energy and food balances, all Austrian farms together have a surplus in their energy balance of meat, milk and grain. The total amount of energy in products (102 Petajoule PJ) is 2.3 times higher than the input energy (45 PJ), but the quality is different. Low quality energy from pastures and fuels is transformed into food. The national food balance has a surplus in energy input – we have a higher import of feedstuffs (Soy bean) in comparison to export in food (meat, cheese). Grass- and cropland in Austrian is well used in 2008. Concerning the expected gain of the total amount of renewable energy in Austria by local authorities it seems their forecasts will fail. We have estimated that there is no farmland to produce the crops for the expected 74 PJ of fuel or energy forests without changing human diet and production systems. AGS is an impressive tool to estimate and plan the national energy and food supply, but also to check the different agricultural production systems. AGS can also be used to plan the location for new biogas plants or to show the area of supply for cities. Some examples are presented in the paper.

Auto-steer navigation profitability and its influence on management practices: a whole farm analysis

Shockley, J.M.[1], Dillon, C.[2] and Stombaugh, T.[3], [1]University of Kentucky, Agricultural Economics, 331 C.E. Barnhart Building, Lexington, Kentucky, 40546, USA, [2]University of Kentucky, Agricultural Economics, 403 C.E. Barnhart Building, Lexington, Kentucky, 40546, USA, [3]University of Kentucky, Biosystems and Agricultural Engineering, 214 C.E. Barnhart Building, Lexington, Kentucky, 40546, USA

GPS-enabled navigation technology has been scrutinized by many farmers for its potential profitability and whether such technology as auto-steer guidance can pay for itself. This study investigates that very question as well as whether auto-steer navigation can influence on-farm management decisions and the potential for managing production risk. A whole farm economic analysis is conducted in order to provide a detailed assessment of auto-steer navigation by encompassing multiple management practices as well as enterprises. Specifically, a mean variance mathematical programming formulation representing a Kentucky corn and soybean producer is modeled. The Decision Support System for Agrotechnology Transfer (DSSAT v4), a biophysical simulation model, is also utilized to generate yields for varying management practices such as cultivar, plant population, row spacing, planting date, and nitrogen application rates for 30 years of weather data and four soil types in Henderson County, Kentucky. Although many management practices can be analyzed, this study will focus on one of those practices, planting date, and how the adoption of auto-steer navigation will impact this decision. Due to auto-steer's ability to provide additional hours in the field, reduce overlaps and skips, and increase efficiency, it is hypothesized that the adoption of auto-guidance will alter planting date therefore increasing net farm income and reducing risk associated with yield variability. Auto-steer's influence on planting date should increase the value of the technology due to yield increase from optimal planting date.

Improving decision support in plant prodution with GIS
Endler, M. and Roehrig, M., ISIP e.V., Ruedesheimer Strasse 68, 55545 Bad Kreuznach, Germany

Numerous models for plant pests and diseases have been scientifically developed in recent years. But their application in extension and practice has been limited. The German Crop Protection Services (GCPS) started a project in 2001 to close this gap: the Information System for Integrated Plant Production (ISIP, http://www.isip.de) was developed as a universal framework to implement weather-based simulation models in the Internet. In general, the models have a regional output to give the user an overview of the current risk potential. The model output is supplemented by data from field monitoring where available and by a written comment by the regional extension officer of the CPS. This 'threefold decision support' gives a comprehensive overview for a defined pest or disease. In addition, the user can use the models interactively by entering own data into the system and thus receive site-specific simulation results. In ISIP, the results for simulation models are shown on static maps without a georeference. On these maps, for example, the infection pressure of potato late blight is given in five levels, represented in differently coloured symbols at the position of the weather station. The first problem is that user has to decide, which weather station is relevant for him. The second problem is that in some regions, the network of weather stations is sparse. In a recently concluded R&D project (see Zeuner & Kleinhenz: 'Use of Geographic Information Systems in Crop Protection Warning Service', this conference) a scientific method was developed to interpolate weather data. This method calculates air temperature and relative humidity for a 1 x 1 km grid covering all agriculturally used area in Germany. Using this interpolated data as input for the prognosis models, so-called 'risk-maps' can be calculated. To make use of this new method, the databases in ISIP have to be georeferenced and a WEB-GIS component has to be implemented. The component will comply to all relevant standards (OGC, GDI-DE) to ensure interoperability with other geoservices. This will improve (a) the regional overview with new functionalities like map zooming and panning and (b) the site-specific simulation results where production factors (e. g. variety or date of emergence) entered by the user are supplemented by weather data interpolated for the nearest grid point. These enhancements will contribute to a higher acceptance and usage of online decision support in plant protection.

ICT and the blooming bloom trade

Fleming, E.[1], Mueller, R.A.E.[2] and Thiemann, F.[2], [1]University of New England, School of Business Economics and Public Policy, Booloominbah Hills, Armidale, NSW 2350, Australia, [2]CAU Kiel, Dept. of Ag. Economics, Olshausenstr. 40, 24118 Kiel, Germany

Globalization results when markets become more integrated because of reduced transaction and transport costs. These costs have fallen because of sustained advances in transport technology and, more dramatically, in digital information and communication technology (ICT). Although communication costs tend to be a minor component of total trading costs, reductions in these costs may strongly stimulate international trade. The empirical evidence in support of this effect is, however, scant and its strength may depend on the composition of ICT. We test the hypothesis of an ICT effect on trade for the case of trade in flowers. We employ a gravity model of international trade which includes 92 countries and which covers the period from 1995 to 2006. The models explains the volume or value of flower trade in terms of the level of internet and mobile phone diffusion, and of a broad range of factors that might also affect bilateral trade. We tested whether a fixed effects model or random effects model best suits the data; results suggest a fixed effects model is appropriate. Preliminary results suggest that the effects on trade in flowers of ICT varies between exporting and importing countries. Mobile cellphone usage in exporting countries is a significant positive factor, especially for countries with a common language, but mobile cellphone usage in importing countries is not. Fixed telephone and internet usage in importing countries are significant positive factors, but their usage in importing countries is not.

The design of a marketing information system to enhance the competitiveness of agricultural commodities: some evidence from the grain sector

Meyer, C., Fritz, M. and Schiefer, G., Institute of Food and Resource Economics, Chair of Business Management, Organization and Information Management, Meckenheimer Allee 174, 53115 Bonn, Germany

Abstract: In the grain market there are a few intrinsic quality parameters that facilitate global trade. However, recent developments like the increasing use of biotechnology, globalisation of markets or changed consumer demands for quality, food safety and process attributes ask for guarantees to that respect. This requires improved communication concepts and information sharing. Many researchers have focused on end-user marketing of food products. This research instead deals with improving the competitiveness of wheat in the mass market through vertical cooperative marketing with a regional focus from a business-to-business point of view. For this reason the research aims at designing and evaluating a communication concept based on a marketing information system. Much information can be derived from current quality systems of companies. This is why information clusters are formed. This study puts emphasis on the visualization and integration of these clusters in a communication concept in order to support relevant success factors of grain trading and processing companies. The cluster integration is based on usability engineering concepts to support quality guarantees like region of origin. It is also assessed in experiments how far organisational grain buyers respond to this integrated approach.

Use of discrete event simulation model to study the logistic of a supply-chain for fresh produce district

Berruto, R.[1], Busato, P.[1] and Brandimarte, P.[2], [1]University of Turin, DEIAFA, Via L. Da Vinci, 44, 10095 Grugliasco (TO), Italy, [2]Polytechnic of Turin, DISPEA, C.so Duca degli Abruzzi 24, 10100 Torino, Italy; remigio.berruto@unito.it

Customers are asking better quality in the fresh produce, and that imply better management of the SC. The high quality of logistic services could be achieved if the partners are aware of the SC system and its behavior in order to have an efficient response to the customer needs. One survey carried out within the partners of a new district for fruits and vegetable production revealed that one of the most important improvements they expect from the new district was related to the logistics. In order to investigate the supply-chain of fresh produce some trials were carried out in the district region with a survey on working times, material flow and information flow for three produce: potato, melon and lettuce. The survey considered all the links in the chain, from production to the consumption. The authors designed and developed the simulation model with an object oriented simulation language, Extendsim®, that implement the supply-chain logistic of the district, in order to improve the logistics operation and the information sharing. Each single box of fresh product is the single item considered for the simulation. The performance of the simulated supply-chain scenarios can be evaluated with many indexes. The most important are: (1) residual shelf-life at the point of sale. This is an index of quality of the product for the final consumer. This parameter is available for the single box of product moved through the supply-chain. (2) percentage of order filled. This index is important to establish the level of customer service and the reliability of the system. (3) time in system and lead-time. These indexes measure the efficiency of the supply-chain. Reduced distribution times allow to lengthen the residual shelf-life of the product. (4) ratio between distribution cost and value of product. The efficiency and the performance of the supply-chain are to be consider together with the distribution cost and the value of the distributed product. The paper will present the main features and the model architecture. The comparison among local, short supply chain for distribution of fresh produce and traditional, longer supply-chain will be presented as an example to understand the usefulness of the proposed system. The model will be used in the spring by some managers in the horticultural district to have a feed-back on the logistic scenarios proposed.

Supply chain optimization of rapeseed as biomass applied on the Danish conditions

Sambra, A., Sørensen, C.G. and Kristensen, E.F., Univeristy of Aarhus, Agricultural Engineering, Schuttesvej 17, 8700 Horsens, Denmark; Ana.Sambra@agrsci.dk

This research is part of an ongoing extended project, BioREF (Biorefinery for sustainable Reliable Economical Fuel production from energy crops). BioREF is intended to develop, in a dynamic way, a bench-mark for future integrated and sustainable bioenergy production systems that will contribute to enhance Denmark's position in the bioenergy production. The success of shifting from fossil fuels to renewable sources lies not only on the process itself but also on the efficiency of the whole production from the supply chain perspective. In order to obtain a sustainable production of biofuels each step is important and crucial for the overall success. The optimization of the rapeseed supply chain is strongly dependent on the available technologies, government policies, and the human factor support, which are characteristic for each country and may also vary between different areas of the same country. This paper intends to look at the supply chain of rapeseed from sowing, harvesting, transportation, drying and storing point of view, and how can these steps be optimized in order to increase the efficiency of the whole supply chain for the Danish conditions. Different scenarios were analyzed to test the different harvesting and transport systems, operational feasibility, under a variety of different logistics constraints specific for Denmark. The influences of crop production schemes, harvesting workability, crop yield, harvesting technology, transporting rules, etc. were evaluated on empirical data sets. Also, the evaluation includes the identification of key parameters and sensitivity analysis on these parameters. In order to obtain a more precise understanding of the process, and its special conditions for Denmark, the scenarios are based on interviews with farmers and biorefinery representatives, statistical analyses and modeling considerations. Given the personalized character of the analysis, the possible fluctuations are minimized and the level of accuracy will be increased. Evaluation of transport is given, for any location from Denmark, along with time of travel, optimum route and distance to be traveled, which is based on an internet application with consideration to the current Danish laws (e.g. tractors are not allowed on motorways). Each country has different procedures which are influenced by geography, climate, as well as local policies and laws. This is why the optimum chain for Denmark may be completely different or even inefficient comparing to another country (for example: the acceptable moisture content of the oilseeds in Australia is 6% while in Denmark it is 9%, which result in higher expenses for the Danish farmers of drying and storing rapeseed). Given the location conditions some steps of the supply chain may require more attention than others, even if the same country is considered. The search for a process optimum and elimination of logistical bottlenecks require a detailed description of the current conditions and dynamic precalculation which takes the specific factors and their interaction into consideration.

Information transfer in the European beet growing countries

Maassen, J. and other members IIRB Working Group, IIRB, working group Communication techniques, p/a P.O. Box 32, 4600 AA Bergen op Zoom, Netherlands

In almost all European countries sugar beet are grown. There is a strong chain involvement in the European beet sugar sector. Sugar industries, growers and independent research organisations work in the different countries closely together to improve competitiveness and sustainable development of the sugar beet crop. Effective information transfer between researchers, advisers, suppliers, beet processors and growers has always been important for efficient, profitable and environmentally friendly sugar beet production. The final goal is to get the beet grower into action. Information transfer in EU countries is managed in different ways and utilises many forms and routes. Face to Face information transfer can be limited by the number of advisors available and the cost of this system. The use of various multipliers is important as these can help to overcomes many of these problems. A range of different, but complimentary methods, are used and each sugar company, growers organisation or research institute has its own communication techniques, information materials and transfer pathways. ICT is more and more an useful tool in that process. In this introductory presentation about the European sugar beet sector, an overview will be given on the communication techniques and strategies used in Austria, Belgium, Denmark, Finland, France, Germany, Italy, Spain, Sweden, Switzerland, United Kingdom and the Netherlands. These countries are represented in the working group 'Communication Techniques' of the International Institute for Beet Research (IIRB). The IIRB is an international, non-governmental and non-profit organisation. In general the aim of IIRB is to study and promote research on beet. Today the IIRB has about 500 members representing 25 beet growing countries. The main objective of the working group 'Communication Techniques' is to improve knowledge transfer by learning from the communication techniques and tools used in each country. Exchange and collaboration in this field saves money for developing communication techniques. Also some examples of international cooperation in internet services will be shown.

UK beet industry experiences in the provision of internet based communications, decision support tools and contract administration facilities

Bee, P., British Sugar, Agriculture Communications, Sugar Way, Peterborough, PE2 9AY, United Kingdom; paul.bee@britishsugar.com

British Sugar has developed a range of internet based services in the UK since 1990, initially working with a small number of 'early adopter' sugar beet growers and advisers. By using growers' survey data and industry usage information, the company pursued a strategy of providing growers with agronomy information and access to campaign delivery results which enabled growers to benefit from accessing internet based information provided by British Sugar. This created a demand for more internet based services and has seen the use of British Sugar Online (www.bsonline.co.uk) grow to a position where 85% of the UK national tonneage is represented by growers using the site in 2008. A range of specialist services have been developed (many in conjunction with the British Beet Research Organisation) including, growers communications area to select their preferred formats (email, fax, post), decision support tools to select best herbicide treatments and access to sugar beet payment documents online. The system also has an ingenious way of managing mass SMS communications to growers, hauliers and all industry partners (advisers, consultants etc.) In 2008 British Sugar has developed and is testing an online sugar beet contract system that allows growers to complete all of their contract documentation (contract tonneage, seed order and field details etc.). The system also allows hauliers to register their delivery vehicles and to view their campaign delivery schedule. The service will be made available to all growers and hauliers in 2009. The UK industry will continue to provide internet based information and communications services to enable growers to access essential information and to be kept as informed as possible when making their crop management decisions. Note: Abstract part of the sugar beet session.

Audiovisual tools as integrated part of information transfer strategies

Kämmerling, B., Pfeifer & Langen KG, Landwirtschaftlicher Informationsdienst Zuckerrübe (LIZ), Dürener Str. 67, 50189, Germany

The character of information seems to be abstract and objective, but to be accepted by the recipients, it has to be transferred and presented in accordance with the claims of personification and some emotional attractive aspects. On this behalf, the sugar beet-growers are no except. That means, also scientific proven information first has to get admission to the growers 'gut feeling' to make him share ideas or not. Direct and permanent personal presence to clients cannot be assured with today´s manpower capacities of consultants. Therefore extension services for thousands of sugar beet growers are in the need of service tools to close the gap between advisory demands and training capacities. The role of print media is well known and adapted, finally there is something to keep 'in my hands' to rely on. But there is still the disadvantage of a one – way communication. Here the internet fills demands and opens the way to interactive communication, because it is always available, always on the latest level of information and it is an inexhaustible source of information, which is sometimes difficult to select in the right way and: it is not a personal contact, at least not speaking to the recipient. Interactive tools like spreadsheets for economic calculations train the user e.g. to find out where leverage effects of any input turn the result. These tools are helpful, incorrupt and often very profitable for the user. It is something to work on and give personal inputs on, but the user feels still alone with himself. Audiovisual tools such as video clips to explain a complex issue create a closer contact and understanding of the senders message. Living Pictures and visualised statements to underline the information can help in the effort to create a personalized sugar beet community. This community is to be seen as a necessary home of interests, exchange and response of ideas. Learning-videos and agricultural short-films with reports, interviews or technical presentations are a further step to personalize information technology used for crop management. Experiences about using audiovisual media in extension services for arable crops in general and in sugar beet advisory services in special are presented. Conclusions for the implementation of audiovisual media in information transfer strategies are given.

The four horsemen of innovation: learning styles, entrepreneurship, attitudes and knowledge network
Miedema, J. and Faber, N., University of Groningen, Faculty of Economics and Business, P.O. Box 800, 9700 AV Groningen, Netherlands

Agriculture is facing major and rapid changes which can significantly affect the sustainability of the European Union. These changes may include intensification of land use, and depopulation and land abandonment. The (new) policy for market price support may lead to minor changes in production but to profound changes in land prices, income and farm structure. Currently, only a small part of beet growers reach a sufficient return on crops to ensure continuation of the farm. Although previous research programs have yielded valuable knowledge that can help beet growers to innovate farming processes, actual transfer of this knowledge to the growers so far is lacking. Currents ways of knowledge transfer do not match learning styles, personal traits or the social environment of previously identified groups of growers. The current research was designed to develop new means of knowledge transfer: using both digital means, e.g., decision support systems, and other means, e.g. study groups, knowledge transfer was re-assessed to form specific inspiring learning environments (Specifieke Inspirerende LeerOmgevingen, SILO's©). A survey study, sent to a random selection of sugar beet growers in the Netherlands, assessed learning styles, attitudes toward innovation, personality traits related to entrepreneurship and the social network growers use to obtain new knowledge. Earlier research has indicted these constructs can predict innovative behavior; however, they have not been investigated at the same time, in one study, in this domain. The current research was also intended to explore the joint effects of these four constructs, and their interactions. These data were linked to the crop yield data over the previous five years, to be able to compare the influence of learning styles, attitudes, network and individual differences on the occurrence and effectiveness of certain types of innovative behavior. Results indicate that different learning styles correlate with different ways of using one's knowledge network: for instance, people who are more prone to seek help, have significantly more contacts and exchange more knowledge within their networks. Growers whom significantly participate more in meetings and interactions with colleagues, produce an above average crop yield, as compared to other groups. The innovation attitude appeared to predict the innovation intention of growers; people with more positive attitudes are more willing to try new ideas and implement not fully tested techniques than growers with less positive attitudes toward innovation. Knowledge networks are comprised of fellow growers, friends, family, but mostly the growers receive their knowledge from advisors, suppliers and study groups. Preferences for learning and innovating correlate with the size of the network, and how intensively it is used. Although this was only the first study on the joint effects of learning styles, entrepreneurship, attitudes and knowledge networks, we feel we may conclude that these four can be labeled the four horsemen of innovation – not just a pony ride on a children's' farm.

Towards teacher competence on metadata and online resources: the case of agricultural learning resources

Manouselis, N.[1], Sotiriou, S.[2], Tzikopoulos, A.[1], Costopoulou, C.[1] and Sideridis, A.[1], [1]Agricultural University of Athens, Iera Odos 75 Str, 118 55 Athens, Greece, [2]Ellinogermaniki Agogi, Dimitriou Panagea Str, GR 153 51 Pallini Attikis, Greece

Today's teachers need to be prepared to provide technology-supported learning opportunities for their students. Being prepared to use technology and knowing how that technology can support student learning have become integral skills in every teacher's professional repertoire. Teachers need to be prepared to empower students with the advantages technology can bring. Schools and classrooms, both real and virtual, must have teachers who are equipped with technology resources and skills and who can effectively teach the necessary subject matter content while incorporating technology concepts and skills. This need becomes even more evident in the case of agricultural, environmental and other life science topics, where the existence of interactive learning resources can greatly facilitate the teaching activities within the classroom. In this direction, a new European initiative (the METASCOOL Comenius project) has been deployed improve the in-service training of school teachers and school ICT staff on topics related to the organisation, sharing, use and re-use of digital learning resources that can be accessed online through learning repositories. Its aim is to develop a practical training framework for improving the quality of teaching and learning in the classroom through the effective use of digital content. The overall objective of the project is not only to improve teacher practice, but also to raise the awareness of teachers across Europe on the need for accurate tagging of resources. It achieves this through a user-friendly approach that motivates teachers to quickly and easily add metadata to resources that they have both used and created. Particular focus is given to agricultural topics, since learning resources on organic agriculture and agroecology will be collected, organised, annotated with metadata, and published in a learning repository. More specifically, METASCHOOL will mainly carry out the following activities: (1) It will adapt, develop, test, implement and disseminate a new training framework that will support the in-service training of (mainly) teachers and (also) ICT personnel of school staff on topics related to metadata, learning resources, and learning repositories. (2) It will adapt, develop, test, implement and disseminate strategies and best practices for organising favourite/useful learning resources into personal portfolios of digital resources, as well as setting up and using learning repositories on a school or regional level. (3) It will suggest and test a variety of teaching methodologies and pedagogical strategies for using digital learning resources in the classroom, for two particular subject areas: science education and agroecology. (4) It will also focus on promoting the creation of a European virtual space for interconnecting school repositories and exchanging/sharing teaching resources. (5) It will organise pilot training and validation activities of both teachers and ICT staff (where possible) from schools all over Europe. (6) Finally the METASCHOOL consortium aims to deliver a structured & reusable set of guidelines and recommendations in all project languages.

Research on open source e-learning tools and agricultural applications

Lengyel, P. and Herdon, M., University of Debrecen, Business- and Agricultural Informatics, Böszörményi 138, 4032 Debrecen, Hungary; szilagyir@agr.unideb.hu

Within the NODES European e-Learning project (Creation of a European network of multimedia resource centres for adult training) we are doing research and developments based on open source tools. The aims of the NODES are to select those open source components which are suitable for creating the network for mainly adult training / lifelong training, the use of multimedia knowledge to improve competitiveness employability and mobility of different target groups, such as rural living people and farmers. In creating the architecture of the Nodes project (Learning Management System, Local – National Repository, Knowledge Databank Management System, Specific and Shared Databases, EU Index, Internet Content Access and Security Rules) there are more tasks which are coordinate and carried out by our expert group. The selection criteria of open source LMS was set up by the consortium. Our studies 'Survey on the existing systems (What are exists)' were consist of 4 parts: The basic elements of e-Learning Systems, The complements and integration of the e-Learning Systems, The specification of interface. The result of our research work was to select the Moodle LMS. The Moodle provides a reliable platform that supports social and collaborative learning, highly configurable and extensible, implements new features and fixes rapidly, is free of licensing costs. The members of project (France, Spain, Ireland, Hungary, Czech Republic and Romania) implemented the Moodle system creating the LMS. To integrate the different contents we had to develop the EU-index, which is the central, common and shared index database (a metadatabase). The EU Index, the merging of each local - national index (based on KD - Knowledge Database - and selected links like available resources / websites / etc.). The EU-index based on the Dublin Core Patent. The Dublin Core Metadata Element Set is a vocabulary of fifteen properties for use in resource description. Another important function is the multimedia content management. One of these parts is the video objects. From the about 200 Moodle modules and blocks we are using for example the AutoView Presenter which allows you to put video on-line with synchronised slides. The NODES system is used in graduate, postgraduate PhD, adult trainings programmes and it is a very successful system as an educational portal system for our faculty too.

Virtual form of education in a lifelong learning: chance for countryside
Jarolimek, J., Vanek, J., Havlicek, Z., Silerova, E. and Simek, P., FEM Czech University of Life Sciences, Information Technologies, Kamycka 129, 165 21 Prague, Czech Republic

Availability of education, including a lifelong learning; is one of the values which are measure of a life quality in the advanced countries. However, there are still significant differences between a town agglomeration and rural regions. Centres of education are mainly situated in big cities; smaller municipalities are separated out of these centres tens or hundreds of kilometers (according to the conditions of the Czech Republic). While educating young people, it is often relatively positively evaluated that they commute towards education; there is a whole range of social and cultural aspects; and above all, they have time for that, it is their main „working' load. Opposite situation is in the lifelong learning, which is conducted in parallel with a full-time employment, but is necessary for an effective and competitive performing of the employment. For participants of the lifelong learning it is impossible to commute big distances, their working load does not allow it. That is why forms, where the so called „education which goes to the students', is chosen. The Faculty of Economics and Management CULS has already established 8 regional consultancy centres in regions of the Czech Republic. Although a commuting distance and availability is shorter, it is still not sufficient for all forms and participants. Information and communication technologies bear an enormous chance to bring education closer to the rural regions. A text form of an eLearning is practically already standard; but a voice and image broadcast give us inexhaustible possibilities of usage. At CULS, a number of projects on usage of virtual education is nowadays solved. Optimal technologies which are area-accessible in the regions, forms and methods of group and individual education are being searched. It is not possible only to substitute personal contact by electronic and paper web; it is necessary to change the deep-rooted procedures and approaches, that is the most difficult task. In the second half of the year 2008 a pilot verification of virtual courses is being procceded (Forestkeeping and Forest crop establishment), providing video lecture; mutual communication; examination of students knowledge and administration of student registers.

Designing e-learning courses in WebCT environment: a case study

Põldaru, R. and Roots, J., Estonian University of Life Sciences, Institute of Economic and Social Sciences, Kreutzwaldi 1, Tartu, 51014, Estonia

The Internet and associated technologies have spurred evolutionary business, including learning and teaching, processes in higher education organizations. Before the last decade, most colleges and universities providing distance education used face-to-face sessions. Recently, broad use of the Internet for conducting instruction has become popular in universities. These online courses are typically conducted in one of three environments: completely online without face-to-face interaction; as hybrid courses where the class frequently meets face-to-face, as well as online; and as face-to face courses with integrated web-based support materials and activities. Successful use of e-learning systems may be attributed to the availability of Learning Management Systems (LMS) or learning environment. One of the most popular learning environments for facilitating the delivery of online education is WebCT. The interface is easy to use and efficient for students taking online courses. In WebCT, the course designer's role is somewhat complicated. However, there are numerous training and support tools for course builders built into the system. The aims of this paper are: (a) to determine the current usage of WebCT teaching and learning in Institute of Economics and Social Sciences of Estonian University of Life Sciences, (b) to identify the factors and requirements affecting the design of e- learning course in WebCT environment, (c) to develop (to design) a e-learning course of 'Operations Research' for economic speciality in WebCT environment, (d) to analyze lessons learned from designing the e-learning course of 'Operations Research'. The efforts reported in this paper show that creating accessible distance learning courses is an ongoing effort, not a one-time project. E-learning using WebCT environment improves learner's skill and knowledge and has the following advantages: space is not needed; learners do not need to wait until a class is available; learners can complete training when it is least disruptive to their schedule; and the methods can increase learner's interesting, deliver contents clearly, and feedback students easily.

Road mapping of curriculum developments and training experiences in agricultural informatics education

Herdon, M., University of Debrecen, Boszormenyi ut., H-4032 Debrecen, Hungary

During the last 6 years the University of Debrecen developed more curricula for training experts on agricultural informatics. The first course was the five year 'informatics agricultural engineer' course which introduced in the 2002/2003 academic year. The first students will graduate this summer, so we have good experience about this education programme. Because of the higher education transformation this 5 year training programme will be closed. The higher education reform started according to the Bologna declaration, so the new Hungarian Higher Education system was introduced in the 2006/2007 academic year. In the new two level (BSc and MSc) system we developed and introduced the BSc curriculum, namely an 'agricultural engineer for informatics and government'. The rate of informatics and specialised informatics knowledge is more than 30% in the curriculum. We made efforts for developing master training programme for information science and advanced applications in agriculture. The newest success in this effort is that the Hungarian Accreditation Committee accepted the proposal of University of Debrecen to start the Master of Business Informatics programme with a 'Informatics for Rural Development' specialization. This specialization takes 1 academic year study within the 2 year master programme.

Agricultural web TV in France

Waksman, G.[1], Burriel, C.[2], Holl, C.[3] and Masselin-Silvin, S.[1], [1]ACTA Informatique, 149 rue de Bercy, 75595 Paris Cedex 12, France, [2]ENESAD, 26 bd du Docteur Petitjean - BP 87999, 21079 Dijon Cedex, France, [3]EP-TE, 71, Avenue du Docteur Netter, 75012 Paris, France

The Web TV, i.e. a TV channel over the Internet, is a new means to inform farmers that is becoming more and more popular since we observe in France and in Europe initiatives of public authorities as well as of agricultural advisory organisations and commercial companies. Our organisations have been involved in two Agricultural Web TV projects since 2005: - the first one named CanalAgri initiated by a local community in Burgundy, where special emphasis was put on interactivity and, - the second one named Agri Web Télé at French national level aimed at developing internal competences to establish sound relationships with Web TV actors (commercial companies or public actors) operating in France and to propose to them relevant videos upon request. In the first part, we present the two above mentioned projects: necessary equipment, investments, training, running costs, organisational schemes adopted, and examples of successful (and less successful) productions with audience measurements. In the second part, we try to present similar initiatives in France: three private commercial channels and two public channels. Many projects are flourishing in EU member states. From our experience and the above description of running projects, we try in conclusion to draw a few perspectives and to examine how the Web TV may modify the information dissemination process in our Agricultural sector.

Interoperable metadata for a federation of learning repositories on organic agriculture and agroecology
Manouselis, N., Palavitsinis, N. and Kastrantas, K., Greek Research & Technology Network (GRNET), 56 Messogion Av., 115 27, Athens, Greece

To further promote the familiarization of consumers with the benefits of organic agriculture (OA) and agroecology (AE) the most dynamic consumer groups have to be properly educated. Young people at all stages of formal education have to be carefully approached through relevant educational programs in tll kinds of educational institutions, from schools to university departments. Such user groups have to be carefully approached through publicly available, quality, and multilingual educational content. In this direction, the Organic.Edunet initiative (http://www.organic-edunet.eu) has been deployed, aiming to facilitate access, usage and exploitation of digital educational content related to OA and AE. Organic.Edunet will deploy an online federation of learning repositories, populated with quality content from various content producers. In this paper, we will present how metadata can be used to support the development of the Organic.Edunet learning repositories, also maintaining interoperability and communication with other learning repositories (such as FAO's Capacity Building Portal and CGIAR's Online Learning Resource project).

'WebInfo': the new website for the national centre of the Danish agricultural advisory service
Hørning, A., Hansen, J.P., Hansen, N.F., Kjær, K.B. and Bundgaard, E., DAAS, WebInfo, Udkærsvej 15, DK-8200 Århus N., Denmark

Over the past 6 years, the Danish Agricultural Advisory Service (DAAS) has gradually built a new internet platform for its entire organisation. A platform which will be completed with the introduction of the new website for the National Centre (www.landbrugsinfo.dk) in the late fall of 2008. The platform contains around 70 separate websites, of which the National Centre's website is by far the most complex, containing 100,000 webpages. The website of the National Centre is a site, which on one hand serves as an extranet for DAAS, but at the same time it is also a commercial site, to which users can purchase a membership. By becoming a member you gain access to the majority of the information on the website. While the main objective has been to transfer the existing information structure and webpages to the new site, substantial effort has also put into improving the site in the following ways: Collaboration: The website is based on Microsoft SharePoint Server (MOSS 2007), which provides advanced workflow functionality. This has provided DAAS with new possibilities for online collaboration. Intensive metatagging: To facilitate optimal retrieval of pages in all relevant contexts, the webpages have been intensively metatagged. The set of metatags is based on Dublin Core standards, and incorporates thesaurus based subject tagging based on Agrovoc terminology. Single Sign On: The platform offers SSO to all of DAAS' websites and applications. Internationalisation: In order to support DAAS' work abroad, the new platform supports the future use of multiple languages on the website.

Agrarian WWW portal

Simek, P., Jarolimek, J., Vanek, J., Silerova, E. and Havlicek, Z., Czech University of Life Sciences Prague, Department of Information Technologies, Kamycka 129, 165 21 Prague 6, Czech Republic

Together with the development of information technologies and also in compliance with worldwide trends information servers (websites) of a number of institutions and companies were gradually emerging on the Internet in the Czech Republic. The agrarian sector was not an exception. As far as agrarian sector is concerned, general features and principles of ICT use are applied also here. However, there are certain specifics definying this environment and clearly also a time lag in development, which is influenced by these conditions. The term „portal' is mentioned on almost all the Internet eZins as the most suitable tool for providing access to information for a particular group of users. It is not possible to give an exact definition of the portal since there is always a certain view of a particular internet application, which could define it as a portal. The most general definition of the portal says it is an entrance to information, which are targeted to a differently wide group of users. From the very beginning the WWW Portal AGRIS project focuses on creating a platform for providing information in the field of agriculture, food industry, forestry and rural development, among many other related sectors. The aim is to provide access to already existing information sources, create its own news service and help to publish information provided by subjects, which have got limited opportunities for electronic (internet) presentation. This leads to producing a unified block of information from the same field, thus offering better access to information for corporal management, public administration, students, teachers, advisors and all the other users. The whole solution stands on general principles, it is modular and open. Therefore, it is possible to implement it in any other sector and field. This portal solution represents a third, so far the highest level of internet services structure. The research and development of the agrarian information system proper has been heading towards this stage already from the beginning. The portal was created in 1999 in cooperation with the Czech University of Life Sciences (formerly of Agriculture) Prague and the Ministry of Agriculture. At present, development of the portal depends on the Council of Experts consisting of the representatives of the following institutions: Czech University of Life Sciences Prague, Ministry of Agriculture, Ministry of Environment, Ministry of Computer Science, Institute of Nutrition and Agriculture Information, Institute of Agricultural Economics, Office for Statistics, Commodity Exchange, Agricultural Journalists and Publicist Club and others. It precisely responds to information requirements in the agrarian sector as well as information and communication technologies (ICT) development – there is already a third programme version at present. At the moment the agrarian WWW portal AGRIS represents an information source with the highest attendance in the sector.

Precision concentrate rationing to the dairy cow using on-line daily milk composition sensor, milk yield and body weight

Maltz, E., Antler, A., Halachmi, I. and Schmilovitch, Z., A.R.O., The Volcani Center, Institute of Agricultural Engineeng, P.O. Box 6, Bet Dagan, 50250, Israel

The dairy industry undergoes structural and economical changes. The number of farms is dropping while the average number of animals in each farm is increasing. Automated recording of animal data is becoming essential for precision dairy farming. Individual feeding enables tailoring a precise ration for each individual cow according to performance. This feeding method is particularly important when concentrate supplementation is needed (e.g. pasture management, robotic milking, fresh cows or other non homogenous groups) and with the global rise in grain prices, the economical importance of individual feeding is increasing. Emerging new technologies has availed acquisition of new data at significantly higher resolution than previously possible. One of the new devices (developed by S.A.E. Afikim, Kibbutz Afikim, Israel and A.R.O, The Volcani Center, Israel) is a sensor (Afilab™) that measures milk composition. The Afilab™ performs real-time analysis of individual cow milk solids (fat, protein and lactose) and gives indication of blood and SCC. The technology is based on spectroscopy, with no interfere with milk flow nor does it alter the milk in any way. The device is installed at each milking stall and analyzes each individual cow's milk at every milking allowing milk component data flow within the same configuration and time frame as milk weight data measured by the electronic milk meter. In order to examining the use of analyzer data (e.g. fat and protein percentage) together with already available data (milk production and body weight) for individual concentrates allocation, 24 multiparous cows kept in a group were all they were fed forages in the common trough and all the concentrates were fed through dual channel computer controlled self feeders. All the concentrates were rationed individually according to performance to. In order to calculate total ration density and crude protein content, the NRC 2001 formula for predicting dry matter intake (DMI) of individual lactating cow, was applied. This could be done on-line by incorporating the performance variables required by the formula by available sensors (milk yield – Afiflow™; milk fat - Afilab™ or periodical milk test; body weight - Afiweight™). The concentrates allocation was calculated for all cows using daily performance of milk yield and body weight and for 11 cows daily milk composition provided by the Afilab™ and for the rest of the cows periodical milk composition recorded by the milk test milk test performed routinely for the Israel herd book. On average milk yield of both groups was similar. However, milk fat, as well as 4% FCM, was significantly higher and less concentrates were allocated to the cows for which decisions for concentrates allocation were done according to on-line milk composition data compare to those for which the decisions were taken in accordance to the milk composition of the periodical milk test. In addition to this, the cows for which concentrates allocation decisions were taken using Afilab™ data, had a higher dry matter intake and also lost less weight during transition time. On individual basis it could be seen that especially during transition time (first 6 weeks of lactation) the periodical milk test failed to provide the needed information regarding milk fat gradually decrease which led to rationing concentrates in a level that most likely caused a greater decrease in milk fat during this period.

Evaluation of a dynamic linear model to estimate the daily individual milk yield response on concentrate intake and milking interval length of dairy cows

André, G., Bleumer, E.J.B. and Van Duinkerken, G., Wageningen UR, Animal Science Group, P.O. Box 65, 8200 AB Lelystad, Netherlands

Automation of concentrate feeding and milking enables application of individual cow settings for concentrate allocation and milking frequency. An adaptive model was developed to estimate the individual dynamic milk yield response on concentrate intake and milking interval. Real time process data were analysed with a dynamic linear model (DLM) to provide parameter estimates that characterise the response of individual cows on concentrate intake and milking interval length in their actual situation. Based on the parameters, a control algorithm calculates daily individual optimal settings, such that the balance milk returns minus concentrate costs is maximised. The whole application, also called dynamic feeding and milking, fits within the concept of precision dairy farming and is implemented on dairy farms in cooperation with industrial companies. This concept for precision dairy farming is an innovative approach to feeding and milking with promising economic results. The existence of individual variation and temporal variation is recognised in common practice and animal science. However, it is difficult to convince nutritionists, animal scientists and end-users that this variation can be utilised for improvement of feeding and milking. Within the dynamic concept the DLM plays a key role and a good understanding of this self-learning model is essential for biometrical engineers to explain the functioning to animal scientists and farmers. The objective of this paper is to provide this understanding. The DLM is presented and some important statistical aspects are explained. The formulation of the DLM is related to existing paradigms about feeding and milking. Furthermore, results of individual cows, collected during the prototyping phase are used to evaluate the nutritional aspects.

Implementation of an application for daily individual concentrate feeding in commercial software for use on dairy farms
Bleumer, E.J.B., André, G. and Van Duinkerken, G., Wageningen UR, Animal Sciences Group, Edelhertweg 15, 8219 PH Lelystad, Netherlands

To assess implementation strategies for precision livestock farming concepts a case study was performed on the introduction of an innovative model for dairy cow feeding in common practice. Daily concentrate allowances for individual dairy cows are usually based on empiric models. These models are generally based on regression equations derived from population data and do not take into account individual and temporal variation. The application which was implemented consists of an adaptive model for estimating the actual individual response in milk yield on concentrate intake using individual real time process data. Before the application was implemented, a prototype was developed by a team consisting of biometricians, animal nutritionists and ICT application specialists. It was tested in an animal experiment and further developed into a proof of principal, which was implemented for testing in a common practical setting on a research farm. Because the results were very promising, a workshop was organised to introduce the concept to software, hardware and feed industries were they were challenged to participate. In the next collaborative phase with industry involvement the further implementation into a management system was stepwise: 91) technical documentation of algorithms, (2) programming, (3) technical testing of algorithms, (4) testing of the integrated software and, (5) on-farm testing. During the implementation it became clear that steps 1 to 3 were not difficult to perform and took not much time. Steps 4 and 5 were more complicated because: (1) correct data must be generated from the management system as an input for the model and, (2) the output of the model has to be interpreted correctly for calculating concentrate allowances in the management system. However, not only technical aspects of the implementation process are important, also the communication with end users and stakeholders is, especially when you introduce a new concept. While testing and implementing the application it was clear that end users and stake holders wanted to use and accept the new concept but are trying to understand and use this concept with common 'population knowledge' they have.

An approach to precisely calculate variable dosing of highly nutritious and energetic animal feed
Ortiz-Laurel, H.[1] and Rossel, D.[2], [1]Colegio De Postgraduados, Apartado Postal 143, 94500 Cordoba, Veracruz, Mexico, [2]Colegio De Postgraduados, Iturbide 73, 78600 Salinas De Hidalgo, SLP, Mexico

Animals in an intensive production industry tend to be managed as part of a group and rations are formulated for the whole herd. Although, the actual trend is toward fewer and larger farms, animals must received healthy feed adapted to their age and species in sufficient quantity to maintain their health and satisfy their nutritional needs. Precision agriculture applied to the animal industries relies more precisely on individual animals instead of specific spot of land and crop. Thus, global concern about energy efficiency both technical and biological pleads for a more precise energetic calculation from feed constituents. Therefore, automated feeding systems where individual animals are given a precise ration based upon their production level are fairly commonplace. These decision support tools aid when feeding livestock, where it is important to supply them with a ration which however, usually has a large variation regarding its nutritious value. An immediate and simple solution is to supply a ration carefully metered on a very precise scale, which can be substituted later for a volume based system. In a 200 beef cattle farm trials were set to provide animals with variable grading rations and a feed dosing plan was carried out. These data were transformed into equations for modeling the available biological energy and its variation. By making use of this technique as a system control and the employed technology including the calculation for the equivalent energy, both technical and biological, it was possible to have an economical evaluation for mechanizing this process. The result of this investigation is to provide with a method able to prepare a concentrate mixing with minimum variation within its biological value, either for a day or a technological period. Also, a better flavoured ration, elaborated by suitable ingredients encourage maximum forage intake by the animals.

A study on the cause and effect of lameness in broiler chickens

Cangar, O.[1], Everaert, N.[2], De Ketelaere, B.[3], Bahr, C.[1], Decuypere, E.[2] and Berckmans, D.[1], [1]M3-Biores, Department of Biosystems, Katholieke Universiteit Leuven, Kasteelpark Arenberg 30, 3001 Heverlee, Belgium, [2]Laboratory for Physiology, Immunology and Genetics of Domestic Animals, Department of Biosystems, Katholieke Universiteit Leuven, Kasteelpark Arenberg 30, 3001 Heverlee, Belgium, [3]Division of Mechatronics, Biostatistics and Sensors (MeBioS), Department of Biosystems, Katholieke Universiteit Leuven, Kasteelpark Arenberg 30, 3001 Heverlee, Belgium

The reasons for the gait problems in broiler chickens are multiple although weight and growth rate are said to be the main reasons for locomotion problems. Other factors that play role are infectious diseases, genetics, sex, weight and growth rate, age, feed conversion, feeding, management and movement. In this study, gait score as a measure for lameness, is thoroughly investigated in relation to the following physiological variables: weight, sex, corticosterone content in the blood plasma, hock burns, chest dirtiness, foot pad dermatitis, tibial dyschondroplasia and femoral head necrosis. A total of 152 birds were selected from a house of 1,500 birds in 3 consecutive growth periods. The chickens were scored for their locomotion by experts, weighed, their sex was determined and blood samples were taken for quantification of stress hormones in the blood plasma. They were then slaughtered and visually scored for hock burns, chest dirtiness and foot pad dermatitis. The birds were dissected to diagnose tibial dyschondroplasia and femoral head necrosis. Between the 8 mentioned parameters, sex, blood stress hormone corticosterone, chest dirtiness and femoral head necrosis had significant relations with the lameness of the birds. Body weight and gait score relation was significant ($P<0.001$) and non-linear. Males were significantly heavier than the females ($P<0.001$). Females and males had similar average gait scores (2.63, 2.75 out of 5). Higher corticosterone concentration in the blood plasma in the birds increased with higher gait scores ($P=0.006$). Between the three visual quality variables (hock burns, foot pad lesions and chest dirtiness) only chest dirtiness showed a significant relation with gait score ($P=0.02$). Post slaughter diagnosis for femoral head necrosis revealed significant relation with the lameness of the birds ($P<0.001$).

Integrated ecological hotspot identification of organic egg production
Dekker, S.E.M., De Boer, I.J.M., Aarnink, A.J.A. and Groot Koerkamp, P.W.G., Wageningen University and Research Centre, Farm Technology Engineering Group, Animal Production Systems Group, Animal Sciences Group, P.O. Box 17, 6700 AA Wageningen, Netherlands

Ecological sustainability in agriculture is a concept that contains various environmental problems, which are caused by emission of compounds during different processes along the food chain. A precise ecological analysis of farming systems and food chains is needed in order to suggest and implement effective measures to improve sustainability. Life Cycle Assessment (LCA) assesses the environmental impact along the entire chain. In this research, LCA was used to locate environmental hotspots within the organic egg production chain and explore options that substantially improve ecological sustainability using sensitivity analysis. The environmental impact was expressed per kg of organic egg leaving the farm gate. Five environmental impact categories were included: 1) climate change i.e., emission of CO_2, CH_4 and N_2O, 2) eutrophication i.e., emission of NH_3, NO_x, N_2O and leaching of NO_3^- and PO_4^-, 3) acidification i.e., emission of NH_3, NO_x, and SO_x, 4) fossil energy use i.e., oil, gas, uranium and coal and 5) land use. In case of a multifunctional process, economic allocation was used. We interviewed 20 out of 68 Dutch organic egg farmers to collect farm data for 2006. Data on transport, feed, rearing and hatching were gathered by the conduction of interviews with suppliers and from literature. The Life Cycle Inventories of electricity, natural gas, tap water, transport and cultivation originated from the Eco-Invent V2.0 dataset. A sensitivity analysis was executed for production parameters from the laying hen farm. To identify hotspots, the relative contribution of transportation, feed production, rearing and hatching and the laying hen farm, as well as the contribution of various compounds to each impact category was determined. We identified a chain-compound combination as a hotspot if it contributed to more than 40% of the total of the environmental impact category. Results showed four hotspots. First, 62% of climate change was caused by emission of N_2O from soils during growing of feed. Second, 57% of acidification was caused by NH_3 emission from the laying hen farm. Third, 47% of energy use was oil used for cultivation of feed and fourth, 95% of the land use was arable land required for feed production. We identified no hotspot for eutrophication, but feed production contributed most with 37% nitrogen leaching and 26% PO_4^- leaching. From the sensitivity analysis it appeared that the most sensitive parameters on an organic laying hen farm are the number of produced eggs, the amount of feed consumed and the housing system. An increase in average egg production from 276 with a SD of 39 eggs per laying hen reduced climate change with 13%, acidification with 15%, eutrophication with 13%, energy use with 12% and land use with 12%. A reduction in average annual feed consumption from 42.9 kg with the SD of 7.2 kg per laying hen reduced climate change with 14%, acidification with 17%, eutrophication with 15%, energy use with 14% and land use with 13%. A shift from deep litter housing to an aviary housing with manure drying reduced climate change with 11%, acidification with 53%, eutrophication with 18% and had no effect on land use. The effect on energy use is still being assessed. We conclude that feed conversion and housing are effective ecological optimization options for organic laying hen farmers. However ecological sound feed production also needs attention.

Automated monitoring of milk meters

De Mol, R.M. and André, G., Wageningen University and Research Centre, Animal Sciences Group, Edelhertweg 15, 8219 PH Lelystad, Netherlands

The milk yield of dairy cows can be measured by electronic milk meters. For ICAR approval the milk meters must be checked after installation and every year. These routine controls of milk meters are labour-intensive and costly. Automated monitoring might be an alternative. A computer model has been developed for this purpose. If there are more milking stands in the milking parlour, one might expect that the averages per milk stand are more or less equal for each milking. A Dynamic Linear Model (DLM) is based on a comparison per milking of the average per stand with the overall average. The model calculates a stand deviation factor per stand after each milking. An alert is given when the stand deviation factor differs significantly from zero. If there is only one milking stand in the milking parlour, e.g. in case of robotic milking, one might expect that the measured milk is more or less equal to the delivered milk. A DLM model is based on a comparison of the measured milk with the delivered milk. The model calculates a separation term and a deviation factor after each delivery to the milk factory. An alert is given when the separation term or the deviation factor is too big. The measured milk yields of the experimental farm in Zegveld, The Netherlands in the period January 1, 2006 till October 8, 2007 were used to test the model for the comparison of milking stands. On this farm there were 82 cows milked twice a day in a milking parlour with eight stands. The measured and delivered milk of the experimental farm 'high-tech farm' in Lelystad, The Netherlands, in 2006 were used to test the model for the comparison of measured and delivered milk. There were 70 cows on this milking robot farm with 170 milkings per day in one milking robot. It was possible to fit the stand deviation factor with the model for the comparison of milking stands. This factor can be interpreted as the relative error per milking stand and per milking. The results correspond globally with the results of two calibrations. One milking meter found defect at the calibration was signalled by the model several months earlier. The results of the model at the high-tech farm for the comparison of measured and delivered milk were hampered by unknown separations: Separated milk, e.g. from cows treated for mastitis was measured by the milk meter but of course not delivered. These separations were not recorded. The model shows that there were deviations between measured and delivered milk, these could be explained by the separated milk. The conclusions of this research are: (1) automated monitoring of milk meters by a comparison of the average per stand with the overall average is possible; (2) validation on more farms is wanted; (3) the measured milk yield should be corrected by the stand deviation factor if it is used for management purposes; (4) automated monitoring of milk meters op milk robot farms by a comparison of the measured milk with the delivered milk seems possible, provided that separated milk is recorded properly.

Masuring and modelling heat production by hen eggs

Hakimhashemi, M., Eren Özcan, S., Exadaktylos, V. and Berckmans, D., Katholieke Universiteit Leuven, Biosystems, Kasteelpark Arenberg 30, 3001 Leuven, Belgium

In this project it has been tried to find an efficient and convenient way to predict total heat production by eggs in incubators which can be considered as a representative for the comfort level of embryos inside the eggs during the incubation period. Two approaches have been applied to 19,200 eggs in an industrial scale incubator. In first approach, carbon dioxide concentrations at 12 positions inside the incubator and in the air inlet and outlet have been measured as well as velocity of incoming air. Carbon dioxide production by eggs has been then calculated by using gas balance equations and discrete time method. Oxygen consumption has been predicted from carbon dioxide production and an average respiration quotient. Respiration equation for embryonic growth in eggs has been then applied to calculate the amount of heat produced by eggs during the incubation period. The results showed an increase in the carbon dioxide and heat production by the growth of embryo. The embryos produced $0-0.03$ m^3/s carbon dioxide and between $0-3,000$ J/s of heat during 18 days of incubation. These results have been then compared to the results of previous studies for a single egg and they had almost the same trend and values during the incubation period. In the second approach the measurements of temperature and relative humidity at 12 positions inside the incubator and in the air inlet and outlet, cooling and heating provided to the incubator, and velocity of incoming air have been used to calculate total heat production from heat balance equations and by considering once a steady state and then discrete time method. The results obtained for total heat production were in the range of $0-2,000$ J/s for steady state and between $0-6,000$ J/s for discrete time, but since the calculations has been disturbed by many unknown parameters and unplanned difficulties (such as problems due to the inaccurate measurements by velocity sensor), so the results were not reliable. In the next step, heat production was calculated from only sensible heat balance and this time the results were in the range of $250-3,000$ J/s and the graph followed almost the same trend that was expected from the literature.

Dust reduction in poultry houses by spraying rapeseed oil
Aarnink, A.J.A., Van Harn, J., Winkel, A., De Buisonje, F.E., Van Hattum, T.G. and Ogink, N.W.M., Animal Sciences Group, Animal Production, P.O. Box 65, 8200 AB Lelystad, Netherlands

Different publications report effects of fine dust (PM10) in the ambient air on human health. For that reason the EU has set limits for PM10 concentrations. These limits are exceeded in parts of The Netherlands. It is estimated that agriculture is responsible for approximately 20% of the primary fine dust emission in The Netherlands, mainly originating from poultry and pig farms. Livestock farmers, especially poultry and pig farmers, are exposed to dust concentrations inside their animal houses that are a factor 10 to 200 times higher than in the outside air. The prevalence of respiratory problems in livestock farmers is a lot higher than in other occupations. The last 15 years a lot of work has been done on reducing dust by spraying a mixture of oil and water. This method proved to be very effective to reduce dust in animal houses at relatively low costs. Little work has been done until now on testing oil spraying systems in animal houses with bedding, despite these houses generally have the highest dust concentrations and the highest dust emissions. The objective of this study was to determine the effect of spraying oil in houses for broilers (floor housing) and layers (aviary system) on dust concentrations and emissions. Important questions within this study were: (1) How much oil is needed to achieve a certain dust reduction? (2) What should be the spraying interval? (3) What is the effect of spraying oil on production and health of the chickens? The study was done in 5 identical rooms for broilers and 4 rooms with aviary systems for layers. In the broiler rooms an automatic spraying system was installed. In the layer rooms oil was sprayed manually. In the broiler rooms rapeseed oil was sprayed at levels varying from 10-40 ml/m^2 at intervals of once every day and once every two days. In the layer rooms 20 ml oil per m^2 was sprayed daily. Concentrations and emissions of PM10 and PM2.5 were measured gravimetrically during 24 h. In broilers personal dust exposure was measured by the light scattering method and ammonia emissions were measured continuously with an NO$_x$ monitor. Daily gain, feed intake and footpad lesions were determined in broilers; egg production was determined in layers. Results in broilers showed a linear relationship between oil application rate and PM10 reduction, varying from 60-90% at oil application rates varying from 10-30 ml/m^2 per day. PM2.5 reduction was less related to application rate and was approx 70% for all treatments. Personal exposure to PM10 was reduced in a similar way by oil application. Dust reduction in layers was 34% for PM10 and 50% for PM2.5. Daily application of oil seems to give higher reductions than spraying once every two days (at the same total oil use). Ammonia emissions and production results were not affected by oil spraying. There was a higher risk of footpad lesions in broilers at the high oil application levels. It was concluded that oil spraying in bedding systems for broilers and layers is an effective method to reduce dust concentrations and emissions. The application rate should be limited to 20 ml/m^2 per day or less to prevent higher risk of footpad lesions.

Simulating the effect of forced pit ventilation on ammonia emission from a naturally ventilated cow house with CFD

Sapounas, A.A.[1], Campen, J.B.[1], Smits, M.C.J.[2] and Dooren, H.J.C.[2], [1]Wageningen UR, PSG, Bornsesteeg 65, 6700AA Wageningen, Netherlands, [2]Wageningen UR, ASG, Edelhertweg 15, 8219PH Lelystad, Netherlands; athanasios.sapounas@wur.nl

Atmospheric NH_3, mainly originates from agricultural sources, can cause serious environmental problems related to soil acidification and eutrophication. Emissions from dairy houses are 15% of total agricultural NH_3 emission. Due to open buildings, existing abatement options are limited. Pit air separation was identified as a potentially efficacious option. In this study a simulation model of a commercial dairy cow building with slatted floor is presented. The model was solved for 12 cases, differing wind speed, direction and both air and manure temperature. For each case three solutions were obtained correspond to a naturally ventilated building without suction system and with a suction system with capacity of 250 and of 500 m^3/h per cow respectively. The results show that due to forced pit ventilation system the ventilation rate was increased 3.1 and 6.2% at capacity of 250 and 500 m3/h per cow respectively. The contribution of the system to the total ammonia removed from the building during winter ranged from 31-35%, 16-19% and 11-8%, for wind speed of 1.0, 4.0 and 8.0 m/s respectively. Correspondingly, during summer, the contribution of the system ranged from 44-48%, 20-21% and 12-9%. Although obvious benefits arise from a forced pit ventilation system, the main mass flow of ammonia still emitted trough the building ventilation openings, specially at high wind speeds.

Development of new methods and strategies for monitoring operational performance of emission mitigation technology at livestock operations

Ogink, N.W.M., Melse, R.W. and Mosquera, J., Wageningen University and Research Centre, Animal Sciences Group, PO box 65, 8200 AB Lelystad, Netherlands

A wide variety of mitigation techniques has been developed for the abatement of gaseous emissions (ammonia, odor and fine dust) from livestock operations. In the Netherlands over the last fifteen years, an increasing number of low emitting housing systems and air purification techniques are applied at intensive livestock farms in order to comply with air quality regulations. A national regulatory system has been set up that makes use of a list of precisely defined low emitting housing systems and their emission factors. An important aspect of this system is the on farm verification by regulatory authorities of the true and effective application of mitigation techniques at livestock farms. In practice, mitigation options are admitted only to the regulatory list if the method allows an effective verification of its application and effective working on farm. Effective verification is mainly based on the principle of visual observation by regulators of installed technical hardware that is part of the applied mitigation technology. This verification prerequisite has determined to a large extent which categories of mitigation tools could be developed and implemented and which could not. As a result the potential of a number of effective mitigation tools, like those based on additives to feed and manure, so far have not been utilized in practice because verification of use cannot sufficiently linked to the installation of hardware. Another result is that the development of energy saving optimizations of air purification techniques by partial air cleaning is obstructed because its verification of proper use is considered to be too complicated by regulators. The objective of this paper is to describe new methods and strategies for the on farm verification of the proper working of mitigation technologies based on routine farm monitoring of process and output parameters. The adoption of these strategies in the regulatory framework should result in a much wider and better utilization of available mitigation options. After describing the evolution and state of the art of current verification strategies, the development of two alternative approaches will be presented in this paper and their potential use explored. Both approaches are based on intensive monitoring and continuous logging of the mitigation process. In the first approach data from key parameters that determine the effectiveness of the mitigation process are logged in a secured database and made available at any time to interested parties, including regulators. As an example, the paper describes the outline of newly developed monitoring and logging systems in air scrubbers. The second main approach is based on continuous monitoring and logging by sensors of air quality parameters in the outgoing ventilation air. This approach enables a fundamental turn in verification strategy by not focusing on technical means but only on outcome, thus leaving to the farmer complete flexibility in how to comply to air quality standards. The technical outlooks and current feasibility of such air quality monitoring systems will be presented. The paper concludes with a discussion of factors and points of views of stakeholders in current verification strategies, and outlines which steps have to be taken to utilize the proposed technical approaches.

Development and evaluation of two ISOagriNET compliant systems for measuring environment and consumption data in animal housing systems

Kuhlmann, A., Herd, D., Gallmann, E., Rößler, B. and Jungbluth, T., Universität Hohenheim, Institute of Agricultural Engineering (440b), Garbenstr. 9, 70593 Stuttgart, Germany

Measuring of environment and consumption data is important to evaluate animal housing systems. Unfortunately, data transmission to and data storage in a central unit as well as soft- and hardware communication within animal housing systems is limited due to incompatible data formats and protocols. Therefore, it is hard to use process data to improve animal health or production pocess. In order to unify and enhance communication between components, data formats and protocols are defined by the international standard ISOagriNET.The aim of the research project, 'Information and Data Collection in Livestock Systems' is the development of an ISOagriNET compliant IT system, a so called Farming Cell. On the one hand in this Farming Cell the appearing data are collected and evaluated to improve animal health, quality assurance or production process. On the other hand data exchange processes between components are defined and implemented. For setting up the Farming Cell commercial products, e.g. feeding or ventilation systems that meet the ISOagriNET standard exist, but components which link up individual sensors to acquire additional environmental parameters or consumption data are not available. Therefore, various sensors, measuring for example temperature, gas-concentration or power consumption have to be connected to the overall system and suitable hard- and software interfaces have to be developed. Furthermore a comfortable web based application to manage connected sensors is necessary. In this paper two different ISOagriNET conforming hardware approaches to link up sensors are described and evaluated: 1. The first one uses the TINI microcontroller from Maxim, which is a low-cost product for public consumers. 2. The second hardware basis is a product for commercial consumers called ILC 150 by Phoenix Contact. Both differ for example in acquisition costs, hard- and software layout, development process, reliability and consequently in the solution. Developing a reliable and robust solution for ISOagriNET compliant data acquisition is very complex. Limited computing power and memory of the used hardware require experienced software developer, whereas the necessary flexibility in terms of sensor connectivity requires knowledge about electrical engineering. As ISOagriNET conforming prototypes the developed solutions are an important contribution for testing this new standard and they demonstrate the possibilities of efficient data collection. The comparison of the developed solutions demonstrates appropriate areas of application, their advantages and disadvantages.

Dust emission factors and physical properties in order to develop dispersion models
Nannen, C. and Büscher, W., Institute of Agricultural Engineering, University of Bonn, Nussallee 5, D-53115 Bonn, Germany

The actual legislation establishes limit values for emissions and immissions of dust particles from animal barns. For the granting of juridical security in proceeding relating to permission with location and protection of animal barns, dispersion models are used increasingly to simulate immissions. To describe the transmission of dust particles it is necessary to know relevant physical parameters. In this project the aerodynamic properties of dust like shape factor, agglomeration and resuspension are determined. The analysis of particle distributions, desities and sedimentation velocities versus growing stages of animals and local climate conditions allows further the examination of physical characteristics of those dust particles. Together with these characteristics the data basis for dispersion modelling can be improved. The dust is classified into different size ranges measured by scattered light. For collection dust an Andersen-Sampler with eight stages is used. The dust samples were analysed by microscope. The shape factor has been calculated using length and area of the particles for different fraction size. In order to make systematic predictions about the sedimentation of dust in animal housing, it is advantageous to use different densities for different size ranges. Therefore, a sedimentation cylinder has been developed in an earlier project, which provides to measure the sedimentation velocity of dust particles. The results show that there are varities between the sedimentation velocity and aerodynamic behaviour of unisize particles from different animal houses and animal species. Furthermore, the other mentioned physicla properties such as agglomeration and resuspension will be determined in laboratory with a wind tunnel.

A wireless network for measuring rumen pH in Dairy cows

Goense, D.[1], Houwers, W.[1], Müller, H.C.[2], Unsenos, D.[3] and Wehren, W.[4], [1]Wageningen University and Researchcentre, Animal Sciences Group, Edelhertweg 15, 8219 PH Lelystad, Netherlands, [2]Fraunhofer-Institut, Mikroelektronische Schaltungen und Systeme, Finkenstraße 61, 47057 Duisburg, Germany, [3]ISIS IC GmbH, Handelsweg 1, D 46485 Wesel, Germany, [4]Landwirtschaftskammer Nordreinwestfalen, Landwirtschaftszentrum Haus Riswick, Elsenpaß 5, 47533 Kleve, Germany

Subacute rumen acidosis is a risk in Dairy production. This increases in high production herds due to high doses concentrate feed. Diagnoses is difficult as early symptoms are vague. Fluid pH is a clear early indication, but the fluid is difficult to sample. Wireless sensor technology provides the possibility to develop a pH measuring device which is permanently present in the rumen of a cow. The objective of a Dutch-German project is to develop a continuous pH monitoring system for dairy cows based on wirless sensor technology. The system includes the measuring device in the rumen, radio communication to pass the body tissue, a radio to reach a collecting point, data storage and real time presentation. The appropriate radio frequency to pass body tissue is found in laboratory experiments with artificial body tissue, as earlier experiments showed that 433 MHz radio's have difficulty with passing body tissue. The developed measuring devices are initially tested in ruminaly canulated cows. These cows were during a one week experiment also equipped with non wireless pH monitoring devices and gave the opportunity to compare pH measurements with those of research instruments. A radio working in the 124 kHz range is able to pass body tissue at low power levels. In theory a half AA size battery will last ten years when sampling every 2 minutes. Tests on cows showed that there is more data loss when the repeater is mounted on the collar in stead of on a leg, so power was set on a critical level. The pH sensor of an ISFET type is equipped with signal processing electronics and amplifiers. together with the processor, radio and antenna it fitted well in a bolus of 3.2 by 15 cm. The bolus was filled with a harsh-sand mixture for sufficient weight and staid well in the rumen. The repeater exists of two radio's, one to receive the 124 kHz signal and one to transmit the data to the collecting device in 2.4 GHz. The collecting device has a GSM modem and transmitted data is stored in a database and a web server provides the possibility for real time observation. The comparison between the one week measurements with the research devices and the wireless devices as well as data on drift will be presented in the full paper. Power must be increased to some extend to realize near 100 percent arrival at the repeater, which will reduce lifetime of the sensor, though this will survive the reliable measuring period of the sensor itself. Modern 2.4 GHz radio's with antenna's integrated on the processor board perform quit well in an animal house environment. Fluctuations in temperature are clear due to drinking of the cows, and fluctuations of pH are observed.

Measuring rumen pH and temperature by an indwelling and wireless data transmitting unit and application under different feeding conditions

Gasteiner, J., Fallast, M., Rosenkranz, S., Häusler, J., Schneider, K. and Guggenberger, T., Federal Reserach Institute Raumberg-Gumpenstein, Animal Welfare and Animal Health, Raumberg 38, 8952 Irdning, Austria

Subacute rumen acidosis is a common and economically important herd health problem of dairy cattle and there is a crucial need for monitoring systems. Therefore an indwelling wireless data transfer system for monitoring rumen ph and temperature was assembled. Measurement times were user selectable, in our trials measurements were taken every 30 minutes. Stored data were transmitted to an external receiver using ISM-Band (433 mHz). The system was controlled by a microprocessor. Data (pH, temperature) were sampled with an Analog to Digital converter (A/D converter) and stored in an external memory chip. The external receiver was connected with a PC (USB) and results were analysed by software and displayed. The indwelling system could be administered orally, but to service the measuring units, described experiments were conducted using 5 ruminally cannulated steers. After calibrating by using standardized pH-dilutions and check for proper operation, rumen-pH und temperature measures were carried out under different feeding conditions. In feeding experiment 1, 100% roughage (hay) ad. lib. was given to the animals for 1 week, measurements with the pH-monitoring unit were taken for 3 days. In feeding experiment 2, animals had daily pasture (from 4.30 a.m. to 16:30 p.m.) and forage ad lib. (from 17:00 p.m. to 4:00 a.m.) for 3 weeks. Measurements of rumen pH and temperature were carried out for the last 7 days. In feeding experiment 3, animals received a diet containing roughage:concentrate 50:50 for 7 days, measurements of rumen pH and temperature were taken for a period of 3 days. Statistical analysis was conducted using GLM (Statgraphic Plus 5.1) and Bonferroni-Holm-Test. In feeding experiment 1, rumen temperature (mean 38.40±0.70 °C) was influenced significantly by drinking water but it is not connected with feeding time. Mean pH was 6.49±0.39 and nadir was pH 6.14. In feeding experiment 2, mean rumen temperature was 38.12±0.80 °C and mean pH was 6.36±0.22. Nadir during pasture was pH 5.34, nadir during feeding roughage was 6.16. Pasture had a significant influence on rumen pH. In feeding experiment 3, mean rumen temperature was 38.55±0.83 °C and mean pH was 6.37±0.24. Nadir was pH 5.29. Decline of rumen pH was significantly related to the feeding of concentrate. When comparing results of measuring standardized dilutions (pH 4, pH 7) prior and after in vivo measurements, coefficient of correlation was 0.9987 (mean for all probes). Drift pH 4 was 0,197 ± 0,070 and drift pH 7 was 0,107 ± 0,088. Time (%) below pH 5.5 was 0 in feeding trial 1, 12.1% in feeding trial 2 and 3% in feeding trial 3. Time (%) below ph 5.8 was 0 in feeding trial 1, 18.9% in feeding trial 2 and 16.6% in feeding trial 3. Time (%) below pH 6.2 was 55.6% in feeding trial 1, 44.2% in feeding trial 2 and 53.5% in feeding trial 3. Maximum duration of measurement is up to 40 days, then a service (energy) of probes is necessary, transmission of data was trouble free. Results indicate, that the present method is a useful and proper tool for scientific applications, especially for a better understanding and defining of rumen acidosis. As the measuring system can also be administered to uninjured cattle, an adapted rumen pH measuring system will also be assembled for practical purposes in future.

Recording of tracking behaviour of dairy cows with wireless technologies

Ipema, A.H.[1], Bleumer, E.J.B.[1], Hogewerf, P.H.[1], Lokhorst, C.[1], De Mol, R.M.[1], Janssen, H.[2] and Van Der Wal, T.[3], [1]Animal Sciences Group of Wageningen UR, BU Animal Production, P.O. Box 65, 8200 AB Lelystad, Netherlands, [2]Alterra, Centrum Geo-informatie, P.O. Box 47, 6700 AA Wageningen, Netherlands, [3]Portolis BV, Spoorbaanweg 23, 3911 CA Rhenen, Netherlands

The trend of growing farm sizes is expected to continue in the coming decades. Increased herd sizes should not lead to less attention for the health and welfare of the individual animal. Moreover societal requirements for food safety and quality and for the production environment increase. Current developments in wireless sensor technology provide good possibilities for real time acquisition of information about the production environment (climate, weather, housing) and production factors (animal, food, pasture). Localization and tracking of individual animals also contains interesting information for operational management decisions. Examples are data about social interactions in a herd (ranking, grouping and seclusion behaviour) and frequency and duration of visits to interesting locations in barn or pasture (milking robot, feeding station, water trough, pasture border). A literature overview of available localization technologies will be presented. The application of wireless acceleration sensors for recording lying, standing, walking, feeding of grazing behaviour will be discussed based on literature and own experiments. In these experiments wireless sensor networks were used for acquiring on-line information of dairy cows kept in a barn with during certain hours at daytime access to a pasture. In a last experiment experiences with combining wireless localization (GPS) and behaviour (accelerometer) information, made available in a GIS application, will be discussed.

Estimating impact on clover-grass yield caused by traffic intensities

Jørgensen, R.N.[1], Sørensen, C.G.[2], Green, O.[2] and Kristensen, K.[3], [1]University of Southern Denmark, Faculty of Engineering (KBM), Niels Bohrs Allé 1, 5230 Odense M, Denmark, [2]Aarhus University, Department of Agricultural Engineering, Schüttesvej 17, 8700 Horsens, Denmark, [3]Aarhus University, Department of Genetics and Biotechnology, Blichers Allé 20, Postboks 50, 8830 Tjele, Denmark; rasj@kbm.sdu.dk

Traffic intensities have a significant influence on a range of crop and soil parameters. For grass and especially clover, the yield response is negative as a function of traffic intensity. During the growing season, conventional grass-clover production for silage experience high traffic intensities due to operations like fertilizing with slurry, cutting the grass, rolling the grass into swaths, and collecting and chopping the grass into trailers with a forage harvester. Normally, the traffic is distributed all over the field area during the growth season. In this way, the track impacts formed by the machines will influence the grass and clover growth and yield differently. As clover is known to have a higher feed value, the evaluation of the quantitative and qualitative affects on the combined clover-grass entity, the individual components must be determined. The objective of this paper was to measure yield affects on clover-grass as a consequence of different traffic intensities. The experiments were carried out in the context of a full scale field trial. A 14 hectare full scale grass-clover field trial with 24 different traffic intensities and 35 replicates was established. Each net parcel measured 9 x 1.3 m and the 24 treatments were randomized onto the 840 net parcels. The grass clover was established in spring 2007 using RTK-GPS auto steered tractors and implements. A Claas Axion tractor equipped with AutoFarm RTK AutoSteer and a 15 m^3 Kimadan slurry tanker on two axels, was used to perform the simulated traffic treatment on the parcels. The different traffic intensities are combinations of different tire pressure (1.0 and 2.5 bar), tire load (3,000 and 6,000 kg), time of year and number of passes (variating from 0 to 8). The harvesting procedure was preformed with a Haldrup plot harvester modified with RTK-GPS. This paper shows the initial results from measuring the yield affects.

Impelementation of herd management system with wireless sensor networks

Wu, T., Goo, S., Goh, H., Kwong, K., Michie, C. and Andonovic, I., University of Strathclyde, Department of Electronic and Electrical Engineering, 204 George Street, G1 1XW Glasgow, United Kingdom

The newly emerged wireless sensor network (WSN) technology has spread rapidly into various multi-disciplinary. Agriculture and farming is one of the industries which have recently diverted their attention to WSN, seeking this cost effective technology to improve its production and enhance animal health care standard. This paper reports on the application of WSN technology to cattle health monitoring. By monitoring and understanding cattle's individual and herd behaviour, farmers can potentially identify the onset of illness, lameness or other conditions which might benefit from early intervention. The sensor node, which is small in size and low in power consumption, shows significant potential in this context. However, WSN solution faces with a number of significant technical challenges before it can be suitably and routinely applied. This paper thus focuses on challenges that closely relate to data transportation from cattle mounted sensory devices, comprises of data protocol, power consumption, mobility, operational range, data transmission volumes and herd size. Challenges imposed from adaptation of WSN in agriculture and farming have been studied and evaluated. The main difficulty of adapting conventional WSN systems into cattle monitoring application is to support node's mobility that is caused by animal's movement. Although there have been diverse algorithms for data fusion and packet routing successfully applied in commercial product and a lot more being proposed by various researchers, majority of these schemes are only applicable to static sensor nodes. The key reason is most of the schemes are based on the assumption that the nodes will stay at a fixed position. Animal movement and herd mobility have been studied with the use of GPS collars in our research. The results are related to the time any individual animal will spend in proximity to a base station. This study considerably embraced the impact it has on the targeted network performance. A novel Virtual Connection Routing Protocol (VCRP) is also proposed. VCRP is a light weight and agile routing algorithm that can satisfactorily support frequent animal movement. It allows a number of data transmission paths generated dynamically and enables sensory data to be relayed to centre management system. In terms of the radio channel itself, the presence of large numbers of cattle can seriously impact on the channel quality. The use of antennae diversity has been studied as a means of optimising access to a base station. We present the results from models and farm based trials.

Could virtual fences work without giving cows electric shocks?

Umstatter, C.[1], Tailleur, C.[2], Ross, D.[1] and Haskell, M.J.[1], [1]Scottish Agricultural College, Sustainable Livestock Systems, Kings Buildings, West Mains Road, Edinburgh EH9 3JG, United Kingdom, [2]AgroParisTech, 16 rue Claude Bernard, 75231 Paris cedex 05, France; christina.umstatter@sac.ac.uk

Virtual fences could have many positive effects on livestock farming and nature conservation: reduction of fencing costs and maintenance costs, feasibility to fence in difficult terrain, increased flexibility in fencing, no adverse effects on wildlife to name just a few. However, the approaches at the moment do all include electric shock as aversive stimulus. This approach might have welfare implications and so-called shock collars for dogs are already banned in Wales and in some European countries. Therefore, our objective was to determine whether a sound cue could replace the electric shock as the aversive stimulus in a virtual fence. We exposed 8 Angus X Limousin cows in a pen to 23 different sounds in a random order and recorded the response of the group on a scale from 0 (low) to 5 (high). From those results we identified 9 sounds which gave the greatest alerting response. In the second experiment we used 7 of the cows for the treatment and one was used as a 'buddy' for the test cows. Each cow was tested individually, with a 'buddy' cow to prevent isolation-induced stress. A collar was put on each cow with loudspeakers attached to it and she was led with the 'buddy' into a test pen. A control sound was played at the beginning and the end of each session. Between this we played three different sounds in a random order with at least 2 minutes in between each sound. Again, the reaction of the cow was recorded using a scale from 0 to 5. Each of the 7 cows was exposed to the 9 sounds in three sessions. From these results we identified 2 sounds which gave the highest response. Sound 1 scored on average 3.7 and Sound 2 scored 3.1. The third best sound scored only 2.3. In comparison, the control sound scored 0.5. These two sounds were then used in a paddock trial on 8 naïve Angus X Limousin cows. We established a virtual fence across the centre line of a 40 X 20m paddock. Cows were habituated to the paddock in 2 trial sessions without any cues. There were two test weeks in which each test cow received a control session and two test sessions, one with each of the two test sound cues. All sessions were at least 30 min and all test cows were accompanied by a 'buddy' cow. Test cows wore a collar in all control and test sessions. In test sessions the cow was played the test cue as she approached the virtual fence. If there was no response the cue was repeated up to 5 times. In the control and test sessions, the position of the cow relative to the virtual fence was recorded. In the test sessions, the number of times the cow approached the line and response to the sound was recorded. The responsive cows either bolted forward or turned round. The success rate was defined by cows turning away from the virtual fence line after the cue was triggered. In this trial, 53% of the cows responded successful to the cues played. There might be a trend that cows with a higher percentage of Limousin genes respond better (61.3% cue success rate) compared to cows with higher percentage of Angus genes (40.3%). Sound 1 had a success rate of 49% whereas Sound 2 had a success rate of 58%. The results of this first pilot study were promising. However, more research needs to be done to develop a non-shock system, which could sufficiently replace fences. It might be, that virtual fencing is more successful with some breeds than others.

A method for managing cow traffic in a pastoral automatic milking system
Jago, J., Bright, K., Jensen, R. and Dela Rue, B., DairyNZ, Private Bag 3221, Hamilton 3240, New Zealand

Automatic milking systems (AMS) are a widely accepted technology in housed dairy systems where minimal grazing is practiced. By contrast, adoption of AMS has been slow in the established pastoral dairying countries of New Zealand, Australia and Ireland with only a few farms adopting the technology. Challenges include generating voluntary cow movement when the majority of feed is pasture (i.e. low supplementation in or near the AMS), achieving 24h utilisation of the AMS and managing clustering of cow traffic which is typical in grazing systems. This paper describes a method for managing cow flow in a pastoral dairy system using automatic milking technology. The system consists of one or more selection units (SU) located up to 500m from the dairy housing the AMS. Access to pasture, water and the AMS act as incentives for cows to enter the SU. Each SU comprises a set of pneumatically-powered drafting gates operated by an embedded controller (EC) with an associated radio frequency ID tag reader. Each EC communicates via radio with a central PC based software program located at the dairy. The SU can be powered either by mains, or solar panel with battery backup. When a cow presents at a selection unit the EC identifies the cow and transmits the cow's ID to the PC based software program 'SMARTY', which in turn determines the destination to which the cow should go, and transmits that destination information back to the embedded controller. The EC manages the operation of the gates in the correct sequence, monitors the movement of the cow through the gates, restoring the gates to their stand-by position after the cow has passed. Any following cow attempting to push through behind a drafted cow is blocked by the EC. The SU drafting gates direct the cow to the dairy or one of two grazing destinations, either the same area that she has just come from or a new grazing area. The cow is drafted to the dairy based on a calculated projection of her accumulated yield compared against a farmer-set target yield. Cows are granted access to the dairy based on a number of decision rules: 1. Does she meet the projected milk yield criteria? 2. How many cows are already queueing for milking? 3. What is the likelihood the cow will return to the SU if rejected for milking? A cow will be directed to the new or old grazing area depending on the time at which a new grazing area is made available relative to when she was last milked. The system is designed to manage cow traffic so as to maximise the efficiency of the AMS by preventing cows not ready for milking from reaching the dairy; minimising walking to and from the dairy; controlling individual cow milking frequency; prioritising cows to overcome issues of hierachy and dominance, and ensuring that cows gain even access to fresh pasture on a 24h basis. Grazing is managed by allocating three new areas of pasture daily, the timing of which is designed to assist with generating 24h use of the AMS. The system is operating at the DairyNZ Greenfield Research Farm. A 54 ha grazing area supports a herd of 180 cows that are milked using two AMS. The diet is 97% pasture and calving is seasonal. Results suggest that cows readily adapt to using the SU and that best practice pasture management can be achieved in a voluntary pastoral milking system.

Stakeless fencing for mountain pastures

Monod, M.O., Faure, P., Moiroux, L. and Rameau, P., Cemagref, TSCF, 24 avenue des Landais BP 50085, 63172 Aubiere, France

Protecting biotopes and maintaining biodiversity are ecological aims highlighted in European policies. Grazing systems which have to adapt to these new constraints require new fencing technologies. An effective livestock fence without stakes or barbed wire has been developed. Based on a battery or solar powered wire that delineates the boundaries of the pasture and works in connection with electrical receptors built into collars placed on the cattle, its use in mountain pastures has been tried and found to be a success. The principle is the following: a generator sends a small current through a coated, light-colored wire easily noticed by the animals. This creates a magnetic field that is sensed by a receiver in the collar of each cow. When it approaches within five feet of the wire, it then hears an audible signal that is entering the 'warning' zone. If it continues its course, it reaches the 'exclusion' zone, where a mild electric shock is generated in the neck region by the collar. This discharge depends on the position of the animal at the time and its previous behaviour. Based on a major behavioural study of cows, the system is able to analyze the reaction of cattle faced with different situations. A microprocessor built into the collar stores in memory the actions of the animal and adapts its response to the present situation. The system is completely self-contained during the high pasture season. The generator is fed from a battery or solar panels, while the collar is fitted with a set of standard batteries that last for one season. The wire can be on ground, vegetation, or even buried. It is insensitive to the presence of water and is no danger to humans. More than 800 days of experiment on 150 cattle of different breeds were carried out in France between 1999 and 2003. Bringing more flexibility in fencing, this new system could lead to more precise management of grazing in protected zones and allows for environmental renovation of wild areas.

Pastures from space: evaluating satellite-based pasture measurement for Australian dairy farmers

Eastwood, C.R.[1], Mata, G.[2], Handcock, R.[2] and Kenny, S.[3], [1]The University of Melbourne, Faculty of Land and Food Resources, Parkville, Victoria 3010, Australia, [2]Commonwealth Scientific and Industrial Research Organisation, Livestock Industries, Private Bag 5, Wembley, Australia, [3]Victorian Department of Primary Industries, Warrnambool, Victoria 3280, Australia; callumre@unimelb.edu.au

Australian dairy farms rely on grazing pastures as their primary and cheapest source of feed. Accurate and timely measurement of pasture biomass is integral for effective grazing management practice, however few Australian dairy farmers record objective pasture mass. A system using satellite imagery has been developed to measure pasture biomass at a paddock scale in Australia. The concept was evaluated through a 17 farm pilot trial over the 3 month spring growth period. The trial was evaluated in terms of technology fit with grazing management practice of participant farmers. Qualitative research methods, including semi-structured interviews and a group workshop, were used to ascertain participant views on issues such as trust, usefulness, and usability, within the context of farming systems. Results from the trial indicate that a technology such as satellite pasture measurement has potential application in Australian dairy farm systems. However the provision of data alone does not guarantee successful technology uptake. Support structures must also be provided to help farmers interpret the information within the specific context of their farm system. These support structures may include use of private agronomists, agriculture extension personnel, or associated software applications.

Cow body shape and automation of condition scoring

Halachmi, I.[1], Polak, P.[2], Roberts, D.J.[3], Boyce, R.E.[4] and Klopcic, M.[5], [1]Agricultural Research Organization, ARO, Institute of Agricultural Engineering, the Volcani Center, P.O. Box 6, 50250 Bet Dagan, Israel, [2]Slovak Agricultural Research Center, Research Institute for Animal Production, Nitra, Nitra, Slovakia (Slovak Republic), [3]SAC Dairy Research Centre, Crichton Royal Farm, Midpark House. Bankend Road Dumfries, DG1 4SZ Dumfries, Scotland, United Kingdom, [4]IceRobotics, Roslin BioCentre, Roslin, Midlothian, EH25 9TT Roslin, United Kingdom, [5]Biotechnical Faculty, University of Ljubljana, Dept. of Animal Science, Domzale, Domzale, Slovenia

The feasibility of including a body shape measure in methods for automatic monitoring of body reserves of cattle was evaluated. The hypothesis tested was that the body shape of a fatter cow probably is rounder than that of a thin cow and, therefore, may better fit a parabolic shape. An image-processing model was designed that calculates a parameter to assess body shape. The model was implemented and its outputs were validated against ultrasonic and thermal camera measurements of the thickness of fat and muscle layers, and manual body condition scoring (BCS) of 186 Holstein-Friesian cows. The thermal camera overcomes some of the drawbacks of a regular camera, the hooks and the tailhead nadirs of a thin cow diverged from the parabolic shape. The correlation between thermal camera's measurements and fat and muscle thickness was 0.47. Mean body condition scorings 2.18, 2.15, 2.23, no significant difference found across the assessment methods. Further research is needed in order to achieve fully automatic, accurate, body condition scoring.

Identifying changes in dairy cow behaviour to predict calving

Macrae, A.I.[1], Miedema, H.M.[1], Dywer, C.[2] and Cockram, M.S.[3], [1]University of Edinburgh, Veterinary Clinical Sciences, R(D)SVS, EBVC, Easter Bush, Roslin, Midlothian, EH25 9RG, United Kingdom, [2]Scottish Agricultural Colleges, Sustainable Livestock Systems Group, Bush Estate, Penicuik, Midlothian, EH26 0PH, United Kingdom, [3]Atlantic Veterinary College, Sir James Dunn Animal Welfare Centre, University of Prince Edward Island, Prince Edward Island, Canada

Dystocia can result in significant economic loss due to cow and calf deaths, as well as reduced milk production and reproductive performance. Experienced stockmen use judgements based on physical and behavioural changes to recognise when cows may be about to calve, and offer assistance when required. With large herd sizes, and large numbers of cows per stockman, individual attention is difficult. This study aims to identify consistent behavioural changes which could potentially be used to predict calving. The behaviour of twenty multiparous Holstein-Friesian cows housed in straw-bedded pens, under continuous fluorescent or infra-red lighting, was recorded for 24 hours prior to the calf being expelled and for a 24-hour control period during late pregnancy. Continuous focal observations from video recordings were used to quantify the frequencies and durations of behaviours during 6-hour periods. For each 6-hour period, Wilcoxon tests were used to examine differences between behaviour during the calving and control periods. The frequencies of lying and tail raising were the most useful indicators of calving, as they showed consistent changes in the final 6-hour period during calving. During this period, lying frequency (number of lying bouts/6-hour period) was significantly higher ($P<0.001$) at calving (median=13, inter-quartile range=9-17) than during late pregnancy (median=4, IQR=3-5), and all cows showed an increase of ≥ 2 bouts. The frequency of tail raising also increased significantly ($P<0.001$) during the final 6 hours before calving (median=35, IQR=27-55) compared to the control period (median=5, IQR=3-7). Durations of lying, walking, eating and ground-licking, and number of walking bouts, did not show consistent changes at the time of calving. This study has shown that counting transitions between standing and lying, or tail raises, could potentially be useful for predicting calving.

Combination of activity and lying/standing data for detection of oestrus in cows

Jónsson, R.I.[1], Blanke, M.[1], Poulsen, N.K.[2], Højsgaard, S.[3] and Munksgaard, L.[4], [1]Technical University of Denmark, Department of Electrical Engineering, DTU, Building 326, 2800 Kgs. Lyngby, Denmark, [2]Technical University of Denmark, Department of Informatics and Mathematical Modeling, DTU, Building 321, 2800 Kgs. Lyngby, Denmark, [3]University of Aarhus, Institute of Genetics and Biotechnology, Faculty of Agricultural Sciences, Research Centre Foulum, Blichers Allé 20, 8830 Tjele, Denmark, [4]University of Aarhus, Institute of Animal Health, Welfare and Nutrition, Faculty of Agricultural Sciences, Research Centre Foulum, Blichers Allé 20, 8830 Tjele, Denmark; rij@elektro.dtu.dk

The objective of this study is to derive an algorithm for detecting oestrus in dairy cows from measurements of activity and duration of lying/standing periods. Early detection of oestrus in cows is very important for modern highly efficient farmers. Visual detection of oestrus is a difficult task and requires highly skilled personnel. Even with experienced personnel, the success rate in visual detection can be relatively low. Modern dairy farms can have several hundred cows and with labour being expensive in most European countries, the trend is that there becomes less and less time for focusing on each individual animal. Therefore there is a need for alternative reliable and cheap methods of oestrus detection. The algorithm derived in this paper analyzes measurements of activity and measurements of lying/standing periods. The cows' activity is measured by a necklace attached sensor that returns an activity measurement consisting of an activity index for each hour. Lying period duration is measured by a sensor attached on the cow's hindleg. Diurnal activity variations in the activity measurements are identified for the ensemble and for the individual cows. An adaptive diurnal filter removes the average daily variation of the individual. Change detection algorithms are designed for the actual probability densities of the activity measurements, which are assumed to be Rayleigh distributed, with individual parameters for each cow. Activity variations that are shared for groups of cows in the herd are detected so as not to affect the oestrus detection. The lying/standing behaviour is modelled by means of a discrete event model. The discrete event model is constructed using automata theory. There are two inputs to the model, time of day and the cow's lying behaviour. The cow's lying balance is a quantity describing how much the cow has been lying for the preceding period. The balance is growing while the cow is lying and diminishing while it is standing. The cows' lying/standing pattern is modelled with two models, one describes the normal behaviour, the other behaviour during oestrus. The consistency of the cow's behaviour with the two models is checked for each observed event. If the behaviour is at some time not consistent with the model describing normal behaviour and at the same time is consistent with the model describing oestrus behaviour the discrete event model assumes oestrus behaviour. The detection results from the two models analyzing activity and lying/standing behaviour, are combined in a decision algorithm for pointing out cows in oestrus. This new approach of combining statistical change detection methods for activity with discrete event models proves advantageous and to give good results with respect to false alarm and missed detection statistics.

Mathematical optimization to improve cows' artificial insemination services

Halachmi, I.[1], Shneider, B.[2], Gilad, D.[3] and Eben Chaime, M.[2], [1]Agricultural Research Organization A.R.O., Agricultural Engineering Institution, P.O. Box 6. The Volcani Center, 50250 Bet Dagan, Israel, [2]Ben-Gurion University of the Negev, Department of Industrial Engineering and Management, Beer Sheva, 84105 Beer Sheva, Israel, [3]Sion Ltd, artificial insemination services, Hafez Haim, Hafez Haim, 79800, Israel

Engineering tools and mathematical optimization are applied in this study to plan the work of the agents of the cow artificial insemination service (inseminator) in Israel. Time is crucial in insemination as the chances of impregnation decline with increasing delay between the start of estrus and insemination. About 1,090 artificial inseminations of cows are performed daily in Israel. They involve 412 farms in 283 villages, and are performed by 29 inseminators; the work plan should balance the work load among the inseminators. To this ends, the working time of an inseminator in each village is required. Thus, a model to predict the working time in a village was developed. Subsequently, a mathematical optimization model was designed and solved, which aims to allocate customers to trips and to determine the itinerary of each trip so as to minimize total distance/time. The main benefits included a 21.4% reduction in total traveling time, and a 55% reduction in the difference between the lengths of the longest and shortest working days. Moreover, the longest delay in reaching an estrous cow is reduced from 7.6 to 5.9 h, i.e., by 1.7 h, which may increase the conception ratio by some 7%. In addition, the trade-off between work balance and total traveling time was studied.

A new generation of fertility monitoring in cattle herds

Balzer, H.-U., Kultus, K. and Köhler, S., Institut of Agricultural and Urban Ecological Projects, affiliated to Berlin Humboldt University, Faculty of Agriculture and Horticulture, Invalidenstrasse 42, D-10115 Berlin, Germany

An increasing problem in modern livestock production is the maintenance of animal health due to the growing livestock in the farm stables and the rising performance of the animals, which is at least connected with lost of money. Alternatives for the time-consuming individual observation of each animal are i.e. the utilization of a pedometer or video-based arrangements. All these systems have one thing in common: they only compass one single parameter and thereby implicate a lack of precision. At IASP – Institute Of Agricultural And Urban Ecological Projects affiliated to Berlin Humboldt University, Faculty of Agriculture and Horticulture – we developed a new methodology for animal monitoring in cooperation with different companies – especially to keep track of fertility and reproduction in cattle. Our system is based on constant measurements of eight psycho-physiological parameters and behavior parameters on animal skin. We created a neckband in which the sensor is implemented. The electrodes are newly developed coat-electrodes, which fix the sensor on the skin. The sensor sends the parameters wireless to a receiver connected with a PC, but it also has a small data memory. With the developed method we are able to identify two important states of female animals: 1. Point of ovulation for optimal insemination and 2. Pregnancy. Additional the farmers get simple information about the status of every animal in a traffic light principle by the software: green – everything is ok, yellow – observation is necessarily and red – call the veterinary. If he needs, the veterinary can get special information about every parameter, i.e. temperature. Every organism reacts on changes in its environment or by itself (= stress) with the modification of physiological regulation functions and tries to adapt. Regarding stress differentiation can be made between vegetative-emotional reactions (parameter skin resistance), vegetative-nerval reactions (parameter skin potential) and muscular reactions (parameter elektromyogram). Four different characteristic changes of regulation are observed by analyzing measured parameters (= stress types). With continuous monitoring of the different parameters it is possible to make evidence of fertility status of one organism and in sum also of the complete herd. The farmer can determine the exact point of insemination or see quite soon, when the conception is gone. This helps him to save a lot of money. Furthermore, it is possible to rate changes and react directly to minimize stress for the animals. Using special stress tests animals can be grouped. We observed that different stress types show varying performance. This monitoring system first time allows the user to give an objective statement about animal well-being - how they feel in special situations or how suitable the housing systems are.

First results of a large field trial regarding electronic tagging of sheep in Germany

Bauer, U., Kilian, M., Harms, J. and Wendl, G., Bavarian State Research Centre for Agriculture, Institute of Agricultural Engineering and Animal Husbandry, Prof.-Duerwaechter-Platz 2, 85586 Poing-Grub, Germany; Ulrike.Bauer@LfL.bayern.de

Results of different studies on electronic identification of sheep and goats showed that the electronic identification of animals is basically practicable but all these studies emphasized, that there are still certain weak points and questions which have to be clarified. These critical points should be solved before a compulsory electronic identification of sheep and goats is introduced in the EU. The aim of this trial was to investigate all available transponders (ear tags and boluses) and readers under typical German husbandry systems with different sheep breeds in order to work out implementation recommendations for farmers and administration. The project started in Sept. 2007, until publication results about tagging of approx. 10,000 animals will be available. During the field trial a total of approx. 10,000 sheep and goats of 27 breeds kept in five different husbandry systems on 28 farms in six federal states of Germany are tagged with 16 different electronic identification media (ear tags & ruminal boluses). Especially the tagging reliability and permanence, loss ratios and health affections of different identification media in different husbandry systems e.g. like herding sheep in mountain regions or in pasture regions with many bushes should be examined. For judging health affections e. g. inflammations caused by electronic ear tags, sheep ears were divided into 16 sections in order to find out if the position of the ear tags has an influence besides the design. Furthermore nine different wand-readers and management programs of different producers are tested. Until August 2008, 5,244 animals were tagged electronically and control readings after four weeks were conducted including judging ear affections by electronic ear tags. According to first results the position of piercing the ear tags has an influence on some health affections e. g. pressure marks on the ears. There are remarkable differences between electronic ear tags of different producers in health affections of the sheep ear. Fold-over ear tags cause less inflammations and provide more space for appropriate wound healing than button tags with trapeze-shaped pin-pieces. Until now ruminal boluses show good results regarding loss rate, health problems and application. First results demonstrate that it is possible to avoid mechanical caused affections of the sheep ear to a very high percentage (93% to 100%, depending on ear tag design) by piercing the ear tag in the correct section of the ear. Moreover ear size and bowing play an important role. Regarding loss ratios of the ear tags and permanence of the identification it has to be clarified if the position which shows less ear affection is also the best regarding loss ratios.

Using a wide electronic pop hole based on RFID-technology with high-frequency transponders to monitor the ranging behaviour of laying hens in alternative housing systems

Thurner, S.[1], Pauli, S.[1], Wendl, G.[1] and Preisinger, R.[2], [1]Institute for Agricultural Engineering and Animal Husbandry, Bavarian State Research Centre for Agriculture, Voettinger Strasse 36, 85354 Freising, Germany, [2]Lohmann Tierzucht GmbH, Am Seedeich 9-11, Postfach 460, 27454 Cuxhaven, Germany

The number of laying hens in alternative housing systems which offer a free-range area is increasing world wide. However, it has been found, that numerous hens of a flock do not enter the free-range area. In order to find the motivation for using the free range area and to optimize the design of the outside area, behaviour data of all hens in a flock should be automatically recorded during their entire stay in the laying barn. So far a narrow electronic pop hole (EPH) was available to accomplish this task. Anyhow the EPH with its height of 27 cm and width of 16 cm is not conform with the EU-regulation 1999/74/EC. Therefore a wide electronic pop hole (WEPH) based on high-frequency (HF-) transponders (13.56 MHz) with at least 35 cm height and 40 cm width should be developed and optimized. A prototype of a WEPH with a height of 35 cm and a length of 100 cm was designed in a way that allowed to change the width in three steps from 40 cm to 55 cm and 70 cm. Furthermore it was possible to use bars and a cavity to slow down the hens during the passage. Two HF-antennas (length: 62 cm; width: 23 cm) were placed 56 cm apart from each other on top of the WEPH. They were connected to a multiplexer who switched the signal from the long-range-reader between the antennas as fast as possible (switching time: 50 to 500 ms, depending on the number of transponders in the reading area of each antenna). All hens (in total 257 hens, whereof 181 hens could be used for further evaluations; age at begin of data recording: 40 weeks; experienced with the EPH) were tagged with a HF-transponder using a wing tag. For 5 months data from all passages were recorded and evaluated using self programmed software tools. The identification reliability was verified using video recordings. The identification reliability for a width of 70 cm was 97.6% (n=3,113 passages), respectively 99.3% for a width of 55 cm (n=606) and 99.8% for a width of 40 cm (n=582). It was not necessary to use bars or a cavity to slow down the hens in order to get good reading results. Only 3 hens of the 181 hens did not use the free-range area. After a fortnight the hens were well adapted to the WEPH and nearly 90% of the hens used the free-range area daily at least for one visit. The identification reliability for all tested widths of the WEPH was similar or higher when compared to the EPH. With a width of 40 cm the hens could not move simultaneously in both directions through the pop hole, while at a width of 55 cm or 70 cm this was possible. There were more hens using the free-range area with the WEPH compared to the EPH (e.g. Thurner and Wendl, 2005: only 50% of the flock with one visit per day). The WEPH proofed to be a reliable system for the automatic evaluation of the ranging behaviour of laying hens. Therefore data for the optimization of free-range areas and for breeding purposes can be collected with a minimum input of labour and at relatively low costs using the WEPH.

Using injectable transponders for sheep identification

Hogewerf, P.H., Ipema, A.H., Binnendijk, G.P., Lambooij, E. and Schuiling, H.J., Animal Sciences Group of Wageningen UR, Animal Production, Edelhertweg 15, 8219 PH Lelystad, Netherlands

A part of the Dutch small scale sheep and goat keepers and some of the professional keepers have, because the risks of inflammation after application and tags tearing out of the ears, objections against the use of ear tags. An alternative for the ear tag is the bolus transponder. However the size of the bolus does not allow application at birth, while breeders want to identify the lambs at the day of birth. The keepers would like to have approval for using injectable transponders as identification device. The Animal Sciences Group was asked to make the animal well being and food safety aspects around the use of injectable transponders comprehensible and to investigate if justified use of injectable transponders can be achieved. In an experiment 559 lambs at an age of about one month were injected with a glass encapsulated transponder (32x3,9mm). Transponders were injected in neck, armpit, groin or abdomen. The application was carried out by two persons (the farmer or a veterinarian). The person doing the application scored the easiness of application in 3 categories: easy, moderate, difficult. One of two observers scored the reaction of the animals during application in 4 categories: no, light, medium, strong. At a weight of about 30 kg the lambs were slaughtered. Once during the fattening period and at arrival in the slaughterhouse the presence of the transponder was checked by reading the ID number. Transponders not read were referred to as losses. During slaughtering the slaughter line position of the transponder recovery was recorded. The time needed for recovery was scored in 4 categories: <10, 10-30, 30-120, >120 s and the carcass damage was scored in 4 categories: no, light, medium, strong. Application in the groin was assessed as easiest, while abdomen and neck were hardest. The animal reaction during injection were strongest for neck and lowest for groin. The check during the fattening period showed losses of 9.0, 0.0, 6.6 and 15.7% for respectively groin, abdomen, neck and armpit, at the slaughterhouse losses were increased till respectively 10.0, 3.8, 8.3 and 17.4%. In the slaughtering process 50.0% of the transponders in the groin were removed from the carcass during skin removal and opening of the abdomen cavity. For the neck, armpit and abdomen these percentages were respectively 40.4, 13.0 and 5.0%. The other transponders mainly were further in the slaughter line recovered. Time needed for recovery was shortest for abdomen and groin and longest for armpit and neck. This was also caused by more cutting in the carcass necessary for the recovery of the transponder. In total 92.1% of the transponders were physically recovered. Losses in the slaughterhouse were caused because some transponders left the slaughter line with the head (neck transponders), the skin (groin and armpit transponders) and the abdominal content (abdomen transponders). At the end of the slaughter line it was possible to ascertain the absence of a transponder in the carcass with an almost 100% certainty. There was a clear effect of the operators on almost all assessments (easiness of application, animal reaction during application, losses and recovery of the transponders). It was concluded that because of the low losses and simple recovery of the transponders application in the abdomen offers the best perspectives. Fine & tuning and further testing of the application procedures and instructions for injecting transponders in the abdomen are necessary.

Application of RFID technology in cow and herd management in dairy herds in Canada

Murray, B.B.[1], Rumbles, I.[2] and Rodenburg, J.[3], [1]Ontario Ministry of Agriculutre Food and Rural Affairs, P.O. Box 2004, 59 Ministry Road, Kemptville, Ontario, K0G1J0, Canada, [2]CanWestDHI, 660 Speedvale Av. West, Suite 101, Guelph, Ontario, N1K1E5, Canada, [3]DairyLogix, 814471 Muir Line Road, RR#4, Woodstock, Ontario, N4S7V8, Canada

Canadian dairy producers have been required to identify all dairy animals with uniquely numbered ear tags since 1999. RFID eartags have been used beginning in 2004, but use of the radio frequency component both in the supply chain and on farm has been very limited. This technology is rapidly being applied in management of larger dairy herds of 1000 cows and up in United States, however there is a need to develop applications for freestall herds in the 100 to 500 cow range for the technology to gain acceptance in Canada. A trial was conducted to explore the application of RFID technology in cow and herd management in Canadian dairy herds. Two test herds were enrolled which were using computerized herd management software but all operations in the barn were carried out with paper and clipboard. Each herd was provided with: Approved RFID tags cow ID sufficient for 100% compliance, handheld RFID readers, pocket computer devices, and associated hardware and software including wireless communication. Before and after measurements of time and labour used for cow management activities were recorded. Tag issues, adaptability of herdsman's routine, and overcrowding were found to be factors in success or failure of application of this technology. Successful use of RFID technology resulted in time and labour savings, a 90% reduction in data entry and retrieval, reduced veterinarian time, and improved accuracy and protocol compliance. Savings indicated a one year payback on investment in hardware and software.

Electronic ear tags for tracing fattening pigs according to housing and production system

Burose, F.[1], Jungbluth, T.[2] and Zähner, M.[1], [1]Agroscope Reckenholz-Tänikon Research Station ART, Tänikon, CH-8356 Ettenhausen, Switzerland, [2]Institute for Agricultural Engineering, University of Hohenheim, 70593 Stuttgart, Germany

The Swiss Animal Disease Act makes it compulsory to label pigs with an ear tag no later than at the time of weaning. To enable complete traceability of the animals from birth to slaughter, it is essential than the ear tag remains on the animal. The animal must still bear the ear tag when it leaves the fattening farm in order to enable it to be positively identified at the abattoir, assigned to its individual slaughter result, and finally, traced back to the farm where it was born. The loss rate and functional reliability of three electronic as well as the official plastic ear tag are analysed on working farms in different Swiss housing and production systems. Piglets were labelled with different ear tags on 18 farms. Pig farmers were assigned to different categories according to their housing and production systems. Farms producing pigs for a label programme or according to the requirements of the Swiss Meat Quality Management scheme were assessed. This classification was deduced from the different housing requirements of the animals during the lactation, piglet-rearing and fattening periods. In addition, the farms could be subdivided into two production systems. We investigated farms working in closed systems (i.e. those which fattened their own piglets) as well as those specialising in at most two of the three production stages (piglet production, piglet rearing, pig fattening) in the fattening-pig production sector. The official plastic ear tag of the Tierverkehrsdatenbank (Animal Tracking Database) and three different electronic ear tags were tested. Besides standardised (ISO) ear tags, a prototype was used whose transponder contained an anti-collision algorithm, allowing the virtually simultaneous identification of several transponders by means of a single reading antenna. Each of the 600 animals in an experimental group had one ear tag applied. The loss rate of the plastic ear tags was checked visually, whilst the loss rate and functional reliability of the electronic ear tags were checked both visually and with mobile reading devices. Documentation took place during the production process, upon ear-tag application during the lactation period, at the beginning of the rearing and fattening stages, and before the animals were sold for slaughter. Initial analyses show the functional reliability of the electronic ear tags to be good. Ear-tag loss rates vary widely over the farms studied. Concrete findings and conclusions will be published in the full paper.

Real-time measurement of pig activity in practical conditions

Leroy, T.[1], Borgonovo, F.[2], Costa, A.[2], Aerts, J.-M.[1], Guarino, M.[2] and Berckmans, D.[1], [1]Katholieke Universiteit Leuven, M3-BIORES, Kasteelpark Arenberg 30, 3001 Heverlee, Belgium, [2]Università degli Studi, Department of Veterinary Science and Technologies for Food Safety, via Celoria 10, 20133 Milan, Italy

A way to tackle the problem of dust concentrations in pig housing is avoiding the peak concentrations caused by agile movements of the pigs. However, there is a need for sensors to measure pig activity continuously in an on-line way. An automatic system for continuously measuring activity levels of pigs from top-view camera images is developed. In parallel, the behaviour of the pigs in video recordings will be labelled by human experts. A relation is identified between the automatic measurements and the visual behaviour labels, enabling the automatic recognition of pig behaviour. For the experiments two adjacent pig pens were used (6.9 x 2.6 meters), each populated by about 15 pigs. An infrared-sensitive CCD camera was mounted 5 meters above the floor of the pen. Images were captured with a resolution of 768 x 586 pixels at 1Hz frame rate. Software was developed to measure the activity level of animals from the camera image. Four zones were defined in the image, each covering half a pen. Every second, the algorithm logged the camera image and the activity index for each zone, defined as the fraction of the floor space in the pen with pig motion in the camera image. The behaviour of the animals in each zone in every recorded video image was visually labelled as one out of five possible behaviour scores: 'no activity', 'fighting', 'biting', 'nuzzling' and 'feeding'. From the analysis of the automatically measured group activity index compared to the manual labelling a relation was found between the activity index and the behaviour types 'no activity', 'nuzzling' and 'feeding'. However, behaviour related to aggressiveness of individual pigs ('fighting' and 'biting') could not be detected from instantaneous measurements of the group activity index.

Active feeding control and environmental enrichment with call-feeding-stations
Manteuffel, G., Research Institute for the Biology of Farm Animals (FBN), Behavioural Physiology, Wilhelm-Stahl-Allee 2, 18196 Dummerstorf, Germany

Our aim was to develop a new technique of precise and animal friendly feeding for pigs in intensive housing. Using this approach, we further intended to supply the animals with an emotionally positive challenge that they can solve, as this can increase the animals experienced control over the environment. Such positive emotional stimuli are putatively able to increase animal health via effects on the immune system. Since pigs have well-developed auditory functions and a comparatively high cognitive capacity the system calls an animal living in a group with an individual acoustic summons to a feeding place. To be able to do so, each animal has to learn its individual jingle and has to be attentive during the feeding times to hear it. As a result the feeding setup serves as a 'cognitive enrichment'. It further allows to feed individual animals at certain amounts of feed on particular times. For the experimental evaluation of this approach groups of eight pigs at an age of 7 weeks were subjected to the test, which lasted until the 20th week. The experiment was repeated with 6 groups. For comparison a control group of equal age and size was kept without the experimental equipment. This consisted of four 'Call-Feeding-Stations' (CFS) integrated into the pen that played the jingles and delivered feed if the called animal entered. The temporal regime, the individual registration of the animals (ear tag transponder, transponder antenna in the CFS) and the protocol of the behaviour in response to calling was computer-controlled. In parallel various physiological parameters and further behaviour data (in the group and in a open-field as a new environmental context) were collected. The animals were pre-trained for their individual summons during 1 week where the jingle was coupled with feeding in a classical conditioning paradigm. When active calling started the pigs had an initial success rate (entering the CFS after calling) of 49% that increased to 83% within 4 days. Since the feed allowance was adjusted to 80% success the animals received their full ration from this time on and compensated for low feed intake within the first days by increasing the following rate to calling to 85%. The animals further displayed statistically significantly less maladaptive behaviour (belly nosing) compared to the controls, some significantly changed immune parameters and, as a direct measure of fitness, a significant better healing of standard wounds (experimental skin lesions). They fed undisturbed by group mates who had never associated the jingle of a called individual with feeding. In the open-field they displayed significantly less indicators of fearfulness. Further, some parameters of meat quality (e.g. drip loss) were improved. The results indicate that the CFS-technique integrated in pig management seems to have the potential to increase animal welfare. CFS can well be integrated in already existing systems, e.g. automatic feeders. They improve the capacity of management as individuals can be called so that competitive fighting when waiting for entrance to an automatic feeder can be prevented. This can even take into account different needs of different individuals. The emotionally positive challenge of acting for being reliably rewarded may also increase animal health and improve general behavioural traits and reduce the general stress level.

Automatic detection of pig vocalization as a management aid in precision livestock farming
Schön, P.C., Düpjan, S. and Manteuffel, G., Research Institute for the Biology of Farm Animals, Behavioural Physiology, Wilhelm-Stahl-Allee 2, 18196 Dummerstorf, Germany

The aim of sound analysis is to decode animal vocalizations by correlations with behaviour and physiology. The interpretation of the sounds of animals can deliver information on their emotional condition and, as a consequence, on well-being in a non-invasive manner. Sound analysis is also a tool for better supervising the conditions of animal housing. For example, incidences of insufficient feed, water, temperature or the occurrence of diseases can be detected. Unfortunately general techniques that produce good models of arbitrary vocalizations are missing because it can be hardly decided a priorily which features will be relevant for the exact characterization of an acoustical event. Hence, systems that are well adapted to a particular species and class of utterances may be the most straight forward approach to the acoustical monitoring of animals. Specific acoustic monitoring of a species' utterances is able to increase the quality of animal supervision and may be used for increasing the process quality. In the course of the further development of acoustic monitoring of pig housings further differentiation of stress vocalization is intended to output different types of stress experienced by the animals. As a result an immediate targeted elimination of stressors will be possible. Starting with a communication model, we developed a sound analysis library for the phonetic analysis of porcine calls. As a practical application we have developed the automatic sound classificator for the detection of aversive vocalization of pigs (Stress Monitor and Documentation Unit-STREMODO) working under practical farming condition since it filters selectively porcine stress utterances from ambient noise. Further laboratory procedures were developed to classify various stress vocalizations in different stressing situations. They allowed finding minute differences in vocalization resulting from acute pain, fear, frustration or competition for feed. STREMODO detects stress vocalizations of pigs with a similar accuracy as expert human listeners. In various applications it found significantly increased stress vocalizations during feeding when the animal to feeding place ratio was high or temperature was too low. We further got the result that, in a poor housing environment without access to enrichment, the area of the pen (large vs. small 0,44 to 1,06 qm per animal) did not result in significantly different stress vocalizations. Vocalizations of pigs where found to differ when the animals were subjected to various stressors. The differences were displayed as small but significant shifts of main energy and altered degrees of noisy components in the spectrum of the utterances. It has been shown that sophisticated automatic systems for vocalization analysis can be developed when suitable sound models and fast analyses are combined. The resulting data can easily integrate into a comprehensive monitoring system. They can be used as components indicating deviations from optimum housing and management. Further, analyzing momentary vocalization is an additional tool to supervise transportation and slaughtering. Extended knowledge on the fine structure of porcine vocalizations and the integration into automatic detection systems has the potential to result in information tools for farmers that may help reaching a better housing, less stressed animals and, as a result, a better process and product quality as well as enhanced animal welfare.

FutureFarm: the European farm of tomorrow

Blackmore, B.S., UniBots, The Cottage, Wardhedges, Flitton, Bedfordshire MK45 5ED, United Kingdom

The European Farm of Tomorrow called FutureFarm is an EC funded 3 year project involving 4 farms with 15 partners from 10 countries. This project aims to address such wide issues of development from public awareness, decision support, compliance, socio-economic impact, energy efficiency and robotics. The core technology will be to take the already extensive experience in Precision Agriculture research and integrate it into a practical farmer based system. In the future European farmers will have to effectively manage information on and off their farms to improve economic viability and to reduce environmental impact. All three levels, in which agricultural activities need to be harmonized with economical and environmental constraints, require integrated ICT adoption: (1) improvement of farm efficiency; (2) integration of public goods provided by farming into management strategies; (3) relate to the environmental and cultural diversity of Europe's agriculture by addressing the region-farm interaction. In addition, the communication between agriculture and other sectors needs improvement. Crop products for the value added chains must show their provenance through a transparent and certified management strategy and farmers receiving subsidies are requested to respect the environment through compliance of standards. To this end, an integrated information system is needed to advise managers of formal instructions, recommended guidelines and implications resulting from different scenarios at the point of decision making during the crop cycle. This will help directly with making better decisions as the manager will be helped to be compliant at the point and time of decision making. In FUTUREFARM the appropriate tools and technologies will be conceptually designed, prototypes developed and evaluated under practical conditions. Precision Farming as well as robotics are very data intensive and provide a wealth of information that helps to improve crop management and documentation. Based on these technologies a new Farm Information Management Systems (FMIS) will be developed.

Management strategies and practices for precision agriculture operations

Fountas, S.[1], Pedersen, S.[2], Blackmore, S.[1], Chatzinikos, A.[1], Sorensen, C.[3], Pesonen, L.[4], Basso, B.[5] and Nash, E.[6], [1]University of Thessaly, Fytoko str., 38446 Volos, Greece, [2]Copenhagen University, Rolighedsvej 25, 1958 Frederiksberg C, Denmark, [3]Aarhus University, Schuttesvej 17, 8700 Horsens, Denmark, [4]MTT, MTT, FI-31600 Jokioinen, Finland, [5]University of Basilicata, Via Ateno Lucano, 10 Potenza, Italy, [6]University of Rostock, Universitätsplatz 1, 18051 Rostock, Germany

Developing codes of good farming practice, diversifying markets and production systems require implementation of more elaborate management strategies. These have to respect specific ecological conditions, demands from the rural regions and those from the value-added chains. On top of that, these strategies have to be simple, but flexible enough to be adapted easily to changing economic or environmental conditions and they need proof of their compliance with standards and other codes of practices. Many farmers claim that their main purpose is to make money, but that is too simplistic, sustainability and environmental issues are becoming more important. Practices are the management options that apply to a particular sector. Operational elements describe the particular field operations. To be able to manage these forms of complex variability, a co-ordinated approach is required, integrating personal strategies, farm practices and field operations. Firstly, the strategy must be identified as one which reflects the values and preferences of the manager. Subsequently, the practices should be chosen that lie within the values set out by the strategy. The field operations then carry out what is required to meet the goals of the practice. There are three levels of decisions that can be made; strategic, tactical (practice) and operational. Strategic decision making will only occur occasionally from year to year and will affect the strategic goal of the farm enterprise and it is likely to be personal in terms of the individual manager. There are many different factors that must be taken into account when formulating a strategic approach, including balancing the economic returns with the environmental impact and amount of risk envisaged. Other strategies could include such concepts as: best management practices; integrated crop management and integrated pest management; minimal crop risk; minimal financial risk and environmental protection. Eight site-specific operations have been explored. These are variable rate (VR) fertilizing; VR spraying; VR cultivation; VR seeding; automated steering and optimized route planning; optimal harvesting in terms of logistics, time and fleet management; chlorophyll content measurement before harvesting to optimize harvest procedures; and finally site-specific management of area subsidies. Within these operations individual strategies and practices have been identified.

Can compliance to crop production standards be automatically assessed?

Nash, E.J.[1], Fountas, S.[2] and Vatsanidou, A.[2], [1]Rostock University, Institute for Management of Rural Areas, Justus-von-Liebig-Weg 6, 18059 Rostock, Germany, [2]University of Thessaly, School of Agricultural Science, Fytokou St. N. Ionia Magnisias, 38446 Volos, Greece

Crop production standards are playing an increasingly important role in agriculture. As well as the minimum standards of good farming practice which are applicable to all farmers and which may be enforced through the principle of cross-compliance (i.e. reduced subsidy payments for failure to comply), there are many voluntary standards to which farmers may subscribe. Legally-regulated standards such as the EU Organic standards or private standards such as GlobalGap may be used variously to increase the market value of the crop as a prerequisite by buyers. Compliance to standards is generally controlled by a mixture of inspections and self-assessment against the criteria defined by the publishing body. These criteria may be published in the form of one or more checklists (e.g. for GlobalGap) or as a text stating the requirements (e.g. for EU Organic). In this paper we will consider the potential for an automated compliance assessment against these published criteria based on data which may be collected and held within the FMIS or made available to the FMIS from third parties, e.g. via a web service interface. Apart from the obvious benefit of providing a consistent and impartial assessment, a further motivation for such automating compliance checking is the desire to be able to optimise decision-support within the FMIS to take account of the applicable standards when planning operations. The implementation of an automatic assessment of compliance has three prerequisites. Firstly, it must be possible to express the requirements of the standard in an unambiguous and consistent logical form suitable for interpretation by computer. Secondly, compliance to the requirements must be capable of being automatically assessed. In practice this means that compliance can be determined using a numerical or logical comparison rather than a value judgement. Finally, the data required to assess compliance and meet documentation requirements must be available to the FMIS. In this paper, the requirements defined in current standards for crop production will be assessed against these three criteria. This assessment will consider not only data that is currently widely available but also data that should become widely available in the foreseeable future such as process data collected using on-board systems, tracking data from RFID chips and secure electronic certification. Where appropriate, suggestions will be made as to where clarifications to the standards may be needed to enable an automated assessment as envisaged in this paper.

Typology of precision farming technologies suitable for EU-Farms

Schwarz, J., Werner, A.B. and Dreger, F., Centre for Agricultural Landscape Research (ZALF), Institute for Land Use Systems, Eberswalder Str. 84, D-15374 Muencheberg, Germany

In the future European farmers will have to increase the efficiency to manage information on and off their farms in order to hedge economic viability and to reduce environmental impact. Precision Farming (PF) is seen as a collection of techniques and specific equipment as well as integrated concepts and is providing a wealth of data and tools for information management. Second generation of PF-technology is available on the market and has a large potential to foster crop production and documentation, especially by using adapted farm management information systems (FMIS). Still missing are standards and basic solutions for FMIS that fit best to the needs of farmers. The EU-funded collaborative research project FutureFarm (www.futurefarm.eu) is aiming to provide suitable background and prototypes for FMIS. Due to the variety of farm types, crops, equipment, growing regions and 'management cultures' within Europe, it will not be possible to have a general standard or one fixed technical solution for information management on EU-farms in the future. Therefore work in FutureFarm on FMIS will be projected on a typology of precision farming settings. Objective of the work at hand is to produce a generic PF-typology considering general patterns and dynamics in the crop production of European farms. The complexity of PF-technologies which possibly can be used for different crops, cropping measures, sites, farm conditions and production goals is structured systematically within a multi-dimensional typology. The typology focuses on archetypes of PF-technologies and their typical variation within techniques and procedures of production patterns for crops in European regions as well as on farm types. Analyses on the existing crop production structures (crops and farm types) in the EU are used to determine the main driving factors and partly the spatial patterns in the EU and their determinants. For this data from EUROSTAT are used, supported by those from literature. Relevant techniques and concepts in the precision farming technology were studied at relevant exhibitions or farmers' fairs and supplemented through expert interviews. In a first attempt, as main discrimination factors of a precision farming typology were used: (a) techniques for PF: (1) exclusively GPS (e.g. soil sampling, auto tracking), (2) offline (e.g. using maps for VRT), (3) online (using sensors for VRT), or (4) hybrid (combination of online and offline); (b) starting levels in equipment: (1) 'mini' (e.g. low cost PDA and GPS), (2) 'compact' (sensors and DGPS) or (3) 'combination' (automatic steering systems with RTK GPS); (c) farm types and farming structures in the EU (sizes, main crops, …). The PF-typology for European farms is a necessary precondition to reduce the number of possible combinations of farming situations in the EU. In the next step of the project it will be discussed with the FMIS-developers, whether a further aggregation of types is necessary. The PF-typology for European farms is necessary to transfer the project's results from one or few prototypes of the FMIS developed for normative farming situation to the actual and especially to future situations of farms in the EU. These situations can not be analyzed directly. The four practical farm cases of FutureFarm serve as a basic framework for generalisation procedures.

Decision tree induction as an automated detection tool for clinical mastitis using data from six Dutch dairy herds milking with an automatic milking system

Kamphuis, C.[1], Mollenhorst, H.[1], Feelders, A.[2] and Hogeveen, H.[1,3], [1]Utrecht University, Department of farm animal health, Marburglaan 2, 3584 CN Utrecht, Netherlands, [2]Utrecht University, Department of information and computing sciences, Padualaan 14, 3508 TB Utrecht, Netherlands, [3]Wageningen University, Business economics group, Hollandseweg 1, 6706 KN Wageningen, Netherlands

This study explored the potential of decision tree (DT) induction for automated detection of clinical mastitis (CM) in an automatic milking system. Sensor data (including electrical conductivity, colour and yield) of almost 600 cows and over 670,000 quarter milkings were collected from December 2006 till August 2007 at six Dutch dairy herds milking automatically. Farmer recordings of quarter milkings that showed clear clinical signs of mastitis were considered as gold standard positive cases (n=100). Quarter milkings that were checked but scored as being visually normal were considered as gold standard negative cases (n=299). Randomly chosen quarter milkings that were not visually checked and that were outside a 2-week range before or after a gold standard positive case and that were not separated automatically or manually were added to the pool of gold standard negatives to end up with 3,000 gold standard negative cases. Several DT algorithms with varying confidence factors and costs matrixes were implemented to study their effect on performance estimates. The time-window in which an alert was assigned as being a true positive one was narrow: only when the algorithm alerted directly before a gold standard positive quarter milking this was considered a true positive one. Using 10-fold cross validation, the detection performance of a DT was estimated. Performance estimates that were evaluated, at a threshold value of 0.50 for probability estimate for CM, included the sensitivity (SN), the specificity (SP), the number of false positive alerts per 1000 quarter milkings (FAR), and the positive predictive value (PV+). Receiver-operating characteristic (ROC-) curves were constructed to visualize all potential combinations of SN and SP of different DTs. Both the total and transformed partial (for a SP range of 97-100%) area under the curve (AUC) were used to summarise and to compare the diagnostic ability of the different models. Detection performances are comparable with the performance of models used currently by AM systems. As costs for false negative alerts increased and confidence factors decreased, SN an FAR increased to a maximum of 54% and 67.2 using a DT with costs for false negative alerts of 100 and a confidence factor of 0.05. SP and PV+ of this model were 93% and 20.4%. However, this model did not show the highest total AUC (0.7779) which was found for a model with costs for false negative alerts of 20 and a confidence factor of 0.05. When looking at the range of SP of interest (≥97%), the highest partial area (0.6457) was found for a DT with equal costs and a confidence factor of 0.05. As it was possible to achieve these results with the use of noisy data and a narrow time-window, results suggest that DT induction shows potential for detecting CM using AM sensor data.

Mastitis detection: visual observation compared to inline, quarter and milking SCC

Mollenhorst, H.[1], Van Der Tol, P.P.J.[2] and Hogeveen, H.[1,3], [1]Faculty of Veterinary Medicine, Utrecht University, Dept. of Farm Animal Health, Marburglaan 2, 3584 CN Utrecht, Netherlands, [2]Lely Industries N.V., Research Dept., Software Engineering, Weverskade 110, 3147 PA Maassluis, Netherlands, [3]Wageningen University, Business Economics, Hollandseweg 1, 6706 KN Wageningen, Netherlands; H.Mollenhorst@ uu.nl

Every farmer using an automatic milking system (AMS) needs equipment to be able to monitor udder health. For the control of mastitis, several indicators are available. Although much research has been done on electrical conductivity (EC), not much knowledge is available on the value of other indicators. The aim of this experiment was to explore several indicators for somatic cell count (SCC) measured during a milking that can be used best for mastitis control. SCC indicators were used alone or in combination with EC. Data was collected at three farms with an AMS. Four indicators were measured for each quarter; two squirts (3rd and 4th) for visual observation, inline SCC (ISCC), quarter SCC (QSCC) and EC attentions. Furthermore, composite milking SCC (CSCC) was measured at cow level. QSCC and CSCC were determined in a laboratory using Fossomatic. ISCC was measured using a method based on viscosity of milk mixed with a reagent. Visual observations were categorized into normal (n=3186) and abnormal (n=22). SCC was categorized in low and high, with thresholds set at 500 kcells/ml for quarter and 200 kcells/ml for composite milking values. SCC data was normalized by log-transformation to calculate correlations. The correlation between ISCC and QSCC was 0.51 (n=1030) and 0.45 (n=372) for the most interesting interval (200-2000 kcells/ml). The correlation between the average QSCCs and CSCC was 0.87 and was not affected by quarter yield correction. The sensitivity / specificity of ISCC, QSCC and CSCC compared to visual observation was 55 / 94, 50 / 92 and 58 / 76%, respectively, probably partly explained by the timing of observation and sample taken during the milking. Combining SCC indicators with EC attentions could increase specificity, however, at the expense of ability to detect visual abnormal milk (lower sensitivity). Changing threshold values for EC attentions and SCC parameters, however, can improve performance considerably, resulting in sensitivities in the same range as SCC alone, while specificity improves to about 97%. Although the correlation between ISCC and QSCC was not very high, ISCC showed the best performance when compared to visual observation, especially outperforming CSCC. When SCC indicators are combined with, e.g., EC indicators, there seems to be potential for improvement of current detection algorithms through using inline SCC.

Inline SCC monitoring improves clinical mastitis detection in an automatic milking system

Hogeveen, H.[1,2], Kamphuis, C.[1], Sherlock, R.[3], Jago, J.[4] and Mein, G.[5], [1]Utrecht University, Department of farm animal health, marburglaan 2, 3584CN Utrecht, Netherlands, [2]Wageningen University, Business economics group, Hollandseweg 1, 6706KN Wageningen, Netherlands, [3]Smartwork Systems Ltd., P.O. Box 36-515, Christchurch 8146, New Zealand, [4]DairyNZ Ltd., Private Bag 3221, Hamilton, New Zealand, [5]Sensortec Ltd., P.O. Box 11004, Hamilton, New Zealand

This study explored the potential value of in-line composite SCC (ISCC) sensing as a sole criterion, or in combination with measurements of quarter-based electrical conductivity (EC) of milk, for automatic detection of clinical mastitis (CM) during automatic milking. Data, generated from a New Zealand research herd of about 200 cows, milked by 2 automatic milking systems during the 2006/07 milking season, included: EC; ISCC; monthly laboratory-determined SCC and observed cases of CM that were treated with antibiotics. Milk samples for ISCC and laboratory-determined SCC were taken sequentially at the end of a cow milking. Both samples were derived from a composite cow milking obtained from the bottom of the milk receiver. Different time-windows were defined in which true positive, false negative and false positive alerts were determined. Quarters suspected of having CM were visually checked and, if CM was confirmed, sampled for bacteriological culturing and treated with an antibiotic treatment. These treated quarters were considered as gold standard positives for comparing CM detection models. Alert thresholds were adjusted to achieve a sensitivity of 80% in three detection models: using ISCC alone, EC alone or a combination of these two. The 'Success Rate' (also known as the positive predictive value) and the 'False Alert Rate' (number of false positive alerts per 1,000 cow milkings) were used to evaluate detection performance. Normalised ISCC estimates were highly correlated with normalised laboratory-determined SCC measurements (r =0.82) for SCC measurements above 200×10^3cells/ml. Using EC alone as a detection tool resulted in a range of 6.9-11.0% for Success Rate, and a range of 4.7-7.8 for the False Alert Rate. Values for the ISCC model were better than the model using EC alone with 12.7-15.6% for the Success Rate and 2.9-3.7 for the False Alert Rate. Combining sensor information to detect CM, by using a fuzzy logic algorithm, produced a 2 to 3-fold increase in the Success Rate (range 21.9-32.0%) and a 2 to 3-fold decrease in the False Alert Rate (range 1.2-2.1) when compared to the models using ISCC or EC alone. Results suggest that the performance of a CM detection system improved when ISCC information was added to a detection model using EC information.

Use of a cow-specific probability of having clinical mastitis to determine the predictive value positive of automatic milking systems
Steeneveld, W.[1], Barkema, H.W.[2] and Hogeveen, H.[1,3], [1]Utrecht University, Department of farm animal health, marburglaan 2, 3584 CN Utrecht, Netherlands, [2]University of Calgary, Department of Production Animal Health, T2N 4N1, Calgary, Canada, [3]Wageningen University, Business Economics, Hollandseweg 1, 6706 KN Wageningen, Netherlands

An automatic milking system (AMS) generates mastitis attention lists that mention those cows and quarters likely to have clinical mastitis (CM). Unfortunately, a general complaint of dairy farmers working with AMS is the relatively low positive predictive value resulting in a high number of false-positive alerts on these lists. The positive predictive value of AMS is based on the sensitivity, specificity and an equal probability of having CM for all cows on a farm (prevalence). However, all cows on a farm are different and will have different probabilities of having CM. This probability is dependent on for instance parity and somatic cell count history. It is expected that the accuracy of detection of CM with AMS can be improved when cow-specific probabilities of having CM are used instead of an equal probability for all cows. The cow-specific probabilities of having CM can be based on prior knowledge of a cow. With the cow-specific probabilities of having CM, an updated positive predictive value can be determined. The goal of this study was to develop a model that can determine the cow-specific probability of having CM based on prior knowledge of a cow. Sensitivity and specificity values of AMS, obtained from literature, were used to illustrate the difference between positive predictive values obtained with equal probabilities and cow-specific probabilities of having CM. A tree-augmented naive Bayesian network (TAN) was developed to determine the cow-specific probability of having CM based on prior knowledge of a cow. The graphical structure of the TAN was based on expertise, while the conditional probability tables were based on a dataset including 22,860 cows from 274 farms with information on having CM and the cow factors parity, month in lactation, season of the year, somatic cell count and CM history. Subsequently, with a sensitivity and specificity of AMS (obtained from literature) and the cow-specific probability of having CM, an updated positive predictive value was calculated. As illustration, using a sensitivity of 50% and a specificity of 99% and a probability of having CM for all cows on a farm of 0.0177 per 30 days in lactation resulted in a positive predictive value of 0.47. While the positive predictive value varied between 0.17 and 0.97 when cow-specific probabilities of having CM were used. With this information the interpretation of sensor outputs is improved. For instance, it is much more important to control a cow with a positive predictive value of 0.97 than a cow with a positive predictive value of 0.17. Results indicate the potential use of adding prior knowledge of a cow to update the positive predictive value of AMS. With the presented method all available information of a cow (sensor information of AMS and prior knowledge of a cow) is used to determine the positive predictive value of AMS. In the future attention lists, generated by the AMS, can be based on the updated positive predictive values to detect CM. Based on these updated positive predictive values a dairy farmer can interpret the attention lists better: cows with the highest priority for controlling for CM will have the highest positive predictive value and cows with the lowest probability of having CM will have a low positive predictive value.

Analysis of external drivers for farm management and their influences on farm management information systems

Charvat, K. and Gnip, P., Wirelessinfo, 784 01 Litovel, 784 01 Litovel, Czech Republic; charvat@wirelessinfo.cz

This report analyses external drivers for farming and their influence on the Farm Management Information System on the base of literature review and stakeholders opinion. The provided analysis analysed a number of external drivers and grouped them according to their importance and potential influence on future development of agriculture. The primary group of drivers, which will be key issues for future FMIS, are: (1) climate change; (2) growing population; (3) energy cost; (4) urbanisation and land abandonment; (5) aging population and health problems; (6) ethnical and cultural changes. These drivers have to be analysed for future system design independently, because they influence each other and it could have in some cases opposite influence on a number of factors. The group of drivers, where there is not clear evident of their influence, because in some cases it is difficult to predict, what will happen are: politicians, press, international organisations. Important driver, which depends on primary drivers, but, which is relatively independent to other drivers is: Food quality and safety. The two groups of drivers, where all drivers in group have strong synergy are: (1) knowledge based bioeconomy; (2) research and development; (3) information and communication; (4) education; (5) investment. And in the second strong synergy group are categorized: (1) partnerships; (2) cooperation and integration and voluntary agreements; (3) development of sustainable agriculture; (4) valuation of ecological performances. Last group with relatively high synergies are: (1) subsidies; (2) standardisation and regulation; (3) national strategies for rural development; (4) economical instruments.

Potential savings and economic benefits in arable farming from better information management

Pedersen, S.M.[1], Fountas, S.[2], Sørensen, C.G.[3], Pesonen, L.[4], Basso, B.[5], Ørum, J.E.[1] and Blackmore, S.[2], [1]Institute of Food and Resesource Economics, University of Copenhagen, Rolighedsvej 25, 1958 Frederiksberg, Denmark, [2]University of Thessaly, CERETETH, Technology Park of Thessaly, 1st Industrial Area, GR 385 00, Volos, Greece, [3]University of Aarhus, Schüttesvej 17, 8700 Horsens, Denmark, [4]MTT, Agrifood Research Finland, Plant Production Research, FI-31600 Jokioinen, Finland, [5]University of Basilicata, Department of Crop Systems, Forestry and Environmental Sciences, Via Ateno Lucano, 10, Potenza, Italy

The economic consequences of implementing new Farm Management Information Systems (FMIS) imply a comparison between a present situation without FMIS and a new situation with FMIS. What causes the economic benefit depends on many parameters that can be difficult to distinguish and separate from each other? Better ICT and sensing systems may not 'in itself' provide better quality, improved yields or reduced factor costs - but it may enable other technologies to be implemented and used in an intelligent way. The objective of this study was to assess farmers' management of different information management systems (incl. precision farming systems) and other ICT systems in conventional cropping rotations for a number of farming systems in Europe. For these farming systems we have established the cost structure and potential benefit for various farm information management systems and operations. Focus has been put on commonly produced crops based on four modern demonstration farms where current advanced information, precision farming and autonomous systems are implemented. The study includes an economic assessment of farmers saving of labour and use of factor inputs to the farmers' different crops and farm operations. The cost of different information-intensive and safety systems (incl. sensors, safety controllers, navigation controllers, software modules etc.) are estimated and compared to conventional systems. In addition, potential benefits related to these systems such as: labour savings, yield response, fuel savings and higher work quality (accuracies) are estimated and used to quantify the factor productivities for specific field operations. This study indicates that there is a economic potential for using advanced information management systems for arable farming in Europe. What causes the economic benefit depends on many parameters that often are difficult to distinguish from each other. However, an intelligent FMIS may enable the farmer to provide better quality, improved yields and reduced factor costs.

Crop models provide the 'desired extra information' to reduce farmer's risk in decision making: the case of nitrogen application rates

Basso, B.[1], Fountas, S.[2], Sartori, L.[3], Pedersen, S.M.[4], Sorensen, C.[5], Pesonen, L.[6], Werner, A.[7] and Blackmore, S.[2], [1]University of Basilicata, Crop Systems Dept., Potenza, Italy, [2]Univ. of Thessaly, Cereteth, Volos, Greece, [3]Univ. of Padua, Landscape & Agroforestry Dept., Legnaro, Italy, [4]Univ. of Copenhagen, Res. Economics, Copenhagen, Denmark, [5]Univ. of Aarhus, Ag. Eng Dept, Aarhus, Denmark, [6]Mtt, Agri-food, Helsinki, Finland, [7]Zalf, Land Use Syst. Landscape Ecol Dept, Munchenberg, Germany

Process-based crop models can play a significant role in the development of alternatives for obtaining sustainable crop production systems. Farmers often express the wish to have future knowledge of weather or the probable outcome of a management strategy. These information are called the 'desired extra information' that are not currently available but which help making better decision. The objective of the study was to demonstrate that crop models can provide this information by simulating the impact of past weather recorded on the site on a specific management strategy. The knowledge of such information would allow the farmer to go beyond the constraints imposed by current practices in describing their decision making process. We report the results of a model of decision making and information flow using crop models as the tools for providing information useful to reduce the risk associated with the management strategy.

Technologies for a standardised information infrastructure to assist compliance to crop production standards

Nash, E.J.[1], Nikkilä, R.[2], Pesonen, L.[3], Sørensen, C.G.[4] and Fountas, S.[5], [1]Rostock University, Institute for Management of Rural Areas, Faculty of Agricultural and Environmental Sciences, Justus-von-Liebig-Weg 6, 18059 Rostock, Germany, [2]Helsinki University of Technology, Department of Automation and Systems Technology, P.O. Box 5500, 02015 TKK, Finland, [3]MTT Agrifood Research Finland, Plant Production Research, Vakolantie 55, 03400 Vihti, Finland, [4]Århus University, Department of Agricultural Engineering, Faculty of Agricultural Sciences, Research Centre Bygholm, Schüttesvej 17, 8700 Horsens, Denmark, [5]University of Thessaly, School of Agricultural Science, Fytokou St. N. Ionia Magnisias, 38446 Volos, Greece

The assessment of compliance to crop production standards requires not only data concerning farm management and crop operations but also information from external sources. Not least amongst this information is the definition of what is required by the standard. Software providers may be reluctant to embed this in their systems due to concerns regarding liability, particularly as standards are subject to temporal and spatial variation which would require a mechanism for customising the software and ensuring timely updates. Many standards also require that inputs to the farm (seed, fertilisers, crop protection agents, etc.) are both documented and certified and that the details of many on-farm operations are recorded. Furthermore, outputs from the farm, both in terms of agricultural products and by-products, must in many cases be recorded. Required information may currently typically be held using paper records, or simple digitised records such as spreadsheets and text documents, and may be exchanged either in paper form or electronically on physical medium or via the internet. However, technologies (e.g. process data collection during operations, RFID tags, web services, wireless networks, electronic signatures, etc.) already exist which could be used to simplify and automate the electronic collection, management and transfer of information. The administrative workload of demonstrating compliance to standards could thereby be significantly reduced by adoption of appropriate technologies. In this paper we will consider some of the requirements of selected crop production standards as to what information is required to fulfil them, based on a systems analysis resulting in a functional model. Based on this, the technical infrastructure required to collect, manage and present this information will be specified and suggestions made as to where existing or emerging technologies may be used to assist in assessing compliance. Additionally, components requiring openness and standardisation will be identified together with, where possible, existing applicable standards which may be suitable. Finally, aspects such as dependability, capacity of data transfer and the special requirements of real-time decision support will be considered.

FutureFarm and the future of precision agriculture in Europe
Lowenberg-Deboer, J.[1] and Griffin, T.[2], [1]Purdue University, International Programs in Agriculture, 615 W. State St., West Lafayette, IN 47907, USA, [2]University of Arkansas, Cooperative Extension Service, 2301 S. University Ave, Little Rock, AR 72204, USA

Knowledge-intensive agriculture is usually economical only for high value crops and livestock. For lower value commodity crops and livestock, embodied knowledge systems are usually more commercially successful. FutureFarm is a team of 18 universities and research organizations in ten European countries with the goal of supporting European agriculture by integrating farm management information systems to support real time precision agriculture decisions. Purdue University in the USA is an outside partner. The hypothesis of this paper is that FutureFarm will be successful for commodity crops and livestock only if it develops embodied knowledge precision agriculture systems. The approach will summarize worldwide adoption of precision agricultural technology, identify trends and outline the implications of those trends for European agriculture. 'Knowledge-intensive agriculture' is an approach to farming in which input use and timing is determined by a manager using observations on the crop or livestock system. Examples of knowledge intensive agriculture include integrated pest management and the first wave of precision agriculture (e.g. variable rate seed, fertilizer and pesticide application). The opposite end of the agriculture knowledge spectrum is 'embodied knowledge agriculture'. Most of the agricultural technology of the 20th Century was embodied-knowledge technology. Hybrid seed, chemical fertilizer and pesticides are good examples of embodied knowledge technology. Each of these embodied knowledge inputs required sophisticated science to develop, but the user does not need to understand that science to use them effectively. Embodied-knowledge technologies are often economically preferred because skilled human management time is expensive. The adoption record of the first wave of precision agriculture technologies has been mixed, but embodied-knowledge precision agriculture has been more successful. The best precision agriculture adoption data is available for the United States where less than 20% of maize area has been managed with variable rate fertilizer in any one year. In that country maize is the crop most commonly managed with precision agriculture technology. In contrast, guidance technologies based on Global Positioning Systems (GPS) have been highly successful. In the US Midwest in 2008 85% of agricultural custom applicators with ground based equipment use some sort of GPS guidance. About 30% use autosteer systems. GPS guidance spinoff technologies such as automatic shutoffs for sprayers and planters are spreading rapidly. GPS guidance is an embodied knowledge technology because users do not need to analyze information and make management decisions. One approach reducing the management time required for variable rate application is the use of proximate crop sensors (e.g. N-Sensor, CropCircle, Greenseeker) to guide VRT nitrogen fertilizer technologies. Precision agriculture adoption patterns have been similar in mechanized agriculture elsewhere in the world. The FutureFarm plans for automated data collection and standardized information flow are essential for development of embodied-knowledge precision agriculture. FutureFarm needs more attention on algorithms for converting that data into input decisions, and it needs to focus on understanding the socio-economic conditions under which embodied-knowledge precision agriculture can be adopted in Europe.

Future GNSS: farmers navigate towards trusted farming

Lokers, R.M.[1] and Van Der Wal, T.[2], [1]Alterra, Wageningen UR, Centre for Geo-Information, P.O. Box 47, 6700 AA Wageningen, Netherlands, [2]Portolis B.V., Spoorbaanweg 23, 3911 CA Rhenen, Netherlands; rob. lokers@wur.nl

Nowadays, European farmers have to comply with numerous rules and directives concerning a diverse range of issues. The EU requires farmers to comply to legislative rules concerning farm management, environmental issues etc. Non compliancy might result in subsidy reductions or even fines. On the other hand, consumers and government demand that farmers produce 'safe food'. More and more they expect that reliable information regarding the production process is available through the production chain all the way to the origin of the primary product. In order to provide the necessary information on the production process, reliable documentation of the business operations performed on the farm is necessary. To automate this process of documentation, the use of GNSS is indispensable. Using GNSS every elementary machine navigation and operation can be labeled with a time stamp and location. This information can be linked to the product batches that go into the production chain. If spatial data available in a Geographical Information System (GIS) can be integrated into GNSS supported machine navigation and operation a farmer can automatically adapt farm management and operational farming to the relevant rules and restrictions. In the FieldFact project, opportunities of Galileo, the new European GNSS, have been examined. Galileo will provide its users in the agricultural sector with key features that will improve the reliability of recorded information. The combination of an authenticated signal and an integrity message will provide a hallmark for the measured location and time. Thus, the generation of authenticated or trusted documentation on the farm will become possible.

New GPS based methods accredited by the EC for area measurement
Grzebellus, M., NavCert GmbH, Hermann-Blenk-Str. 22, 38108 Braunschweig, Germany

Agricultural subsidies paid by the European Union add up to approx. € 50 billion every year. How much of this amount is paid to the individual farmer depends on the size of the area cultivated and the crops. In a first step, the farmers declare the area of the land cultivated. As the areas cultivated are very large, even minor non-conformities may result in high repayments. This information is then verified on a case-by-case basis by the individual Member States – in Germany, this is the task of the individual German Länder – and the EU, which selects a sample for on-the-spot checks. In the previous years quit a huge amount of payments were under discussions. Farmland area has been calculated by interpreting satellite images and aerial photographs and physical surveying which is quite time consuming. Now the EC has for the first time allowed to use GPS surveying equipment for measuring areas of farmland. In its Regulation 972/2007/EC, the European Union generally approved various approaches to farmland area calculation, provided certain tolerances are not exceeded and the measurement methods are set forth in appropriate EU standards. The Joint Research Centre (JRC) of the EU Commission recommends further for the first time the use of the GPS satellite navigation system for farmland calculation. On the basis of the criteria specified by JRC a test plan was developed together with a voluntary certification mark which manufacturers can use to qualify their GPS surveying equipment for this particular application. NavCert as part of the TÜV SÜD group has been selected to contribute to the requirements and apply for the accreditation. The set up of the validation scheme and the first experience certifying measurement equipment will be presented. High emphasis was put on the proper design of the reference acres to reflect real life conditions with shadowing, irregular shapes, narrow but long acres and multipath. The accreditation is done by the EU. In parallel the underlying technical requirements will be entered into the European standardization scheme to achieve later on a harmonized standard. This new approach for certification provides at an early stage already to the manufacturers a reliable process to get a certification and the confirmation that their equipment can be used for calculation of agriculture subsidies. The usage of EGNOS today may provide additional benefits due to a higher accuracy and integrity. Later on Galileo will be used as well for further elimination of errors. The presentation will provide an overview what has been achieved so far and which improvements are expected.

URM as tool for Shared Environmental Information System (SEIS)

Charvat, K.[1], Kafka, S.[2] and Cepicky, J.[2], [1]Wirelessinfo, Cholinská 1048/19, 784 01 Litovel, Czech Republic, [2]Help Service Remote Sensing, Černoleská 1600, 256 01 Benešov u Prahy, Czech Republic; charvat@ wirelessinfo.cz

The paper describe concept for Sharing Environmental Information System - Uniform Resource Management (URM), which support validation, discovery and access to heterogeneous information and knowledge. It is based on utilisation of metadata schemes. The URM models currently also integrate different tools, which support sharing of knowledge. The URM concept was introduced by NaturNet Redime project as tool for managing of educational context and now is modified for general sharing of information inside of community in c@r project. Uniform Resource Management (URM) provides a framework in which communities can share information and knowledge trough their description, which is easy understandable inside of the community. In order to effectively share information and knowledge, there has to be a standardized scheme, which will support uniform description of information and knowledge including common vocabularies. The main objective of URM will be easy description, discovery and validation of relevant information sources. URM will ensure that any user can easily discover, evaluate and use relevant information. The free text engine (eg. Google) can't be used due to the fact that in many cases a user obtains thousands, if not millions, of irrelevant links. This happens because the free text engines does not fully recognise the context of researched information. The context characterises any information, knowledge and observation. Context strongly influences the way how the information will be used. There are different definitions of context in existence. The important issues for the context are: * to identity of an entity; * to profile of an entity; * spatial information * temporal information * environmental information * social relation * resources that are nearby * availability of resources; Many context attributes characterize the environmental information or knowledge. From the point of view of context, the information or knowledge could be divided into different parties: * Information or knowledge provider i.e. a party supplying the resource; * Custodian, accepts accountability and responsibility for the resources and ensures appropriate care and maintenance of the resource; * Owner of the resource; * User, who uses the resource; * Distributor who distributes the resource; * Originator who created the resource; * Point of Contact to be contacted for acquiring knowledge about or acquisition of the resource; * Principal investigator responsible for gathering information and conducting research; * Processor who has processed the data in a manner such that the original resource has been modified; * Publisher, i.e. party who published the resource; * Author, i.e. party who authored the resource.

System analysis of management information systems for the future

Sørensen, C.G.[1], Fountas, S.[2], Pesonen, L.[3], Basso, B.[4], Pedersen, S.M.[5] and Nash, E.[6], [1]Aarhus University, Agricultural Engineering, Blichers Alle, 8830 Tjele, Denmark, [2]Centre for Research and Technology, Technology Park of Thessaly, GR 385 00 Volos, Greece, [3]MTT Agrifood Research Finland, Plant Production Research, MTT, FI-36000 Jokionen, Finland, [4]University of Bassilicata, Crop Systems, Forestry and Environmental Sciences, Via Ateno Lucano, 10 Potenza, Italy, [5]Copenhagen University, Food and Resource Economics, Rolighedsvej 25, 1958 Frederiksberg C, Denmark, [6]Universität Rostock, Agrar- und Umweltwissenschaftliche Fakultät, Justus-von-Liebig-Weg 6, 18059 Rostock, Germany

Future European farmers will effectively manage information on and off their farms to improve economic viability and to reduce environmental impact. An integration of information systems is needed to advise managers of formal instructions, recommended guidelines and document various decisions making processes. In the project FutureFarm, a new Farm Information Management Systems (FMIS) will be developed. Specifically, the objectives of this paper are to define and analyse the system boundaries and relevant decision processes for such a novel FMIS as a prerequisite for a dedicated information modelling. The boundaries and scope of a system is described in terms of actors and functionalities, where actors are entities interfacing with the system (e.g. managers, software, databases). In order to analyse the complex and soft-systems situations of how to develop an effective FMIS, the soft systems methodology (SSM) was used. This approach involves, identification of the scope of the system, identification of user requirements, conceptual modelling, identification of actors and decisions processes, and information needs as a preliminary step before the detailed information modelling A holistic view and scope of the system is presented together with the system constraints. The results include the analyses of 4 demonstrations farms in the project and representing adverse conditions across the EU. The systems components are depicted and linked to the subsequent conceptual model of the overall system. This research has shown the benefit of using dedicated system analysis methodologies as a preliminary step to the actual design of a novel FMIS. The soft systems methodology targets organisational process modelling and identify unstructured problems as well as identifying non-obvious problem solutions. Specifically, the approach provides the possibility of more clearly capturing the change that is necessary for a current system to transform into a proposed system that will fulfil the user requirements of tomorrow.

The implementation of the European Common Agricultural Policy and spatial data infrastructures
Van Der Wal, T.[1], Kay, S.[2] and Devos, W.[2], [1]Portolis, spoorbaanweg 23, 3911 CA, Netherlands, [2]JRC-ISPC, GeoCAP, Via Fermi 2749, I 21027 Ispra (VA), Italy

The European Common Agricultural Policy (CAP) is already for many years relying on geo-spatial applications in administering, managing and controling farmers declarations. The CAP is under constant change and over the years it has become more spatial explicit with more emphasis on cross compliance to environmental regulations and the inclusion of landscape features in the subsidy system. The implementation of the CAP regulations is left to the member states. National choices has lead to different implementations of the same regulation. The INSPIRE directive provides a regulatory framework for harmonising the reference datasets of these national implementations and this leads to a reference model for land parcel information systems. Also Europe's own satellite navigation system Galileo is an opportunity for the agricultural domain to harmonise methods and datasets for rapid field visits and On The Spot Control. the authors present the opportunties these european geo-spatial initiatives have to offer to the CAP and indicate how the agricultural domain for regulated applications adopts the INSPIRE directive and Galileo programme in both the single payment and the rural development schemes. The CAP can benefit and contribute to the European SDI. These developments and challenges will be presented.

Recording and analysis of locomotion in dairy cows with 3D accelerometers
De Mol, R.M.[1], Lammers, R.J.H.[2], Pompe, J.C.A.M.[2], Hogewerf, P.H.[1] and Ipema, A.H.[1], [1]Wageningen University and Research Centre, Animal Sciences Group, Edelhertweg 15, 8219 PH Lelystad, Netherlands, [2]Wageningen University, Farm Technology Group, Bornsesteeg 59, 6708 PD Wageningen, Netherlands

Lameness is a big problem in modern dairy farming. Lameness causes welfare problems and economic losses for the farmer. Lameness can be detected by regular observations of the locomotion of the cows. But this method is not reliable and more difficult in larger herds. An automated method for lameness detection can solve this problem. Accelerometers attached to a leg of the cow can be used to record the locomotion of a cow. Appropriate data processing can result in differentiating the locomotion of lame and non-lame cows. In an experiment at the experimental farm 'De Ossekampen' the 3D acceleration during walking of non-lame dairy cows was measured and analyzed. Nodes with a 3D accelerometer in a wireless sensor network were applied to track the movement of the leg with a frequency of 50 Hz. The data was analysed in two steps: first the steps were detected, followed by the determination of step parameters. Suitability of both variance analysis and Fast Fourier Transformation for step detection was evaluated. Both gave good results in the detection of steps. For each step the following parameters were calculated: step length, step time and a step index (related to the firmness of a step). The calculated step length was not useful as a step describing parameter. On the other hand, results for step time and step index gave reliable results. For a part of the recorded acceleration periods video recordings were available to confirm the calculated results. The application of the locomotion analysis for lameness detection will be discussed.

An intelligent wireless accelerometer system for measuring gait features and lying time in dairy cows

Pastell, M.[1,2], Hakojärvi, M.[1], Hänninen, L.[2,3] and Tiusanen, J.[1], [1]University of Helsinki, Department of Agrotechnology, P.O Box 28, FI-00014 University of Helsinki, Finland, [2]University of Helsinki, Research Centre for Animal Welfare, P.O Box 57, FI-00014 University of Helsinki, Finland, [3]University of Helsinki, Department of Production Animal Medicine, P.O Box 57, FI-00014 University of Helsinki, Finland; matti. pastell@helsinki.fi

Lameness is a crucial welfare issue in modern dairy production. Lameness is defined as an abnormality of gait and it has been shown that lameness is strongly related to daily lying time. Lame cows suffer discomfort and pain of long duration if the problem remains untreated. The aim of this study was to design and construct a wireless accelerometer system for measuring gait features and daily lying time in dairy cows. The measured data was used together with behavioral observations to develop software for calculating lying time and deriving step features from the data. A new wireless 3-dimensional acceleration measurement device which registers movement and inclination of the device in three axes was developed. We used a license free wireless 869 MHz radio channel which enabled highly reliable transmission in barn conditions. The transmitter ran on an 8051 compatible programmable micro controller. The orientation of the device was calculated with the accuracy of 2 degrees and the measurement rate is adjustable up to 400 Hz. The constructed device was less than 5 cm long and 3 cm thick. This means that the device is small enough to be used also with newborn calves. The sensor was mounted on the leg of four dairy cows and data was collected at 250Hz during the cows' standing, lying and walking on rubber and concrete floors. The test measurements were used to design an intelligent algorithm to be embedded in the sensor node to reduce the radio channel load. The algorithm consisted of two separate rules preventing the node from transmitting values that were within the sensor noise and still enabling fast response to essential acceleration changes. The performance of the algorithm was evaluated by simulating the protocol on the test datasets measured at 250 Hz. We obtained average transmission frequency of 2.7 Hz without losing information about cows' activity. The method for reducing data flow shows successfully the great possibilities of adding simple intelligence to sensors. The algorithm is especially suitable for barns with automatic milking where the activity of the cows is divided over 24 h and the cow behavior is not simultaneous. The data reduction algorithm gives the device a battery exceeding one year. The test measurements were also used to develop algorithms for calculating cows' daily lying, standing and walking time. The lying time measured with the sensor agreed with over 99% of the lying time observed from over 200 h of behavioral recordings. Furthermore and ad-hoc algorithm for segmenting steps and deriving gait features from the raw acceleration data was created. The algorithm calculated the step duration, maximal acceleration in horizontal and vertical direction, the acceleration-deceleration ratio and the stance and swing phase durations. We have successfully developed a new wireless system with intelligent battery saving features for measuring lying time and exact gait features in dairy cows. The next step is to apply the device and the developed algorithms in lameness detection and further validate the algorithms with more animals and different walking surfaces.

Recording of dairy cow behaviour with wireless accelerometers

De Mol, R.M., Bleumer, E.J.B., Hogewerf, P.H. and Ipema, A.H., Wageningen University and Research Centre, Animal Sciences Group, Edelhertweg 15, 8219 PH Lelystad, Netherlands

The daily behaviour of dairy cows reflects the health and well-being status. Tools that measure the cow's behaviour can help the farmer in his daily management. The behaviour can be monitored with wireless accelerometers. For this application, the accelerometers are used as a tilt sensor to measure the angle. The angle of a leg reflects the lying or standing behaviour, the angle of the head reflects the eating behaviour. An experiment was carried out at the experimental farm 'Nij Bosma Zathe', in the North of the Netherlands and lasted 50 days in May/June 2008. The cows were indoors during the first 36 days and had access to a pasture during the day on the last 14 days. Six cows were equipped with two 2D accelerometers, one attached to the head and one attached to the right hind leg. The accelerometers were attached to wireless sensor nodes. Measurement data were transmitted to a gateway directly or via repeaters. The acceleration of the head and leg was recorded every halve minute (average of seven measurements with 1 Hz measuring frequency). Based on calibration measurements, the acceleration was transformed to the angle. A cow was standing when the angle was more than 45°, otherwise lying. Data occasionally appeared to be missing when cows are lying because communication in the WSN was hampered by the body. The method to transform the acceleration to angle and behaviour appears to be appropriate, it is possible to monitor the cow's behaviour with a wireless sensor network equipped with accelerometers. The data from the nodes attached to the leg were used for this. Results will be given.

Approach to model based motion scoring for lameness detection in dairy cattle

Pluk, A.[1], Bahr, C.[1], Maertens, W.[2], Vangeyte, J.[2], Sonck, B.[2] and Berckmans, D.[1], [1]Catholic University of Leuven, Measure, Model & Manage Bioresponses (M3-BIORES), Kasteelpark Arenberg 30, B-3001 Leuven, Belgium, [2]Institute for Agricultural and Fisheries Research Technology & Food Unit, Agricultural Engineering, Burg. van Gansberghelaan 115, B-9820 Merelbeke, Belgium; daniel.berckmans@biw.kuleuven. be

Lameness is an increasing animal welfare problem. It has a negative impact on milk production, body condition and reproductive performance in dairy cows. Since early identification and treatment can reduce the cost of lameness and improve animal wellbeing, more objective measures to quantify lameness need to be used. In the development of an automatic on-line tool for lameness detection several image parameters useful to detect changes in the motion of a cows body are addressed. To avoid the need to redevelop algorithms for every image parameter used in lameness detection, we propose a model based motion scoring system. The model of the cow is derived from an image stream. First of all the cow is extracted from the background using image processing techniques. Now the derivation of the cow model is performed in a top-down approach: The overall position of the body of the cow in the image is detected. This is performed by fitting a rectangle onto its body. This provides us with information about the position and orientation of the body; In the next step the position and orientation of the head of the cow is detected in a similar way; And finally the positions and orientations of the individual limbs are detected. Repeating this procedure gives us a mathematical model of the body position of the cow within each image and it provides us with an objective description of the motion of the cow. The model and motion description can be used for derivation of the image parameters needed for lameness detection.

ICT adoption trends in agriculture

Gelb, E., Center for Agricultural Economic Research, 9 Hagalil st., Rehovot, 76601, Israel

By the mid 1990s the agricultural sector had accumulated substantial experience and adoption proficiency with ICT applications - even as they were constantly innovated, improved and upgraded. Regardless, in many cases, ICT adoption remained considerably short of universal, leaving significant agricultural potential and rural development unrealized. Both are of major concern with worldwide impact. Understanding and alleviating the constraints for adoption of ICT constitute strategic issues. To remedy this shortcoming the EFITA questionnaire data sets were initiated and collated at EFITA conferences and similar professional forums since 1999. These forums representing a very wide range of respondents identified and ranked critical ICT adoption constraints and their consequences for farmers, extension and research. The data sets consistently suggested over time that ICT adoption constraints are universal and of major significance. For example in 2001 72% of questionnaire respondents replied that unsuccessful ICT Adoption is a major concern. Contrary to the improvement predicted by a learning curve an increased, significant 94% of the 2007 replies still identified ICT Adoption as a major issue. The following main impediments cited in the data-sets suggested why: (1) lack of ICT applications 'tailored' to end-user needs; (2) ever increasing application sophistication which in turn imposes enhanced human capital requirements; (3) lack of ICT synchronization with production, market and environment dictates; (4) lack of essential, ongoing end-user, extension and research training. The data sets ranked these and additional significant constraint trends over time. Replies overwhelmingly suggested that providing ICT Services for farmers is a public concern. Public funding of Information Technology services for farmers was consequently universally justified and recommended. In time the questionnaire rankings became a baseline for comparison of ICT adoption constraints gleaned from comparable country and sector specific data sets. Interpretation of the commonalities, trends and insights can substantiate public policy priorities, promote effective measures to alleviate ICT Adoption constraints and focus current ICT adoption programs for agricultural production and rural development.

The impact of ICT on the food economy

Bunte, F.[1], Groeneveld, R.[2], Hofstede, G.J.[3], Top, J.[2] and Wolfert, S.[1], [1]LEI - Wageningen UR, Plant Systems Division, Alexanderveld 5, 2585 DB Den Haag, Netherlands, [2]AFSG - Wageningen UR, Fresh Food and Chains, Bornsesteeg 59, 6708 PD Wageningen, Netherlands, [3]Social Sciences - Wageningen UR, ORL, Hollandseweg 1, 6706 KN Wageningen, Netherlands

The purpose of the paper is to outline the implications of developments in ICT for the food economy. Following Kinsey, the paper presumes that ICT is one of the key technologies driving a wide range of product, process and organizational innovations in Food Supply Chain Networks (FSCNs). Because of the growth of ICT applications, information becomes a competitive factor in itself. Companies increasingly derive value added from the ability to exploit information rather than controlling natural resources. The paper outlines the impact of ICT on the food economy as follows. On basis of a literature review from four disciplines – knowledge management, management information systems, operations research and logistics, and economics - the paper identifies the demand for new ICT applications, the supply of new applications and the match between demand and supply. Subsequently, the paper discusses the impact of new ICT applications for the food economy. The paper relates the development of new technologies to innovation and adoption processes and economic growth, and to concepts of open innovations and living labs. The demand for new ICT applications is due to increases in global competition as well as economies of scale and scope, changes in consumer demand (higher quality, more variety and more convenience), growing consumer and regulatory demands with respect to sustainability and changes in logistics and sourcing. Information plays a key role in organizing product and monetary flows in FSCNs. Information is increasingly shared between companies. Two emerging technologies play a key role in this respect: the Service-Oriented Architecture and Software as a Service. In this framework, the paper identifies ten disruptive ICT technologies. Companies increasingly match demand and supply of new ICT applications by applying business process management. This is a innovative method of efficiently aligning an organization with the wants and needs of clients. Promising applications include traceability, open innovations and living labs. At the market level, ICT applications promote competition processes by reducing transaction costs and increasing transparency. At the economy level, ICT applications turn out to be key factors in explaining productivity differences between the US and the EU. This holds in particular for general merchandising. The information economy offers great opportunities for employees worldwide by raising both responsibilities and payments. On the other hand, it may very well be the case that not everyone profits from the coming about of the knowledge economy. The digital divide may continue or sharpen already existing divisions in society. The main conclusion of the paper is that information is increasingly shared in FSCNs, even though both individuals and organizations still have limited skills of combining and exploiting information. The potential benefits of information sharing for the economy and society are enormous.

Spatial variability of soil properties and the occurrence of soil-borne pests in sugar beet

Scholz, C., Patzold, S. and Welp, G., Institute of Crop Science and Resource Conservation, Division Soil Science, Nussallee 13, 53115 Bonn, Germany

The small-scale spatial variability of soil properties within fields offers varying conditions for the diverse development of crops as well as many soil-borne pests (e.g. pathogens, parasitic nematodes). *Heterodera schachtii* (Schmidt), *Ditylenchus dipsaci* (Kühn), and *Rhizoctonia solani* (Kühn) are economically important nematode species and pathogens, respectively, in sugar beet production. The objective of this study is to elucidate the coincidence of soil properties and soil-borne pests in order to create risk maps at the field scale and to optimize site-specific crop protection. In several investigations, the patchy appearance of nematodes within fields has been linked to soil heterogeneity. In North America, the population density of soybean cyst nematodes was higher at sandy sites than at adjacent loamy or clayey sites. The same relationship with soil texture was found in Western Africa and in Spain. The occurrence of Rhizoctonia solani obviously is related to soil moisture and bulk density. We hypothesize a causal relationship between the above-mentioned soil-borne pests and the spatial variability of soil properties (soil texture, soil organic carbon, pH, lime, and nutrient contents). In April 2008, we monitored 26 sugar beet fields in North Rhine-Westphalia with georeferenced EM38-measurements, a non-invasive method to determine the apparent electrical conductivity (ECa) of the soil, which is highly correlated with the clay content of the soil. Depending on the ECa maps and on the development of the pests during the vegetation period, soil samples will be taken on selected heterogeneous fields in September 2008. Besides various physical and chemical soil properties, population parameters of *Heterodera schachtii, Ditylenchus dipsaci*, and *Rhizoctonia solani* will be monitored. The spatial distribution of soil properties and pests within the fields will be evaluated with geostatistical methods.

Ground-based integration sensor and instrumentation system for measuring crop conditions
Lan, Y.[1], Zhang, H.[2], Lacey, R.[2], Huang, Y.[3] and Hoffmann, W.[1], [1]APMRU, USDA-ARS, 2771 F&B Road, College Station, TX77845, USA, [2]Texas A&M University, Biological and Agricultural Engineering, College Station, TX77843-2117, USA, [3]USDA-ARS, 141 Experiment Station Rd., Stoneville, MS38776, USA

Precision agriculture requires reliable technology to acquire accurate information on crop conditions. Based on this information, the amount of fertilizers and pesticides for the site-specific crop management can be optimized. A ground-based integrated sensor and instrumentation system was developed to measure real-time crop conditions including Normalized Difference Vegetation Index (NDVI), biomass, crop canopy structure, and crop height. Individual sensor components has been calibrated and tested under laboratory and field conditions prior to system integration. The integration system included crop height sensor, crop canopy analyzer for Leaf Area Index, NDVI sensor, multispectral camera, and hyperspectroradiometer. The system was interfaced with a DGPS receiver to provide spatial coordinates for sensor readings. The results show that the integration sensor and instrumentation system supports multi-source information acquisition and management in the farming field.

Differentiation of several leaf rust (*Puccinia rocondita*) severity classes in winter wheat (*Triticum aestivum*) with *in situ* hyperspectral data

Mewes, T.[1], Franke, J.[1] and Menz, G.[2], [1]University of Bonn, Center for Remote Sensing of Land Surfaces (ZFL), Walter-Flex-Str. 3, 53113 Bonn, Germany, [2]University of Bonn, Institute for Geography, Remote Sensing Research Group (RSRG), Meckenheimer Allee 166, 53115 Bonn, Germany

The reduction of fungicide use for more sustainable crop production demands fast and precise pathogen identification on wheat stands. Sensor based detection is a promising approach for site-specific stress control. Several studies have shown possibilities and limitations for the detection of plant stress using spectral sensor data. Hyperspectral data provide the opportunity to collect spectral reflectance in contiguous bands over a broad range of the electromagnetic spectrum. Individual phenomena like the light absorption of leaf pigments can be examined in detail. The precise knowledge of stress-dependent changes in certain spectral wavelengths provides great advantages for the detection of fungal infections. This study focuses on the estimation of disease severity of leaf rust in wheat stands by spectroradiometer data. In a field experiment, 84 randomly sampled points of a 5 ha field were multitemporally measured with a ASD Fieldspec® 3 spectroradiometer. The field was subdivided into four different variants with different application strategies: A customary variant, one with minor fertilizer application, one with minor fungicide application and one with minor fertilizer and minor fungicide application. The four types could be well differentiated and classified using decision tree algorithms. Several classes were defined with different disease severity values. The limitation of disease severity differentiation was tested using an increasing amount of classes. The possible number of classes was limited due to low variation of severities within one variant. Hence, the exact possible number of different severity classes needs to be proven in further studies. However, remote sensing data with high spectral resolutions provide the possibility for accurate discrimination of certain disease severity classes.

Optimization of radiation mode for plants with different geometrical arrangement

Rakut'ko, S., Far Eastern State Agrarian University, Electrification dep., Blagoveschensk, Politechnitcheskaya st., 86, 675005, Russian Federation

For the energy saving purposes the optimal choice for additional radiation treatment of plants is such a structure of light field that corresponds to geometrical structure of plant. Existing characteristic methods of geometrical structure of phytocenosis are not quite suitable for purposes of optimization of radiation mode parameters. The objective of the paper is to formulate parameters of geometrical structure of plants which are suitable for practical application; to develop method of optimization of parameters of plants radiation mode. The computer controlled photometric unit for definition spatial pattern of plant crone was designed. A series of pictures of test plant crone were made under different angles with the help of digital camera. Special software written on Visual Basic analyzes the areas of pictures with the given chromaticity coordinates in RGB system using thresholding method. The research was conducted based on specimen of single ornamental plants (38 types of plants have been tested) with different geometrical structure of crone. For characterize the structure of single plant crone the midlength section curve (MSC) that represents the dependence of crone section area in the given secant plane from the secant plane tilt angle is suggested. The paper shows that it is possible to decrease photometric losses by maximizing of the product of layout factor, which is the characteristic of radiation plant parameters, on the midlength section, which is determined by spatial pattern of irradiated plant crone. Hence, it is possible to optimize the radiation environment for plants with different geometrical arrangement by increasing the conformity of radiation plant layout scheme to the spatial pattern of plant crone.

Analyze of temporal variations in NDVI within-field spatial variability as a possible basis for precision farming

Nováková, M., Halas, J. and Scholtz, P., Soil Science and Conservation Research Institute, Remote Sensing and Informatics, Gagarinova 10, 827 13 Bratislava, Slovakia (Slovak Republic)

Theoretical principle of Precision Farming (PF) resides in application of complex and local variable approach to various agriculture activities supported with relevant scientific knowledge and information technologies. The idea of spatially variable farming is based on simple fact - existence of within-field variability related to soil and landscape properties (with respect to crop yield and soil production). Therefore there is a requirement on identification of an effective and simple method one can utilize for preliminary detection of existing detail-scaled spatial variability (before high-priced analyses are applied). The task seems to be crucial since PF methods are not recommended to be followed anywhere if its economical and ecological benefits should remain sustentative. The objectives of this study were: (a) to analyze NDVI vegetation index as an indicator of within-field variability in relation to soil and landscape properties; (b) to analyze conditions when/how/why NDVI fit within-field variability of soil and landscape properties, as well the crop within-field variability; (c) to test NDVI images as a source data for production zones (PZ) delineation and (d) to outline an example of PZ theoretical application - arrangement of variable irrigation. The experiment was carried out in 2005 and 2006 at pilot area Žihárec (120 ha) located at south-west part of Slovakia (fluvial relief of Váh river). In experiment existing georeferenced data on soil physical and chemical properties (content of soil organic carbon, soil nutrients and soil texture fractions; measurements of soil mechanic resistance, soil moisture and soil electrical conductivity) and other data on NDVI (in 2005 and 2006), DEM, weather (rainfall) and winter wheat yields (in 2005 and 2006) we had at our disposal. Statistical analyses (multiply regression, correlation analyse), spatial and geostatistical analyses (radial basis function, kriging; intersection tools, zonal analyses) were used to investigate relations in agro-ecosystem (defined as system weather-relief-soil-crop) and to explain the position of NDVI as indicator of system spatial variability in 2005 and 2006. On the base of NDVI images (2005) and assumption of high correlation between NDVI and winter wheat yield (2005), production zones were delimited. Two vegetation seasons (in 2005 and 2006) were evaluated in respect to average winter wheat yield and yield differences in individual zones. The results of experiment pointed out two different positions of NDVI as indicator of within-field variability related to soil and landscape properties and crop yield forming: a) as significant indicator (in 2005) and b) as poor indicator (in 2006). In the conditions of pilot area, rainfall (the amount and distribution) was identified as a main factor which defines NDVI actual position. The evaluation of winter wheat yield in 2005 and 2006 based on delineated zones enabled us to specify zones with relatively high production or with relatively low production in two years and zones with different response on different weather conditions (in dry weather low yield, in wet conditions high yield). On the basis of the results of production zones response analyse, we propose to consider additional, spatially variable irrigation to be applied before another variable PF technology (as fertilizer application).

Small scale spatial variability of soil physical properties of a ferralsol with three different land uses

Paz-Ferreiro, J.[1], Pereira De Almeida, V.[2] and Alves, M.C.[2], [1]University of Corunna (UDC), Soil Science, Facultad de Ciencias, 15071, Coruña, Spain, [2]University of the State of São Paulo (UNESP), Soil Science, Faculdade de Engenharia, 15385-000 Ilha Solteira, SP, Brazil

Neglecting spatial heterogeneity pattern in soil nutrient status may result in reduced yield and in environmental damage. In order to analyze the impact of land use on the pattern of spatial variability at the stand scale a study was conducted in Ilha Solteira, São Paulo state, Brazil, on a Ferralsol. Land uses were pasture, mango orchard and corn field. Soil samples were taken at 84 individual points following a nested sampling design. Texture, soil penetration resistance and soil water content were analyzed. Classical statistics revealed a characteristic ranking in the variability of the studied soil properties. Geostatistical analysis showed contrasting patterns of spatial dependence for the different soil uses, sampling depths and studied properties. The nugget variance showed various degrees of spatial dependence from no spatial pattern i.e. pure nugget effect to weak, moderate or even strong spatial structure. For most of the study properties the range was between 30 and 60 m. Semivariograms of properties such as penetration resistance had a relatively high proportion of structural variance. Land use effects may have contributed to differences in variability between the experimental plots. Geostatistical analysis allowed maps to be produced, which may be useful for site specific management.

Spy-See: advanced vision system for phenotyping in greenhouses

Polder, G.[1], Van Der Heijden, G.W.A.M.[1], Glasbey, C.A.[2] and Dieleman, J.A.[3], [1]Wageningen UR, Biometris, P.O. Box 100, 6700 AC, Wageningen, Netherlands, [2]Biomathematics & Statistics Scotland, JCMB, The Kings Buildings, Edinburgh EH9 3JZ, Schotland, United Kingdom, [3]Wageningen UR, Plant Research International, P.O. Box 16, 6700 AA, Wageningen, Netherlands

The EU project SPICY (Smart tools for Prediction and Improvement of Crop Yield) is aimed to develop a suite of tools for molecular breeding of crop plants for sustainable and competitive agriculture. These tools help the breeders in predicting phenotypic response of genotypes for complex traits under a range of environmental conditions. Pepper will be used as a model crop. Molecular breeding will not completely replace large scale phenotyping. Hence, automated and fast high-throughput tools to reduce the amount of manual labour necessary in phenotyping experiments are called for. In the SPICY project an image analysis tool and a fluorescence tool will be developed to measure large numbers of phenotypic traits automatically. This poster will describe the technical details of the vision system we designed. Important features for phenotyping include number and size of leaves, flowers and fruits, time of flowering, and color of leaves and fruits. The plants grow in rows with a row distance of 1 meter. In between are heating pipes. The maximum height of the plants is about 3 meter. We designed a system on a trolley which can travel over the heating pipes. The distance to the plants is very short, therefore to cover the whole 3 meter height we use high resolution cameras with a very large field of view lens at four height positions. Three different cameras are used: 1) high resolution color cameras; 2) high resolution infra red cameras (750-900 nm); 3) low resolution range cameras, based on the TOF (Time of Flight Principle). For illumination xenon strobed flashlights are used. For image capture all cameras and flashlights are triggered simultaneously using an encoder on one of the wheels of the trolley. This way images are captured automatically on a fixed interval when the trolley moves over the heating pipes. By capturing a large number of overlapping images, and with the low resolution information from the range cameras we expect to be able to do high resolution 3D reconstruction. The Spy-See vision system described here will produce a huge amount of detailed image information which can be used as input for the image analysis tool which will be developed in the SPICY project.

Modelling crop root growth and biomass accumulation

Hautala, M. and Hakojärvi, M., University of Helsinki, Dept of Agrotechnology, P.O. Box 28, FIN-00014 Helsinki, Finland

During the last few decades technology used in crop production has developed noticeably. The work needed to be done by humans becomes easier and finally decreases by means of technology and automation. The aim of this research project is to find out requirements for methods and automated machines needed in automated crop production. Modelling and simulation will be used as research methods. A simple plant growth simulation model was established that enables statistical study of biomass growth as a function of any of the variables, e.g. mass of seeds, intensity of radiation, amount of rain and effective leaf area index. Statistical rain is created by three distributions using Monte Carlo method; distribution of dry periods, rain times and amount of rain. Typically hundred yearly rains are worked out for each combination of parameters. In order to make the simulation feasible, crop growth and water transport models were kept as simple as possible taking into account the objectives of the simulation. Therefore the number of parameters was also held small. Soil water transport model has four parameters: SWC, FC, PWC and saturated hydraulic conductivity. Soil is divided into four layers and water moves only when water content is above FC. Crop growth model has only two crop specific parameters; mass of seeded grain and seeding density. Seed mass is divided into leaves and roots. Further dry mass growth comes from the experimental fact that C3-plants produce 1.4 g HCO_2/MJ solar energy when radiation is above 100 W/m^2 and there is no lack of water and nutrients. An experimental (and physical) fact is that about 500 moles of water is transpired when 1 mole of CO_2 is used in photosynthesis. This water is taken from the root volume if available. If there is no water available for the crop, the growth ceases. In this paper results of simulations based on various root growth strategies are presented. Biomass increase is calculated as a function of various root parameters like the number of main roots, maximum rooting depth and lateral root spreading. Results (for example root to shoot ratio and total biomass distributions) are calculated as a function of rain, soiltype and soil temperature.

Variable-rate lime application for Louisiana sugarcane production systems

Johnson, R.[1] and Viator, H.[2], [1]USDA-ARS, Sugarcane Research Unit, 5883 USDA Road, Houma, LA 70360, USA, [2]Louisiana State University Agricultural Center, Iberia Research Station, P. O. Box 466, Jeanerette, LA 70544, USA

Precision agriculture may offer sugarcane growers a management system that decreases costs and maximizes profits, while minimizing any potential negative environmental impact. Variable rate (VR) application of lime and fertilizers is one area in which significant advantages may be realized. A series of experiments were initiated in the summer of 2006 to determine the utility of VR lime application to Louisiana sugarcane production systems and to investigate alternate methods to estimate lime requirements for sugarcane. Soil electrical conductivity (EC_a) mapping techniques were also evaluated as potential tools to develop management zones for VR lime application. Soil samples (0-15 cm) were collected from a nine hectare field on a 0.3 ha grid to map variability in soil pH and lime requirement. Both shallow (0-30 cm) and deep (0-90 cm) soil EC_a data were also collected at the same time using a Veris soil EC_a mapping system. Soil samples were analyzed for soil pH and lime requirement was determined by the standard Shoemaker-McLean-Pratt (SMP) procedure and two additional methods (Adams-Evans (AE) and Woodruff (WD) lime requirement methods). VR application maps were prepared utilizing variogram analysis and kriging of the grid soils data. Treatments compared the SMP, AE and WD lime requirement estimates in both VR and uniform scenarios along with a no-lime control. Plots were seven rows wide (15 m) by ~85 m, and there were six replications. Lime treatments were applied with a Newton Crouch VR lime applicator equipped with a Mid-Tech VR controller. Soil pH was found to vary from 4.1 to 8.1 prior to lime application and the corresponding lime recommendations were also variable with the calculated lime rates ranging from 0-7.8, 0-7.4 and 0-3.4 Mg/ha for the Shoemaker-McLean-Pratt, Adams-Evans and Woodruff lime requirement methods, respectively. Soil EC_a measurements were correlated with soil pH levels from grid soil samples with pH increasing with soil EC_a levels. Sugarcane yield results from this study showed a significant advantage in the theoretically recoverable sugar (TRS) levels (kg/Mg) with VR lime application. The Adams-Evans VR treatment resulted in the highest TRS of all methods (237 kg/Mg) and the no-lime control the lowest (227 kg/Mg). Sugar yield (kg/ha) was significantly greater ($P=0.18$) with the Woodruff VR method (9,,450 kg/ha) as compared to the conventional SMP method (8,800 kg/ha). These combined data suggest that sufficient variability exists in both soil properties and cane and sugar yields to justify a precision management approach because if similar yields can be obtained with the VR system while actually applying fewer inputs, then Louisiana sugarcane producers can show an overall increase in profitability.

Use of aerial imaging and electrical conductivity for spatial variability mapping of soil conditions

Lukas, V., Neudert, L. and Kren, J., Mendel University of Agriculture and Forestry in Brno, Department of Agrosystems and Bioclimatology, Zemedelska 1, 61300 Brno, Czech Republic

Mapping within-field variability of soil conditions using traditional practices such as soil sampling are not too sufficient for economical and organizing reasons. The measurement of soil electrical conductivity and aerial imaging was verified to optimize soil sampling. Verification was carried out at two pedologically and topographically different locations in the Czech Republic. In the years 2004 – 2008 soil samples in regular grid (50 x 50 m) were taken for analysis of nutrient content (P, K Mg and Ca), organic matter (OM), pH value and soil texture. At the same time, apparent soil electrical conductivity (ECa) using EM38 and aerial imaging of bare soil using multispectral camera Duncan Tech MS3100, thermocamera Fluke Ti-55 and DSLR camera Nikon D80 were measured. The results of indirect methods and soil sampling were compared using correlation analysis on the points with 10 m buffer zone. Results show that between indirect methods (ECa, aerial imaging in visible, NIR and thermal spectrum), OM content and soil texture there were moderately strong correlations while in the case of P and K nutrient content and pH values the correlations were on lower level. But these relationships were not the same at both locations due to different soil moisture conditions. This was obvious especially in ECa. Similar direction of correlations as in the case of ECa was found also in thermography but mostly on lower level. On the other hand reflectance in visible and NIR spectrum showed different directions of correlations. Based on the results of correlation analysis, two different sampling designs – regular with high density of samples 50 x 50 m (totally 214 samples on the 53 ha field) and irregular with total number of 40 samples were compared. The prediction error of interpolation process of OM content from irregular grid was lower using cokriging method with ECa layer than without it and it was on the same level as ordinary kriging interpolation from regular grid. Obtained results lead to these conclusions: (1) indirect methods based on geophysical or spectral characteristics can be used for the mapping of spatial variability of chosen soil characteristics, (2) results of indirect measurements are in contrast to laboratory analysis less accurate but data are provided with much more higher density and cover the whole field, (3) combination of indirect methods with soil sampling can lead to reduction of soil samples and to lower costs of the whole process of mapping. This work was supported by the Ministry of Education, Youth and Sports of the Czech Republic as project No. 2B06124 'Reducing of impacts and risks on environment and information acquisition for qualified decision-making by methods of precision agriculture'

Evaluating water status in irrigated grapevines and olives using thermal imaging

Alchanatis, V.[1], Cohen, Y.[1], Sprinstin, M.[2], Naor, A.[2], Meron, M.[3], Cohen, S.[4], Ben-Gal, A.[5], Agam, N.[5], Yermiyahu, U.[5] and Dag, A.[5], [1]ARO - Volcani Center, Institute of Agricultural Engineering, P.O. Box 6, Bet Dagan 50250, Israel, [2]MIGAL Galilee Technology Center, Agro Innovation Migal, Southern Industrial Zone, P.O. Box 831, Kiryat-Shmona 11016, Israel, [3]MIGAL Galilee Technology Center, Crop Ecology Laboratory, Southern Industrial Zone, P.O. Box 831, Kiryat-Shmona 11016, Israel, [4]ARO - Volcani Center, Institute of Soil and Water, P.O. Box 6, Bet Dagan 50250, Israel, [5]ARO - Volcani, Gilat Research Center, Gilat, M.P. Negev 85280, Israel; victor@volcani.agri.gov.il

Irrigation management of grapevines and olive orchards should optimize both yields and quality of wine and oil, respectively. Achieving this goal depends on the ability to maintain mild to moderate levels of water stress during the growing season. Spatial in-field variability can lead to variable stress levels in different parts of the vineyard/orchard resulting in variability in the quality of grapes/olives at harvest. This can be prevented by precision irrigation at the tree level. The ability to monitor water status at the plant level is the first step towards precision irrigation. We report on the progress of an ongoing multi-year investigation of the use of thermal imaging for monitoring crop water status of individual plant. Irrigation treatments were applied in a wine-grape vineyard (vitis vinifera cv. Merlot) in northern Israel and in an olive orchard (Olea europea cv. Barnea) in southern Israel. Images in the thermal and the visible ranges were acquired from a 22 meters high crane. Images in the visible range were used to mask soil and shadowed leaves. Maps of crop water stress index (CWSI) were created using artificial and theoretical references of Tbase and Tdry. Tbase is related as the temperature of a totally transpiring leaf while Tdry is related to temperature of non-transpiring leaf. Results from two seasons will be presented (2005 and 2007 in grapevines and 2007 and 2008 in olives). In general CWSI based on both artificial and theoretical reference surfaces was well correlated with stem water potential and stomatal conductance.

The effect of dew on reflectance obtained from ground active optical sensors

Povh, F.P.[1], Gimenez, L.M.[2] and Molin, J.P.[1], [1]University of São Paulo (ESALQ), Department of Rural Engineering, Av. Pádua Dias, 11, 13.418-900 Piracicaba, SP, Brazil, [2]ABC Foundation, Farm Machinery, Rod. PR-151, km 288 - P.O. Box 1003, 84.165-700 Castro, PR, Brazil

A lot have been discussed about the use of optical sensors for nitrogen management. Considering the active ones, they can be used any time of the day, including night readings. However, there is a period of the day that the leaves are covered with dew, which is the condensation of atmospheric moisture onto a surface, forming droplets. The presence of dew can change the values of vegetation indices compared to the data acquisition in the same area without dew. This work had the objective of evaluating the effect of dew on NDVI collected with two commercial active optical sensors. The experiment was conducted in Itaberá, SP – Brazil, in an experimental field of ABC Foundation. Three conditions of soil covering were considered, one black oats crop with 10 days after seeding and two wheat crops with 23 and 35 days after seeding. The data acquisition with a GreenSeeker and a Crop Circle was realized in four different morning times, 7:30, 8:30, 9:15 and 11:30 am, which corresponded to four amounts of dew on the leaves, and three replications. Besides the NDVI, the Crop Circle sensor allowed to analyze the reflectance of visible and near infrared wavelengths. Statistical analyses were done using SAS with Tukey test at 5% of significance. The results confirm that dew influences the NDVI values for both the sensors, with significant differences in different times of the morning. The NDVI values increased 12% for GreenSeeker and 27% for Crop Circle, from the first (7:30 am) to the last reading (11:30 am). Near infrared showed small differences in the readings, not significant, unlike the results of the visible data. These results lead to conclude that the visible wavelength is more affected by the presence of dew than the near infrared. Then, the presence of dew should be considered for nitrogen recommendation based on active optical sensors.

Yield-SAFE for water management: calibration and validation for maize in Mediterranean and Atlantic regions

Mayus, M.[1], Palma, J.[2], Topcu, S.[3], Herzog, F.[4] and Van Keulen, H.[5], [1]University of Hohenheim, Intsitude of Crop Production and Grassland Research, Fruwirthstrasse 23, 70599 Stuttgart, Germany, [2]Universidade Tecnica de Lisboa, Instituto Superior de Agronomia, Tapada da Ajuda, 1349-017 Lisboa, Portugal, [3]Cukurova Üniversitesi, Ziraat Fakültesi, Tarimscal Yapilar ve Sulama Bolumu, 01330 Adana, Turkey, [4]Eidgenössische Volkswirtschaftsdepartment EVD, Forschungsanstalt Agroscope Reckenholz_Tänikon ART, Reckenholzstrasse 191, 8046 Zürich, Switzerland, [5]Wageningen University, Department of Plant Science, Haarweg 333, 6700 AK Wageningen, Netherlands

Optimisation of water and nitrogen utilization is required to increase crop yields in a sustainable and economic way. In particular in the Mediterranean region, nitrogen leaching, as a result of inappropriate fertilizer and irrigation management is a problem. Prediction models ease the development of site -specific optimum management strategies and for generating background knowledge for taking related policy decisions. Yield-SAFE, a parameter-sparse biophysical model for long-term predicting of crop and tree yields, is based on resource capture and resource use efficiency. The model takes into account irrigation. Nutrient dynamics are not included. As an alternative we propose application of the so-called 'Triple Quadrant-approach', which estimates the effect of fertiliser application on crop yields, based on two relations: 1) that between nitrogen application and uptake and 2) that between uptake and crop yield (invariable per crop species). Nitrogen leaching is quantified from model predictions on nitrogen recovery and water flow to the groundwater. The model has been evaluated in terms of biomass production of summer and winter cereals and four tree species in Atlantic and Mediterranean regions. This paper presents a refined calibration and model improvements with respect to the fluxes of water, based on data from an irrigation/fertilisation experiment with maize at the research station of Cukurova University, Adana, Turkey. The experimental data, obtained during the seasons of 2001, 2002, 2004 and 2005, comprise: daily weather, weekly soil water contents, and relevant soil physical characteristics, plant physiological measurements, crop yields and leaf area development. In a next step the belowground module was calibrated and validated for the Atlantic region with data from the Wageningen University (the Netherlands) and the University of Stuttgart (Germany). Yield-SAFE satisfactorily describes the water dynamics in the soil. Improved quantification of water capture and losses allows more accurate yield predictions. The estimates of nitrogen leaching provide information of the magnitude of the problem.

Using vegetation indices to determine peanut maturity

Vellidis, G.[1], Ortiz, B.[2], Beasley, J.[3] and Rucker, K.[4], [1]University of Georgia, Biological & Agricultural Engineering, P.O. Box 748, Tifton GA 31793-0748, USA, [2]Auburn University, Agronomoy & Soils, 202 Funchess Hall, Auburn AL 36849-5412, USA, [3]University of Georgia, Crop & Soil Sciences, P.O. Box 748, Tifton GA 31793-0748, USA, [4]University of Georgia, Tift Co. Ag Service Center, 1468 Carpenter Road South, Tifton GA 31793-7548, USA

Peanut plants develop pods which contain the desirable peanut kernels in the soil. Consequently, like root crops, it is difficult to gauge maturity by observing the above-ground plant. To prepare for harvest, peanut plants are dug, the pods shaken free of soil, and the whole plant inverted before being laid back on the soil surface to dry before harvest. Digging too early or too late can greatly reduce both the quality and yield of a peanut crop. Harvesting mature peanuts is critical to financial success. The best tasting and best paying peanut is a mature one. Maturity affects flavor, grade, milling quality and shelf life. Mature peanuts have quality characteristics consumers want and are, therefore, worth more to the producer. Determining when to harvest peanuts is always a difficult decision for peanut producers because there are always maturity differences in any peanut field. The hull-scrape method allows us to objectively determine maturity. However, to properly represent the maturity of the entire field, this method requires manually digging up peanut plant samples weekly as maturity approaches and then performing the labor-intensive test. It is cumbersome and time consuming to collect samples from many locations in a field and to do that repeatedly. Consequently, most producers may not be harvesting their peanuts at the optimal time. Harvesting decisions could be greatly improved if maturity maps of the field were available. Research has shown that vegetation indices, or VIs, derived from remotely sensed data may be used to measure plant vigor, biomass, canopy stress and plant maturity. VIs are mathematical ratios of reflectance at specific wavelengths. Reflectance is a measure of the percentage of sunlight reflected by plants at those wavelengths. Reflectance is measured with multispectral cameras or similar sensors. Two of these sensors are the GreenSeeker and the Crop Circle. Although dozens of vegetation indices have been developed, the one most commonly used is NDVI or the Normalized Difference Vegetation Index. Another commonly used index is the RVI or Ratio Vegetation Index. In this poster, we will present the findings of a study using NDVI and RVI to identify peanut maturity. We are using the GreenSeeker and Crop Circle sensors to do this because both of these instruments are commercially available and commonly used by growers in other commodities (corn and cotton are good examples). In the process, we are also evaluating each of the instruments for adaptability and ease of use with peanuts. We are evaluating our ability to predict peanut maturity with NDVI and RVI by collecting peanut samples from many locations in the field and by conducting maturity assessments on these samples. Maturity assessments are conducted weekly using the hull-scrape method. The poster will present data linking peanut maturity to NDVI and RVI values, maturity maps created with data from the hull-scrape method showing the spatial variability of maturity, comparisons between the spatial patterns of maturity maps and NDVI and RVI maps, and an assessment on whether instruments like the GreenSeeker and Crop Circle can be used as tools to help the peanut producer determine optimal maturity.

Mapping available soil micronutrients in an Atlantic agricultural landscape

Paz-Ferreiro, J.[1] and Vieira, S.R.[2], [1]University of Corunna (UDC), Soil Science, Facultad de Ciencias, 15071 Coruña, Spain, [2]Instituto Agrômico de Campinas (IAC), Soil Science and Environmental Resources, Barão de Itapura, 1481, 13020 - 902 Campinas, SP, Brazil

Taking into account the spatial variation in soil nutrient status is a prerequisite for an optimized use of precision agriculture techniques. The potential of site-specific management to increase crop yield and to reduce inappropriate fertilization has been demonstrated, however a lack of field-proven decision criteria for variable rate fertilizer application remains, which is mainly true for micronutrient amendments. Nutrient maps based on intensive soil sampling are useful to develop site-specific management practices. Geostatistical methods have been increasingly used to determine the spatial correlation and the range of spatial dependence at different sampling scales. If spatial dependence occurs, the modelled semivariograms can then be used to map the investigated variable by kriging, an interpolation method that yields unbiased estimates with minimal estimation variance. Moreover, the indicator kriging approach may be used to model the uncertainty of mapping soil properties. Until now, most of the research on nutrient spatial variation focussed on macronutrients due to its economic and ecological importance. However it is well-known that an adequate and balanced supply of all plant nutrients is required for high yields. The objectives of this paper were to examine and to map the spatial distribution of the micronutrients Cu, Zn, Fe and Mn on an agricultural area in Galicia, Spain, under Atlantic climatic conditions. The surface of the study area is 25 ha at it comprises three major soil units with a difference of about 35 m between the lowest and the highest point. The coordinate position of seventy-three randomly located sampling points was recorded with a topographic total station. Soil samples were collected from the 0-30 depth and analyzed for Cu, Zn, Fe and Mn with DTPA and Mehlich-3 extraction. All four elements analyzed showed spatial dependence. The strength of spatial dependence as assessed by the values of nugget effect and range of spatial dependence decreased in the order Cu>Mn>Fe>Zn. Spherical models were fitted and validated with the jack-kniffing technique, which showed that they were all satisfactory. The maps of individual nutrients constructed with the values estimated by kriging showed some similarity in the spatial distribution, suggesting the delimitation of uniform management zones. These similarities neither were identified with topography nor with any other single soil property such as organic matter content, pH and clay content. This is an indication that the factor o factors that affect nutrient contents changes across the study area.

Comparison between the tillage force and the cone penetrometer data

Csiba, M., Milics, G., Barthalos, P. and Nemenyi, M., University of West Hungary, Faculty of Agricultural and Food Sciences, Institute of Biosystems Engineering, Var 2., 9200 Mosonmagyarovar, Hungary

Using precision soil tillage techniques in a sustainable management both research and field work users are comparing several times the cone penetrometer data to other different measurable values, for example: yield, tillage force, soil water content, etc. Our trial was based on results of Desbiolles *et al.*, where they predicted the tillage implement draught using a cone penetrometer data. The goal of our investigations was to redeem the penetrometer measurements - but don't loose the valuable information – with an on-line method (using a tractor), which provides much more data. To measure the cone index we used 3T System penetrometer up to depth 40 cm. As a matter the soil moisture (the most influencing parameter of cone index) content was recorded of course. From these data we calculated the cone energy Pe [kPa × m] with the mathematical integration of the cross sectional area of the cone. These values were compared to the soil strength data S [kN/m] calculated from tillage force, what was measured based on our earlier development. We also modified the dimensionless numbers in the denominator of the given equation specifically to our tine. According to earlier researches in our institute it was found a correlation between tillage force and the yield, so it should be a logical conclusion that there is a connection also between penetrometer and tillage force data, as the plant growth is mostly affected by soil conditions. Also tillage draft prediction is important for proper tractor-implement matching, due to the rising cost of fuel.

The software for the mobile information-measuring complexes
Petrushin, A.F., Agrophysical Research Institute Russian Academy of Agricultural Sciences, 14 Grazhdansky prosp., 195220 Saint-Petersburg, Russian Federation; apetrushin@agrophys.ru

The software for mobile information-measuring complexes is considered. The simple complex configuration consists of the GPS-receiver, on-board computer with the software and various data acquisition sensors. Program filling provides connection between all components of the complex and enables to record to the computer memory geographical co-ordinates from the GPS-receiver, information from measuring sensors and to represent graphic information. The main advantage of the software is cross-platform for creating modular information-measuring complexes for specific problems. The cross-platform software ensures independence from on-board computer hardware. Another advantage is an open API of the software allows to independent developers connect their own units which are responsible for data exchange with measuring sensors.

Mobile measuring-calculating agro-meteorological complex

Efimov, A.E., Danilova, T.N., Kozyreva, L.V., Kochegarov, S.F. and Uskov, I.B., Agrophysical scientifically research institute, Saint-Petersburg, Grazhdanskij avenue, 14, 195220, Russian Federation

Modern technologies of precious agriculture requires availability of authentic and sufficient information for time-space differentiation of plant condition's parameters, root-dwelled soil layers and landed atmosphere layers on specific agricultural lands. Mobile measuring-calculating agro-meteorological complex was designed and manufactured for applications to route measurement of microclimate parameters in divided landscapes and agro landscapes, field crops, landscape architectures and landscape designs. With GPS transducer complex is applicable for agro-meteorological mapping of agricultural holdings and other territories. Complex supplied with original embedded software which allows controlling parameters of retrieval, storage and primary processing of agro-meteorological information. Complex consists of blocks: primary converters block; block of data acquiring and data storage; block for primary processing of experimental data. 1. Primary converters block consists of: temperature, moisture, air pressure, air speed, temperature of soil surface, radiation balance, direct and dispersed solar radiation sensors; micro-controller for input/output operations and wireless transmission of acquired information through radio channel. 2. Block of data acquiring and data storage mates with micro-controller through radio channel. This block is presented as a laptop with specialized data acquiring and storage software. 3. Block for primary processing of experimental data is presented by a collection of programs: programs to control physical parameters sensors; server program for complex settings and configuration control. Complex is running under control of Linux operating system and main programs developed with Java programming language. Example of resulting information for gradient measurements of temperature, moisture, pressure of air layers near surface and radiation flows at local points and along displacement route of sensors system (along fields and crop area) is shown. Competitive advantages: 1. Mobility and stand-alone work 2. Complexity measurement of heat and moisture flows 3. Gradient measurements of heat and moisture flows on fields, crops and landscape structures 4. Specialized mathematical software for complex configuring and configuring of desirable output data format.

Development of a yield mapping system for precision agriculture

Zhang, M., Li, M., Liu, G. and Wang, M., China Agricultural University, Key Laboratory of Modern Precision Agriculture System Integration Research, CAU 125, Qinghua Donglu 17, Haidian District, Beijing, 100083, China

In practice of precision agriculture, most important thing is to realize the spatial and temporal variability of the field conditions, yield, soil fertilizer, crop growing status, and so on. Then, it is needed to analyze the reason influence crop growing according to above information. And finally, it is necessary to input fertilizer, pesticide etc based on the crop demand. A yield map is the basis for understanding the yield variability within a field, analyzing reasons behind the yield variability, and improving management according to the increase of the profit. In order to obtain correct grain yield information and yield map, except for grain weight sensor or grain bulk sensor, some other sensors are necessary. It includes six sensors (grain flow sensor, header height sensor, elevator speed sensor, ground speed sensor, grain moisture and temperature sensor) and some other components, such as a GPS receiver, a liquid crystal display (LCD), a touch screen, and an intelligent controller. The yield monitor collects four analog signals (grain flow, header height, grain moisture, and grain temperature), two pulse signals (ground speed and elevator speed), and DGPS data at the same time. After a series of process, the system stores all data on a CompactFlash® (CF) card once per second. The yield monitor also includes a liquid crystal display (LCD) and a touch screen as the Man-Machine interface. The system was installed on the combine harvester and began to harvest test. The crop harvested was wheat, and the harvesting combine used to equip the monitor was JL1603, a typical machine in northern China with 4 m of header width. After harvest, the data that collected by the yield monitor system were output and be analyzed by some software. The maps were generated using Surfer6 via IDW (Inverse Distance Weighted) interpolation. According to this yield map, it is possible to understand the yield variability within a field, analyze reasons behind the yield variability, improve management, and then increase the profit. Field test showed that the system had the capability to predict accumulated grain mass with an error less than 3% under the given harvest conditions. More filed tests are needed to improve the calibration method and to reduce the effect of the harvest conditions on measurement accuracy of the monitor system. This study was supported by National High Technology Research and Development Program of China (863 Program): 2006AA10A305.

Method of determination of thermal properties of soil
Ivanova, K.F., Agrophysical Research Institute, 14 Grazhdansky Prospect, 195220, Russian Federation

The analytical method for thermal characteristics definition and their change with time in the arable layer of derno-podsolic soil was offered to estimate soil thermal properties for precision agriculture technologies. To solve the problem of soil temperature profile measurements, heat flow in the top soil and daily dynamics of these two characteristics received at automated agrometeorological station. The thermal parameters determination was carried out in two stages. Method includes consecutive solution of two equations: Fourier's law for the heat conduction; the equation of thermal conductivity for a one-dimensional case. The algorithm of calculation of the thermal parameters was build up on approximation of experimental temperature measurements as a function of depth and time and thermal characteristics as a function of time with help Fourier series. The offered technique of determination of thermal conductivity coefficient and soil bulk heat capacity allowed to reveal harmonious character of the coefficients change during the growing season. The calculated average values of the parameters are in the range of changes of similar characteristics for different soil types from peat to sand.

Real-time analysis of soil parameters with NIR spectroscopy
Li, M. and Zheng, L., China Agricultural University, Key Laboratory of Modern Precision Agriculture System Integration Research, CAU 125, Qinghua Donglu 17, Haidian District, Beijing, 100083, China

Precision agriculture is a management approach to variations in the field. Describing the variability of soil parameters is an important step for promoting precision agriculture. NIR spectroscopy appears as a rapid, convenient and simple nondestructive technique used to quantify several soil properties in many research. Thus, the grey-brown alluvial soil in the Northern China was selected as research object, and the estimation models of soil moisture content, SOM, and TN were developed with NIR spectra of raw soil samples. Soil samples were collected from an experimental farm of winter wheat in China Agricultural University. After the soil samples were taken into the lab, NIR absorbance spectra and soil moisture were rapidly measured under their original conditions. After measurement of NIR spectra, each soil sample was divided into two parts. One was used to directly analyze soil moisture. The other was used to analyze soil parameters after air-dried. Moisture was measured by 105 °C-24h method with an electric fan heater. Air-dried soil samples were analyzed using chemistry method. The soil parameters analyzed were soil organic matter (SOM) and soil total nitrogen (TN). The soil moisture content has high correlation with original absorbance at every waveband. 150 spectral data sets were randomly divided into two groups: calibration group and validation group. 90 data sets were used as calibration group to build an estimation model, and the other 60 data sets were used as validation group to check the model. First the correlation coefficient between the soil moisture content and spectral absorbance was calculated in every waveband. All correlation coefficient were higher than 0.93. It would be better to use single linear model to evaluate soil moisture. The correlation coefficients between soil TN content and absorbance spectra at every wavelength were calculated. Result showed that the soil TN content had low correlation coefficient with the absorbance spectra. It was necessary to seek a new means of preprocessing NIR spectroscopic data of raw soil samples, and analyze the correlation between absorbance spectra and soil TN content. Thus, 150 data sets were divided into 30 groups based on the interval of 0.055%, and 30 new data of soil TN content were obtained by calculating the average of each group of data set. Correspondingly 30 new NIR absorbance spectra were also obtained by the same calculation. The correlation coefficients of the new data sets were calculated at every wavelength. Result shows the correlation coefficient was high in main absorption bands of TN. 15 new data sets were obtained by the same algorithm to validate estimation model. Regression model was established using 30 new data sets preprocessed from original NIR absorbance spectra as above. And high correlation coefficient was obtained. The NIR absorbance spectra of raw soil samples were measured by FT-NIR analyzer. And correlation analysis and regression were conducted on three soil parameters, soil moisture, SOM and TN content. High correlation coefficients were obtained for all parameters. This study was supported by 863 Program (2006AA10A301).

Development of an online measurement system to soil EC

Zhao, Y., Li, M., Zhang, J. and Zhang, M., China Agricultural University, Key Laboratory of Modern Precision Agriculture System Integration Research, CAU 125, Qinghua Donglu 17, Haidian District, Beijing, 100083, China

It is reported that soil electrical conductivity (EC) can serve as a proxy for soil physical properties such as organic matter, clay content, cation exchange capacity, and depth of topsoil. Hence, we developed a an online measurement system to soil EC, in order to satisfy the demand of fast measuring soil properties in the farm, and support decision-making in precision management of crop. This on-the-go measuring system mounted on the tractor not only can measure soil EC in various kinds of soils, but also can perform weeding and plowing of farm works in the period of crop growth season. Furthermore, this system was easy to be operated by farmers to make field management decision. The measuring system is based on four-electrode method, which was thought to cover a big area with ease and be less susceptible to outside interference. The measuring system with six electrodes employs two arrays to investigate deep soil and surface soil respectively, and soil EC data can't only be saved in flash disk, but also be downloaded to field computer for making the soil EC map. The device consists of three parts: the sensor unit included six electrodes, the control and display unit, and the data processing software. The frame of the sensor unit adopts three-points hanging structure to connect with a tractor, thus the sensor unit will be controlled to rise or fall by the device of tractor's hydraulic pressure. Using this structure, the sensor unit has well flexibility. The tip of the electrodes is designed to be shovel shape so that the electrodes have good capability of cutting soil and weeding. Moreover, because the tip of electrodes inserts soil deeply, the tips will contact soil much morn closely. This will improve measuring stability and precision for our device too. As illuminating above, the developed sensing system has a especial advantage that the device could not only measure soil EC for various kinds of soils in no-crop condition, but also can perform weeding and plowing of farm works in the period of crop growth season. Test The performance tests of the device were carried out in a laboratory and a trail field. The results showed the output of detecting voltage electrodes and the input of alternating current electrodes had a linear relationship. With the value of alternating current increasing, the value of detecting voltage would increase in term of direct proportion relationship. This study was supported by 863 Program (2006BAD11A10-001) and (2006AA10A301).

Using X-ray method for prediction of field seeds germination in precision agriculture
Arkhipov, M.V., Gusakova, L.P., Velikanov, L.P., Vilichko, A.K., Alferova, D.V. and Zheludkov, A.G.,
Agrophysical Research Institute of the Russian Academy of Agricultural Sciences, 14 Grazhdansky prosp.,
195220 Saint-Petersburg, Russian Federation

Precision seed-growing using new biophysical methods for assessment of seed quality, is one of important trend of precision agriculture. The X-ray method developed in Agrophysical Research Institute allows to realize the early prediction of laboratory and field germination. It is determined, that the pinpointed part of latent defects in a germ of barley seeds correlates with presence in a lot of blind seeds. It is shown, that direct x-ray increase (20-30 times) allows to detect structural damages of seed that is technically impossible by 2-5 times increase. The databank of latent defects of lots of barley seeds received with the help of various technologies is created.

A new self-excited oscillator method for measuring soil water content and elecrtical conductivity in technologies of precision agriculture

Ananyev, I.P., Agrophysical Scientific Research Institute, Measurement Instruments R&D Laboratory, Grazdansky prospect, 14, 195220, Saint-Petersburg, Russian Federation

The new measurement method for material real ε' and imaginary ε' components of complex dielectric permittivity, and also electrical conductivity σ, based on usage of a self-excited oscillator with inertial stabilization of oscillation amplitude is considered, offered in Agrophysical Scientific Research Institute. The instrumentation for on-contact and noncontact measurement of soil water content and electrical conductivity, including means for in situ measurement and continuous measurement in movement, developed with use of the new method are considered. To these means concern: the instrument with a two-component complex permittivity transducer (TCPT) utilizing the self-excited oscillator for routing inspection of the fields, provided with a rod capacitance probe for measurements on depths up to 50 cm; the TCPT with a short four-pin capacitance probe for installation on various depths in soil and incorporation into structure of automatic agrometeorological stations; the mobile measuring device for on-contact measurement in movement of soil water content, electrical conductivity, temperature and horizontal penetration resistance of a soil arable layer; the device for noncontact nondestructive measurement of soil water content and electrical conductivity with the help of an inductive ring loop antenna included into a resonant oscillatory circuit of the TCPT. Developed instrumentation application enables operative measurement of soil water content and electrical conductivity for efficient evaluation of water and total mineral nutrition requirements of plants and fields during the vegetation period. The received measuring information is necessary for support of the differentiated technologies in precision agriculture and quality monitoring of agricultural assignment grounds.

The analysis and application of navigation system for autonomous vehicle in agriculture

Zhang, M., Liu, G. and Zhou, J.J., China Agriculture University, Key Laboratory of Modern Precision Agriculture System Integration Research, No.17 Tsinghua East Road, Beijing, 100083, China

This paper presents an overview on navigation system for autonomous vehicle in agriculture. The meanings, the application, and the evaluation of the navigation system for autonomous off-road vehicle are discussed. There are two key technologies of the navigation system. One is positioning, which determines vehicle position and velocity to a known reference coordinates. The other is controlling, which controls and operates the vehicle to move from the actual position to the desired location. Various position sensors, such as encoder, gyroscope, accelerator, inertial measurement unit, machine vision and Global Positioning System are introduced. In order to improving accuracy of positioning, the data obtained from different sensors need to be fused. The fusion algorithm, such as Kalman Filter, is discussed. For controlling, the block diagram of supervisory preview navigation control is analyzed. Based on a market demand analysis for different applications in agriculture, two different navigation system schemes with different cost and performance are designed. For soil sampling, the demand accuracy is about 1 m, and a low-cost navigation system is designed. For fertilizing or weeding, the accuracy is about 1-5 cm, and a high-accuracy system is designed. An evaluation method of navigation system to accuracy, cost and security is also introduced, especially to accuracy. The safety look-ahead distance and the path curve are analyzed according to different traveling speed. This study was supported by National High Technology Research and Development Program of China (863 Program: 2006AA10A304) and Beijing science and technology new start plan (2007A098)

Positioning and control technologies of automatic navigation system of agricultural machinery
Gang, L. and Xiangjian, M., China Agricultural University, P.O. Box 125 Qinghua Donglu 17, Haidian District, Beijing, 100083, China

Positioning and Control technologies are important parts in the automatic navigation system of. A Kalman filter was designed for fusing the information from the DGPS and the electronic compass to improve the navigation positioning accuracy. A fuzzy self-adjusting PID controller was designed for deciding the steering angle of the agricultural machinery to improve the control accuracy. A lot of experiments were carried based on an electro-mobile to test the performance of the whole system. The results show that the average and the range of the electro-mobile Cross Tracking Error are all reduced. The dynamic positioning accuracy and control accuracy of the electro-mobile are improved above 10% and 20% respectively.

Spatial prediction of wheat yield and grain production using terrain attributes by Artificial Neural Network (ANN)

Ayoubi, S.[1], Nouruzi, M.[1], Jalalian, A.[1], Deghani, A.A.[2] and Khademi, H.[1], [1]Isfahan University of Technology, Department of Soil Science, College of Agriculture, Isfahan, 8415683111, Iran, [2]Gorgan University of Agricultural Sciences and Natural Resources, Department of Water Engineering, 8615642, Iran

Crop production systems are very complex in semiarid regions with variable topography. The knowledge about the effects of terrain attributes on strategic crops such as wheat production can provide valuable information for sustainable management of landscape. Spatial variability of yield production and affecting factors within the hillslopes could be applied for precision farming. Black box models such as the artificial neural network (ANN) provide a mathematically flexible structure to identify the complex non-linear relationship between inputs and outputs withoutattempting to explain the nature of the phenomena The objectives of this study were to: (1) compare of the prediction capabilities multiple linear regression (MLR) models to artificial neural network(ANN)models for rainfed wheat yield production and (2) determine important terrain factors that influence rainfed wheat yield components in Ardal district of Iran. Wheat yield data collected from 100 selected points at 1 m^2 plots in late June of 2007. Sampling points were distributed randomly stratified at the given geomorphic surfaces in the study area including: summit, shoulder, back slope, foot slope, and toe slope. Digital elevation model (DEM) was applied to calculate primary and secondary terrain attributes (Elevation, Slope, Aspect, Curvature, Plan Curvature, Profile Curvature, Specific catchment area, Wetness index, Stream power index and Sediment transport index). Artificial neural network analysis was performed using Feed Forward Back Propagation method by Neural Solution software. ANN models for the study area resulted in R^2 and RMSE of 0.81, 0.031 for biomass and 0.72, 0.048 for grain yield respectively. Whereas, MLR models could just explained 62% and 22% of biomass and grain yield of wheat variability with 0.76 and 0.34 RMSE respectively within the study area Sediment transport index (LS factor) and Stream power index were identified as the top terrain attributes for explaining variability of biomass and grain yield, respectively. These results indicated that hydraulic processes and soil sediment transport through the hillslope had a significant effect on distribution of moisture, soil colloidal portions, and soil nutrition, which can affect directly on crop production. Overall results indicated that ANN models had more performance to predict of wheat yield components in the hillslopes in the semiarid region using terrain attributes.

Diagnosing crop growth and nitrogen nitritional status around panicle initiation stage of rice with digital camera image analysis

Lee, K.J., Choi, D.H. and Lee, B.W., Seoul National University, Department of Plant Science, Kwanakgu Shillim-9-dong, 151-921 seoul, Korea, South

Nitrogen top dressing management at panicle initiation stage (PIS) of rice is important and the last chance for fine-tuning the plant growth and leading to high yield and high quality production of rice. Model prescribing nitrogen topdressing rate at PIS was constructed and validated for its effectiveness through a series of previous field experiments. The model uses shoot nitrogen density calculated as shoot dry weight x shoot N concentration at PIS of rice and soil N supply from PIS to maturity for calculating the nitrogen top dressing rate for securing target yield and/or target protein content. However, the monitoring of shoot N density is destructive, time- consuming, and expensive. We tried to employ digital camera, available universally to farmers, for non-destructive, timely, and cheap monitoring of shoot dry weight and N concentration. Field experiments consisting of four rice varieties and five N fertilization levels were conducted in 2006, and color images of paddy field over the experimental plots were taken at a constant height of 2 m with the digital camera between 11:00 and 15:00 around panicle initiation stage of rice. Just after taking color images, rice plants were sampled for measuring shoot dry weight and N concentration. Digital image analyses were done with a simple program developed in Visual Basic. The percent rice canopy cover that was calculated as the ratio of the number of rice plant pixel to total pixels of digital camera image taken over rice canopy around PIS, showed good correlation with shoot dry weight of rice canopy. And also normalized R, normalized difference index and hue value that were calculated from RGB values of plant pixels showed highly significant correlations with shoot nitrogen concentration. These results indicate that digital camera image analysis could be applied to diagnosing crop growth and N nutrition status for recommending N topdressing rate of rice.

Environment monitoring system based on the wireless sensor network

Li, L. and Haixia, L., China Agricultural University, Key Lab on Precision Agriculture System Integration Research, P.O. Box 125, Qinghua Donglu 17, Haidian, 100083, Beijing, China

Wireless Sensor Network (WSN) is increasingly popular in the field of environmental monitoring due to its promising capability. WSN is a system comprised of radio frequency (RF) transceivers, sensors, microcontrollers and power sources. Instrumented with a variety of sensors, such as temperature, humidity and CO_2 detection, WSN can operate in a wide range of environments and provide advantages in cost, size, power, flexibility and distributed intelligence compared to wired ones. In recent years, energy saving technology becomes more and more important in precision agiculture in the world, due to the higher price of energy. Many producers changed their produce mode and made the strategies more flexible. WSN offers permanent online access to the environmental condition. In a network, if a node cannot directly contact the base station, the message may be forwarded over multiple hops. By auto configuration set up, the network could continue to operate as nodes are moved, introduced or removed. A greenhouse environment monitoring system was designed first taking the advantages of the WSN, which integrated wireless sensor network, embedded technologies and GPRS. A field controller was designed based on the ARM9 and embedded Linux operating system. The data of greenhouse environment could be received, transformed, displayed, and saved by the controller. The remote database server could receive the data and send command to the field controller device. The large amount of data of greenhouse environment acquired, and then communicated by the wireless sensor network. The design gives a new way to collect the data of environment instead of the tradition way using wires.

A new sensor investigation for in-field soil nitrate nutrient rapid detection

Zhang, M.[1], Ang, S.[2], Wang, M.H.[1] and Nguyen, C.[2], [1]China Agricultural University, Key Laboratory on Modern Precision Agriculture System Integration Research, Ministry of Education, Tsinghua East Rd. No.17, 100083 Beijing, China, [2]University of Arkansas, Department of Electrical Engineering, University of Arkansas, 72701 Arkansas, USA

In the past few decades, excessive fertilization caused higher production cost, lower product quality and serious environmental pollution and resource waste in China. The soil testing for fertilization recommendation had become wide spread nationwide since 2005. Soil nitrate test was a key part for guiding site-specific fertilization. Conventional testing methods are either time consuming or too expensive. Simple, rapid, robust and reliable soil nitrate detection methods, suitable for in-field application, are urgently needed. Many soil parameters were tested and analyzed based on commercially available equipments. However, these systems are quite complex and expensive for soil nutrient detection in lab. Ion selective electrode (ISE) technology has been widely employed in environmental analysis. But, the inconsistent repeatability of ion selective electrodes limited their extensive use in practical systems. In this paper, a platform for rapid detection of soil nutrients with automatic fluidic control system was developed. Micro pumps were used for solution delivery. Custom controller software running on the laptop was programmed to automatically control the fluidic system. The polypyrrole modified glassy carbon electrode was investigated as the sensing unit, which were mounted onto a sealed test chamber. Potentiometric response of the nitrate-doped polypyrrole ($PPy(NO_3^-)$) electrode followed a quasi-Nerstian response over a linear range from 10^{-4} M to 10^{-1}M nitrate. The lower detect limit was $(4\pm1)*10^{-5}$M. Because of the morphology change of polypyrrole film, potential drifts occurred when detections were conducted at the interval of more than 1 hour. As such, Nernst slope was used in favor of the potentiometric measurement. After 4.5 minutes of self-calibration, only 90 seconds are needed to perform a sample test. The potentiometric slopes obtained were (50 ± 2) mV/decade from 10^{-1}M to 10^{-2}M $NaNO_3$, (45 ± 3) mV/decade from 10^{-2}M to 10^{-3}M $NaNO_3$, and (20 ± 5) mV/decade from 10^{-3}M to 10^{-4}M $NaNO_3$. Compared to the conventional method, the nitrate sensor with automatic fluidic control system enormously increases soil nutrient test efficiency with acceptable test accuracy.

Effect of organic manure and nitrogen level on sugar beet (*Beta vulgaris*) yield and root nitrate content
Abd El Lateef, E.M., National Research Centre, Field Crop Research Department, Tahriir Street, Dokki, Cairo, 12611, Egypt

Sugar beet is a relatively new important crop in Egypt and it demands moderate quantities of organic manure and nitrogen and lower water requirements compared with sugar cane; however, due to some environmental problems like nitrate leaching in the desert lands and the lack of organic manure, it is important to find alternative organic resources. Field trial was conducted in the winter season of 1997/98 on a private farm,Nubaria district; Behaira Governorate (84 km Alex-Cairo desert road); in a newly reclaimed desert soil. The objective of the trial was to investigate the effect of organic manure (composted sludge) and nitrogen on sugar beet yield. Another objective of the trial was to evaluate the fertilizer equivalency value of the organic manure types. The area of the trial was 0.25 ha (0.59 feddan), the physical and chemical analysis of the soil was⁻ (pH 8.5; EC 0.24 dsm^{-1}; OM 0.73; N 1,400 ppm; P 132 ppm; K 826 ppm; Fe 3694 ppm; Mn 56.8 ppm; Zn 17.8; Cu 3.78; Cd 0.02 ppm; Pb 1.36 ppm; Ni 2.9 ppm). The experiment included 20 treatments which were 5 nitrogen fertilizer levels i.e.; 0,15,30, 45 and 60 kg fd^{-1} and 4 compost levels i.e. 5, 10, 15 and 20 m^3 fd^{-1} with and without adjusted N fertilizer rate (30 kg N fd^{-1}). The experimental design in the trial was complete randomized block. Root and shoot yields fd^{-1} were determined. Sap nitrate concentration was determined by Nitra- check meter. The N equivalency value was estimated The data were statistically analysed using software pckage (Cohort 2). Highly significant effects of the treatments were found by ANOVA ($P<0.001$). Nitrogen fertilizer application on its own did not result in any significant effects on yield or sap nitrate concentration. Highest root yield and sap nitrate was observed at 30 kg N fd^{-1}, although shoot yields were progressively increased by fertilizer addition up to the highest rate applied (60 kg N fd^{-1}). Compost applied at 5 m^3 fd^{-1}, with or without N fertilizer (30 kg N fd^{-1}), did not significantly increase yields over the control. Higher application rates of compost, with or without the adjusted rate of N fertilizer, significantly increased yields over the lower rate of compost or. Thus there were substantial yield benefits from compost addition, although there was also an eight-fold increase in sap nitrate concentration which could potentially detract the commercial value of the crop to the sugar beet processor. Nitrogen fertilizer equivalency value of the compost for sugar beet production was determined and indicated that the regression coefficients for the yield responses to mineral and compost N were similar and suggested there was equality in N value between the fertilizer and the compost. Crop yield increased in linear relation to N addition although much higher root production was observed with increasing rate of N application in compost compared with the range of N application rates supplied for inorganic fertilizer. Sap nitrate concentration in beet root was also raised significantly by compost application. The measurement of plant nitrate status confirmed the significant plant available N content in the compost, but in the case of sugar beet a high nitrate content may reduce the economic value of the crop to the beet processor. 1 fd = Feddan (4,200 m^2).

Radio-controlled complex of agricultural crops aeromonitoring for information support of precision agriculture technology

Slinchuk, S.G., Petrushin, A.F. and Aivazov, G.S., Agrophysical Research Institute Russian Academy of Agricultural Sciences, 14 Grazhdansky prosp., 195220 Saint-Petersburg, Russian Federation

The radio-controlled complex of agricultural crops aeromonitoring for information support of precision agriculture technology has been developed in Agrophysical Institute to collect heterogeneous data about crops state on the basis of an aerial photography. The complex consists of radio-controlled model of the glider airplane equipped with a photocamera and a field computer with the software to process findings. The information system is based on digital photographs accessing from the radio-controlled complex of aeromonitoring and following processing by object-oriented decoding methods in ERDAS Imagine software environment. The vector objects (polygons) with certain attributive data are generated as a result of the classification. GIS-analysis is made on the basis of the obtained information and required subject maps and schemes for decision making on agrotechnological procedures management are developed.

In situ and at real time characterization of the heterogeneity of soil and the dynamical behavior vehicle

Chanet, M.[1], Marionneau, A.[2] and Seger, M.[1], [1]Cemagref, UR TSCF, 24 Avenue des Landais, BP 50085, 63172 Aubiere, France, [2]Cemagref, UR TSCF, Domaine des Palaquins, 03150 Montoldre, France

To assess the real comportment of the soil-machine interaction during the displacement of vehicles, a tractor wheel, named 'the sensor-tyre' (patent no. 05-11455), was tooled up with electrodes measuring the electrical impedance of the soil under the wheel and linear transducers providing the vertical, longitudinal and radial deformation of the tyre. The electrical impedance was used to characterize the soil state (i.e. the water content and the bulk density of the soil). Shaped by the wheel load and the soil bearing capacity, the deformations of the tyre offer additional information about the soil-wheel interaction. Since the mechanical properties of soil are strongly influenced by moisture and bulk density, we suppose that the combination of the four measurements provided by the sensor-tyre (the tyre deformations along the x, y and z axes and the soil electrical impedance) may be used to evaluate (1) the wheel load, (2) the area of the tyre-ground contact, (3) the torque applied to the wheel and (4) the soil strength. This work aims at proposing simple models to evaluate these parameters for sandy situation. It was intended to provide, in situ and at the real time of field operation, the key information required by current models to estimate the risk of soil compaction. The deformations of the tyre (traction tyre Kleber 16.9 R 38) are given by the displacement along the x, y and z axes (calculated after trilateration) of a point P located inside the tyre (in the middle of the tyre width) and related with strings to three transducers fixed on the rim. The x and y axes belong to the plane tangent to the tyre at P point, the first is oriented to the wheel rotation direction and the second to the transversal direction. The z axe is orthogonal to the previous axes. For the electrical impedance of the soil, the electrodes were fixed at the top of 23 lugs, on each side of the tyre. The parameters which we made vary during the experiments are the load and the couple applied to the wheel, the pressure in the tyre, as well as the humidity and the nature of the ground. For all the experiments, the latitude and the longitude were supplied by a Differential Global Positioning System (DGPS). The signal treatment and statistical analyses were achieved with Matlab and are presented and discussed in this paper. This paper presents original solutions to evaluate, in situ and at the real time of field operation, several parameters which can for example achieved the follow-up of the grounds compaction, or also the security of vehicles and operators during the field operations.

Participatory-decision support system for irrigation management based on earth observation methodologies and open-source webGIS tool: a case study from Italy

Osann Jochum, M.A.[1], Belmonte Calera, A.[1], Nino, P.[2], Lupia, F.[2] and Vanino, S.[2], [1]Universidad de Castilla-La Mancha, Instituto de Desarrollo Regional, C/ Bachiller Sansón Carrasco, 15, E-02071, Spain, [2]National Institute for Agricultural Economics - INEA, Servizio 5, via Barberini, 36, 00187, Italy

The PLEIADeS project (funded by European VI Framework Program) addresses the improvement of water use and management in agriculture through innovative Information Technologies (IT) and the most recent Earth Observation (EO) methodologies. Within this framework a tool, based on 'open source' softwares, which aims at helping water managers to optimize the water consumption, has been realized. One of the key features of the system is the delivering of a near real-time irrigation schedule to farmers produced through the integration of EO–derived product and field data inside a GIS environment that provides a reliable crop requirement estimation at farm level. The paper describes the approaches and the first results obtained during the project activities in the CugaRiver basin (Sardinia–Italy), a site with relevant irrigated areas and with a strong conflict for water target (industrial, agricultural, civil, etc.). System development has been based upon two research parts devoted to the analysis of both technical and social aspects. The technical part consists in the SPIDER software development through user requirement analysis and EO-based models validation by field measurement survey. During this phase, satellite image acquisition, field data collection (crop phenology and Kc, height, fractional cover and agrometeorological station data) and generation of multi-level EO-assisted products (colour combination image, NDVI map, map of crop coefficient Kc, map of crop evapotranspiration ETc) has been performed. The non-technical part gives a strong emphasis to the active participation among all stakeholders involved in water management (legislators, planners and users) in order to capture all those characteristics useful for tailoring the system to the investigated area.

Assessment of weed cover: a study of the consistency and accuracy of human perception
Andujar, D.[1], Ribeiro, A.[2], Fernandez-Quintanilla, C.[1] and Dorado, J.[1], [1]ICA-CSIC, Serrano 115B, 28006 Madrid, Spain, [2]IAI-CSIC, Carretera de Campo Real km 0.2, 28500 Arganda del Rey, Madrid, Spain

Visual estimation of relative leaf area of weeds (weed cover) is commonly used in weed science to quantify weed infestation level in crops. This methodology is especially indicated when a large surface or a large number of samples must be assessed. However, the visual estimation carried out by an observer may have several sources of imprecision: lack of experience, lack of attention after a period of repetitive assessments, influence of the previous cases assessed, influence of the infestation level present (it is easier to discriminate between 0 and 5% weed cover than between 50 and 55% weed cover). Automatic image analysis systems for monitoring weed cover may provide a good alternative to visual estimation. In order to design accurate and consistent methods for automatic weed assessment it is necessary to know which are the major factors affecting this process. A perception study was conducted in order to evaluate the reliability and consistency of visual estimation of weed cover. A total of 747 sampling points were sampled during three years in two maize fields located in Central Spain. Digital images were obtained in a regularly spaced grid of 12 ×12 m. Photographs were taken in each point with a 35-mm camera from a height of 1.6 m. The images were obtained in May, at the three-to-four leaves growth stage of maize. Percent weed cover was determined in two different ways: (1) four observers visually evaluated the digital images, estimating the area covered by weeds using a scale from 0 (no weeds) to 100% (complete cover); and (2) hand drawing (in black) on a transparent sheet the area covered by weeds in the image, digitizing the black and white images with a frame scanner and analyzing the produced pixel values for weeds by using a computer software. The 'errors' in the visual estimations were estimated by comparison with real weed cover data obtained by digitalization of manual drawings. A reliability analysis was conducted to assess the differences in the visual estimation values recorded by the four observers. Repeatability analysis were conducted to assess the differences in the assessments performed by an observer over time or when the images were presented using a different order. The results of this study have shown a poor agreement between the assessments of the various observers as well as significant errors in their visual estimation of weed cover. In general, visual perception tended to overestimate weed cover at low weed densities and underestimate it when densities were high. Other factors responsible for this poor agreement will be discussed in the paper. The results obtained justify the need of automatic image analysis systems for monitoring weed abundance.

An inexpensive process for 350-1100 nm wavelength aerial photography for precision agriculture

Pudełko, R.[1] and Igras, J.[2], [1]Institute of Soil Science and Plant Cultivation – State Research Institute (IUNG-PIB), Department of Agrometeorology and Applied Informatics, Czartoryskich 8, 24-100 Puławy, Poland, [2]Institute of Soil Science and Plant Cultivation – State Research Institute (IUNG-PIB), Department of Plant Nutrition and Fertilization, Czartoryskich 8, 24-100 Pulawy, Poland; rpudelko@iung.pulawy.pl

The aim of this work was to present the possibilities of obtaining four-channels of aerial photographs (blue, green, red, infrared) and their use in agriculture research. Two synchronised and gyro-stabilised digital cameras FujiFilm IS-1 sets were used, which were installed on a Cessna airplane. The digital cameras used in research were factory set and adapted to perform photos in 350-1,100 nm wavelength. To separate the photographs content into spectral channels, selective filters were used and installed on the lens as well as combining the photographs from two cameras in a GIS computer program. The high-resolution multispectral photographs were the source material for the spatial analyses. For example, it can be used in precision agriculture - e.g. vegetation index (NDVI, SAVI) map drawing. In this work, an evaluation of the cost for obtaining this type of data and the cost comparison of an alternative manner for obtaining the high-resolution multispectral aerial and satellite photos were presented.

Low altitude aerial photography used in remote sensing for monitoring the expansion of diseases

Nieróbca, A., Pudełko, R. and Kozyra, J., Institute of Soil Science and Plant Cultivation – State Research Institute (IUNG-PIB), Agrometeorology and Applied Informatics, Czartoryskich 8, 24-100 Pulawy, Poland

The aim of this research was to present the practical application of remote sensing methods in order to monitor the expansion for focal points of fungus diseases in winter wheat cultivation. This was performed with the use of non-metric aerial photography, taken at a low altitude (approx. 600 m) from a Cessna airplane over different dates. The photography was taken over fields from the Research Farm IUNG-PIB located in southern eastern Poland. The aerial photography was analysed with the spatial method, based on spectral properties recorded in the RGB channels, with the Definiens 7.0 program. The index of infected roots, index infected stem base diseases, and two vegetation indexes (NDVI, SAVI) were performed on selected places on the field by the use of ground observations. A comparison between the remote control data with infections and the vegetation index, allowed a classification of the spatial analysed results from the aerial photography, which enabled obtaining a map of infection for the field. A comparison was made between the fungus disease infection maps from different dates, and this allowed the infection spread rate to be accessed for the vegetative period, as well as defining a calendar for carrying out plant protection processes for a precision agriculture system.

Use of RADAR measured precipitation data in disease forecast models
Jung, J. and Kleinhenz, B., ZEPP, Rüdesheimer Str. 68, 55545 Bad Kreuznach, Germany

Infection of leaves by plant diseases depends in high degree on leaf wetness duration. Availability of leaf wetness data depends on the number and location of met. stations and is low in agricultural regions. A second method to obtain leaf wetness data is the simulation by using standard met. data (temperature, rel humidity, radiation, precipitation). A new method to receive precipitation data with high resolution is measurement by radar. German Meteorological Service (GMS) provides such data in a resolution of 1 km^2 thru a System called RADOLAN. In an project the usability of such data for leaf wetness simulation and disease forecast is researched. The validation of precipitation data took place in special chosen validation areas. For the verification of radar data the GMS grid was accomplished with the grid of the meteorological stations. In this way it was possible to relate spatial identical values. These pairs built the basis for verification, which was done by statistic methods. In addition it was checked, if the results from the models used by ZEPP (Central Institution for Decision Support Systems in Crop Protection), gave the same value calculated with the various measured precipitation values. Therefore the leaf wetness was calculated with both methods and compared. Additional other aspects like the distance to the radar system were analysed. Results showed an improvement of spatial disease forecast given by the DSS.

Geophilus electricus: a new soil mapping system

Ruehlmann, J.[1], Lueck, E.[2] and Spangenberg, U.[2], [1]Institute of Vegetable and Ornamental Crops, Theodor-Echtermeyer-Weg 1, 14979 Grossbeeren, Germany, [2]University of Potsdam, Institute of Geosciences, Karl-Liebknecht-Strasse 24, 14476 Golm/Potsdam, Germany

Digital soil mapping becomes more and more popular and necessary in precision agriculture. This trend is rooted in the common requirements to decrease the input of energy and costs to the agricultural production processes as well as to protect the environment e.g. by minimizing the losses of nutrients from the plant-soil-system. In this context, the knowledge of spatial variability soil properties becomes essential. The indirect methods of geophysical measurements allow rapid and cost-efficient acquisition of soil related data. Therefore, we want to present a new soil mapping system – Geophilus electricus which is based on soil electrical conductivity (EC_a) measurements. The Geophilus system is based on rolling electrodes moved by jeep or tractor and is combined with a measurement instrument developed by T. Radic. The current electrode configuration is an equatorial dipole-dipole array with 1 m dipole width and 0.5 m dipole spacing which allows measuring electrical soil parameters in 5 different depths levels up to about 1.5 m. The measurement instrument measures the complex resistivity (amplitude and phase angle) for a frequency range between 1 mHz and 1 kHz. Four frequencies can be measured simultaneously. The resulting spectral information can provide additional information for an enhanced characterization of the soil. Simultaneously, GPS-data are recorded for geo-referencing the electrical measurements. The measurement velocity of the rolling electrode system is about 15 km/h. Using a distances of about 18 m between the rows, the daily mapping capacity is about 100 hectare. The data taking rate is about 1/s. Besides the presentation of the technical details of the Geophilus system, we want briefly to overview the results of current system application.

Using optical remote sensing for estimating canopy water content
Clevers, J.G.P.W., Wageningen University, Centre for Geo-Information, P.O. Box 47, 6700 AA Wageningen, Netherlands

Hyperspectral remote sensing has demonstrated great potential for accurate retrieval of canopy water content (CWC). This CWC is defined by the product of the leaf equivalent water thickness (EWT) and the leaf area index (LAI). In this paper, in particular the spectral information provided by the canopy water absorption features at 970 nm and 1,200 nm for estimating and predicting CWC was studied using a modelling approach, in-situ spectroradiometric measurements, and airborne hyperspectral data. The relationship of spectral derivatives and water band indices with CWC was investigated with the PROSAIL radiative transfer model and tested for field spectroradiometer measurements on both a grassland site and a natural vegetation site. For the latter also an airborne hyperspectral image was used. PROSAIL simulations showed a linear relationship between derivatives at the slopes of the 970 nm and 1,200 nm absorption features, hardly sensitive to variations in model input like leaf dry matter content, leaf structure parameter, leaf angle distribution, hot-spot parameter, leaf angle distribution, soil reflectance, illumination geometry and viewing geometry. At 942.5 nm the derivative yielded an R^2 of 0.98 with CWC. For 20 plots at a homogeneous grassland field spectral derivatives obtained with an ASD FieldSpec spectroradiometer yielded an R^2 of 0.71 for the derivative at 942.5 nm with CWC. For 12 plots at a heterogeneous natural vegetation site ASD spectral measurements yielded an R^2 of 0.43 for the derivative at 942.5 nm with CWC. The derivative at 946.5 nm, obtained from the airborne hyperspectral sensor (AHS), gave an R^2 of 0.51 with CWC. The relationship obtained with the PROSAIL simulations could be used as a predictor for CWC at both sites with promising results for mapping CWC over larger areas, giving them a physical basis and more general applicability. Derivatives outperformed water band indices. Since the spectral information at the left slopes of the water absorption features may be confounded by atmospheric water vapour absorption, our findings suggest derivatives at the right slope of the 970 nm absorption feature also to be useful in estimating and predicting CWC.

Assessment of soil properties on cropping farms with *Geophilus electricus*

Spangenberg, U.[1], Lueck, E.[1] and Ruehlmann, J.[2], [1]University of Potsdam, Institute of Geosciences, Karl-Liebknecht-Str.24, 14476 Golm/Potsdam, Germany, [2]Institute of Vegetable and Ornamental Crops, Theodor-Echtermeyer-Weg 1, 14979 Großbeeren, Germany

To date electrical conductivity surveys have been used in agriculture to identify heterogeneous pattern in the field without differentiation of these pattern with depth. However, the combination of electrical survey with electrical sounding based on multi electrode technology is state of the art in geophysical exploration methods. It allows the delineation of conductivity structures in the three spatial dimensions and the transformation of the measured structures in a subsurface model. Numerous lab series of complex electrical conductivity measurements provided evidence that beside that absolute conductivity value the phase angle and the frequency dependence of complex electrical conductivity allows a more reliable interpretation of the measured data. Based on these experiences the University of Potsdam together with the IGZeV developed a multi-electrode soil sensor called „Geophilus electricus'. This multi-electrode system allows the measurement of the complex electrical conductivity of the underground with 4 defined frequencies in 5 depth intervals. The project „Precision Farming With A High Resolution Electrical Soil Sensor System' aims on the verification of practical applicability of the system under routine conditions in two cropping farms. We will present data series measured over the period of two years together with their interpretation with respect to soil properties of agricultural interest.

Kopernikus inside of EarthlookCz project

Charvat, K., Horak, P. and Vlk, M., Wirelessinfo, Cholínská 1048/19, 784 01 Litovel, Czech Republic;
charvat@wirelessinfo.cz

Global Monitoring for Environment and Security is one of 4 ranges of solutions in the framework of ERA-STAR Regions project. GMES is one of the most important issues of European Space Policy and is focused on the building of an Earth monitoring system from which the data will be used mainly for environment and security. Project EarthLookCZ (www.earthlook.cz) is one of pilot projects developed in the Czech Republic in the ERA-STAR regions framework. Support the implementation of GMES in the Czech Republic is one the main goals of the EarthLookCZ project. An Analysis of the GMES Situation in the Czech Republic was completed in 2007 as the first project output. The analysis presents the importance of GMES within a European policy framework for Space research and refers current GMES activity on the both the European and national levels. In the Czech Republic the analysis is focused on current GMES solutions projects and on the activities currently taking place in the EarthLookCZ project with the main goal the implementation of the solutions. An overview of current data sources in public administration is another part of the study. From a technical point of view, the main outputs of the project are presented by The Design of Technological Infrastructure on the Basis of ISO and OGC Standards and also Design and Implementation of the Prototype of GMES Spatial Data Infrastructure Solution. The solution is designed in distributed system form, which will provide the connection to metadata about GMES data and services. This early prototyping solution will test the principle of GMES catalogue services on a national level which should be used for the later international EarthLookCZ context. A GMES national catalogue portal is one from independent components of EarthLookCZ complex system for raster and vector spatial data sharing. The catalogue portal will provide data source searching on the basis of their metadata records through structure queries. The portal will also contain edit functionality for new metadata records creating or editing. The metadata and catalogue system corresponds to ISO 19115/19119/19139 standards and provides for cascade searching on the other standardized catalogue systems, GeoNetwork for instance. The main requirement is creation of a mobile field-data collecting system with possibility to update it in future. User-programmable system should be based on open formats and structures and should not be strictly bound to one device type. System may be developed on both commercial and OpenSource platforms and should contain both common mapping tools and special utilities aimed clearly to problems solved within the Forest Management Institute. This system should allow the first of all: • usage of existing data sets for data collection; • integration of Earth Observation data to the process of data collection; • data input to user-defined structure; • connection system to other special purpose devices supporting data collecting; • connection to map servers – query system and update systems. Module architecture will allow building different applications from the simplest (for example just for GPS data collecting) to the applications with maximum functions.

Interconnection of altitude of stationary GPS observation points and soil moisture with formation of winter wheat grain yield

Lapins, D., Dinaburga, G., Plume, A., Berzins, A. and Kopmanis, J., Latvia University of Agriculture, Institute of Soil and Plant Science, Liela iela 2, Jelgava, LV-3001, Latvia

Field trials were carried out at the Research and Study farm 'Vecauce' of Latvia University of Agriculture during the years 2006 and 2007 to investigate factors influencing formation of winter wheat grain yield. Researches have been carried out in the stationary observation points. Results show tight negative correlation between altitude of observation points and soil moisture. Correlation is significant ($P<0.05$) in both trial years but coherence is more tight in year with reduced amount of precipitation, like it was observed in period April–July 2006. Significant negative correlation established between altitude and winter wheat grain yield. In year 2006, when lack of precipitations was observed, this coherence is with increased probability $P<0.01$. Correlations between altitude and number of wheat plant per square meter in autumn, fresh weight of plant in tillering stage, coefficient of tillering and area of flag leaf was not significant. Soil moisture at the depth of 40-45 cm were below optimum – 25% - in both trial years and also in both stationary observation levels with average altitude 95 and 102 m above the see level. Also it was significantly lower in the top points of terrain. Significant higher grain yield of winter wheat were obtained in field points with an average altitude 95 m above the see level. Also this coherence is more explicit in year with less precipitation like it was observed in year 2006. Analysis of correlation established that soil moisture at the depth of 40-45 cm has significant positive impact on winter wheat grain yield and on the area of flag leaf. Increase winter wheat yield levels increase also soil moisture significance in yield formation process.

Modelling spatio-temporal weed population dynamics for the development of strategies in site specific weed control

Nordmeyer, H.[1], Richter, O.[2] and Sandt, N.[2], [1]Julius Kühn-Institute, Institute for Plant Protection in Field Crops and Grassland, Messeweg 11-12, 38104 Braunschweig, Germany, [2]Technische Universität Braunschweig, Institute for Geoecology, Langer Kamp 19c, 38106 Braunschweig, Germany

Weeds have a patchy distribution within fields. This is due to heterogeneity in soil and environmental factors and to weed biology. The concept of site specific weed control take into account this variability. This concept is intuitively attractive but the adoption of this concept in agricultural practice has been slower than initially expected. A reason for this delay is rooted in missing automatic weed recognition systems and in the assessment of consequences of site specific weed control for following crops. Weed population simulation models provide the possibility to estimate the effect of different weed control strategies and to give informations about weed infestation in following crops. Herbicide application strategies are investigated to minimize the amount used without significantly yield reduction. Site specific weed control considering economic thresholds is a promising strategy for reducing herbicide use. A mathematical model has been developed to study spatial dynamics of different weed populations. Spatial dynamics were modelled by coupling a local population model with a cellular automaton model. The model was applied to several weed species, Apera spica venti, representative for a grass weed, as well as Stellaria media, representative for a broad-leaved weed. Different scenarios of site specific weed control were simulated over different time periods and the weed population were calculated and evaluated. This approach also allowed for the consideration of the dependence of weed dynamics on habitat characteristics. By means of this cellular automaton, the spatial spread of weed species can be modelled under consideration of heterogeneous distribution of soil properties on a field. Validating the model, the actual spatial distribution of weed species is compared with the results of simulation taking into account the soil properties and starting from a series of different initial weed distributions. Based on the actual distribution infestation forecasts for the following years are presented. The use of this model for site specific weed control is discussed.

Spatial and temporal variability of N, P and K in a rice field

Morales, L.A.[1] and Paz-Ferreiro, J.[2], [1]University Nacional del Nordeste (UNNE), Soil Science, Sargento Cabral, 3400, Corrientes, Argentina, [2]University of Corunna (UDC), Soil Science, Facultad de Ciencias, 15071, Coruña, Spain

Rice plays an increasing important role in Corrientes and neighbouring provinces of Argentina, where lime amendment is becoming a current practice because of the acid nature of the soil. Soil spatial variability is a natural occurring and or management induced feature, which is important for site-specific management. The objective of this study was to compare the effect of lime additions on the spatial variability of macronutrients (NH_4^+-N, P, K) measured by routine methods at a rice soil in Corrientes (Argentine). The study soil was a typic Plintacualf. Field trials were set up involving three treatments, control plus two dolomite applications of 625 and 1250 kg.ha^{-1}. Soil was sampled at three different stages, first in aerobic conditions and also two more times in anaerobiosis. Ninety-three samples per plot were taken during each of the three sampling periods. Since the estimated range of spatial dependence relies upon sampling area size, a nested sampling strategy was used. The basic sample grid was 11.9 x 20 m and the study area 5.1 ha. The used scheme provided sufficient numbers of pairs over a wide range of distances and thus allows identification of short- and long-range variations. The statistical variability was rather important. Significant nutrient availability differences between dolomite treatments were found. Moreover, the spatial variability of the studied soil properties was assessed using semivariogram analysis and examination of the kriged maps. All the three elements analyzed exhibited a rather strong spatial dependence with small nugget effect and this all over the three different study periods and for the three lime treatments. Geostatistical techniques were used to examine de data as follows: (1) spatial structure of macronutrient contents within and between crop stages, (2) stability and persistence of patches with high and low nutrient content. Geostatistical analysis also provided insight into possible processes responsible of the observed spatial variability patterns within the rice soil.

Modeling case study of expected maize yield quantity

Boksai, D.B. and Erdélyi, É.E., Corvinus University, Department of Mathematics and Informatics, Villányi út 29-43., 1118. Budapest, Hungary

Serious consequences of global climate change are expected in agriculture. The aim of our experiment was to study the effect of different sowing times based on two weather scenarios (transient UKTR 3140 developed by U.K. Hadley Centre and a scenario modeling the current weather conditions in Debrecen) on maize grain yield, biomass quantity and phenophase starting days. Our aim was to find out whether extreme climate effects can be buffered by earlier or later sowing times. Our baseline was sowing on 25th April. We applied further four sowing times, pushing the sowing day one week earlier and one week later. Therefore, our earliest date was 11th April and latest date was 9th May. We used the most up-to-date, 4.1 version of a simulation model system called 4M, which is based on the CERES model and which has been further developed and updated by Hungarian scientists. 4M calculates soil-plant-weather connections on a daily level. Our results show that yield quantity does not change significantly with different sowing times. But if sowing times are two weeks earlier, yield and biomass quantity and yield safety increases, resulting in a lower risk. With earlier sowing times, ripening is more likely, as the plant avoids the warmest time.

Decision making for cattle using movement-based pattern recognition
Godsk, T.[1] and Kjærgaard, M.B.[2], [1]Danish Agricultural Advisory Service, National Centre, Web & It, Udkaersvej 15, 8200 Aarhus N, Denmark, [2]University of Aarhus, Department of Computer Science, Aabogade 34, 8200 Aarhus N, Denmark; tbg@landscentret.dk

Galileo is the name of a European satellite-based positioning system, similar to the American GPS, which is expected to be fully developed and running in the year 2013. The system will provide position measurements outdoor as well as indoor, and the measurements will be more accurate than what is provided by GPS. A Danish consortium supported by The Danish National Advanced Technology Foundation in Denmark, aim to develop a Galileo-based platform for pervasive positioning. As a part of the consortium the Danish Agricultural Advisory Service, National Centre whishes to develop a system for supported decision making. Through position measurements it is possible to track each animal in the herd, and thereby represent the individual animal's movement pattern. Preceding research has shown that such movement patterns also holds information about the health condition as well as the welfare of the animal, i.e. when a movement pattern diverges from the animal's 'normal' movement pattern, it might be a symptom of a health state that requires attention. Using a suitable pattern recognition technique will help detect such abnormal movement patterns in the herd. By making use of Galileo-based position measurements, it is possible to track the cattle indoor in the stable environment as well as outdoor on the fields, and also a sufficient accuracy will be achieved. In order to find a suitable pattern recognition technique for the above mentioned cattle case, one focus point of the research is on the use of one or more known pattern recognition techniques to detect movement abnormalities amongst the cattle. Another focus point of the research is that once a suitable technique is found, it will be integrated in the Galileo-based platform for pervasive positioning as a generic software component and used to satisfy the need for pattern recognition in the decision supporting system. The research will follow an iterative process over experimental development, prototyping and evaluation. So I would like to present the current state and the immediate result of my research: Which pattern recognition techniques have been evaluated and what is the outcome? How does the evaluation take place and what are the evaluation criteria? While waiting for Galileo to be fully functional, other technologies like GPS, RFID, activity measuring devices and more comes in use; how does these technologies cope with the indoor/ outdoor scenery? And what is my experience on fusing these different sensor technologies? Would it be a relevant subject for future work?

Prototype system to recognize agricultural operations automatically based on RFID

Sugahara, K.[1,2], Nanseki, T.[2] and Fukatsu, T.[1], [1]National Agricultural Research Center, Field Monitoring Research Team, 3-1-1 Kannondai, Tsukuba, Ibaraki, 305-8666, Japan, [2]Kyushu University, Faculty of Agriculture, 6-10-1 Hakozaki, Higashi-ku, Fukuoka, 812-8581, Japan

Efficient data collection methods of agricultural operations are required for process control, quality control, labor management and cost accounting in agriculture. The purpose of this study was to clarify the possibility and problems of a method to recognize and record agricultural operations automatically based on RFID (Radio Frequency Identification) as a innovative automatic identification technology. A prototype system to collect data of agricultural operations using small RFID tags and wearable RFID reading devices was developed, and basic operation tests of this system were carried out in order to verify its effectiveness. The wearable device used in this system reads the identification data (ID) of the RFID tags attached to various agricultural objects, such as implements, machinery, facilities, materials, plants and fields, during the agricultural operation. The wearable device continuously sends data of the object ID and the reading time to the database server via wireless LAN. The server detects each task of the agricultural operation by matching the 'task pattern' as combination of the specific objects to the time-series data which are collected by reading the RFID tags. The prototype system applied two types of RFID tags, the microwave type (2.45 GHz) and the electromagnetic type (13.56 MHz). As a result of the operation tests of this system, it was possible to detect the tasks from the time-series data collected during the simple operations such as pesticide preparation. However, it was demonstrated that there were problems to be solved for further development of a practical system. This system might be able to cover various operations, not only in agriculture. It is necessary to study what kind of workplaces and operations the system can be applied for.

Controlling of silage crop compaction for validation of silage quality and aerobic stability in case of silage bagging technology

Maack, C. and Büscher, W., Institut für Landtechnik, Livestock technology, Nussallee 5, 53115 Bonn, Germany

The conservation of feed crops in silage bags has been developed as an alternative operating technique to clamp silos. Distinguishing is the flexible applicability without buildings and structures which reduces the fix costs significantly. In times of high prices of feed crops the minimal losses in the bag gives the opportunity of secondary saving potential. Providing transfer rates higher than 150 t/h the technology may be employed in harvesting systems with high performance, which has already made the concept of silage bags relevant in the field of biogas crops conservation. The crop compaction is one of the main factors with an important effect on silage conservation and the aerobic stability after silo opening. In order to compact the silage crop, it is pressed under constant mass flow into the tube using a rotor. During the process the tube should be seated on solid ground. While pressing more raw material into the tube the pressure inside increases and a force in driving direction pushes the machine forward. The compaction process is regulated by a brake system controlling the forward movement. In order to regulate the braking pressure the tube material's expansion has to be controlled at fixed measuring points on the tube. The right braking pressure depends on feed crop, its DM-content and the ground the machine is used on. Therefore the rate of compaction depends on the operator's experience which makes intensive training necessary. As part of a PhD project a small bagger was constructed in order to perform different measurements during the pressing process. Instead of locating the tube on the ground it is used on a rolling carriage. This construction allows pressing under constant conditions and gives the opportunity to move the whole tube by a forklift. The aim of the present project is to develop suitable control loops for optimised forage compression and for the reproducibility of the silage success by means of bagging machines. During the presentation the model of the bagger, the used sensor technique and results of the measurements will be shown.

Prospects for the application of genomic markers in precision livestock farming
Smits, M.A., Calus, M.P. and Veerkamp, R.F., Animal Sciences Group - Wageningen UR, Animal Breeding and Genomics Centre, P.O. Box 65, 8200 AB Lelystad, Netherlands

Animal sciences have arrived at the threshold of the genomic revolution. The growing sequence information and the fast increase in the availability of many thousands of single-nucleotide polymorphisms (SNPs) scattered across livestock genomes provide realistic opportunities for the improvement of animal traits by precision livestock breeding. This explosion of genomic data and the constantly reducing costs for mass genotyping, allows the identification of thousands of SNPs in individual farm animals and the use of this information for genome-based selections. Such approaches will accelerate the genetic selection process and lead to more precise genetic improvements. Several breeding industries already apply genomic selection procedures using >10,000 markers. Also the steadily increasing knowledge on genotype–phenotype relationships will furnish opportunities for precision livestock farming, including the design of precision mating systems and the sorting of animals into specific management systems. The latter offers the possibility to optimize production environments, for example by the use of genotype-based healthcare programs. Functional genomic studies fill the gap between genotype and phenotype and aim to get insight into biological processes that underlie animal traits. Understanding the effect of nutrients on gene expression may provide promises to optimize the nutritional needs of animals. Similarly, improved knowledge on host–pathogen interactions may be used to develop new tools and strategies for prognosis, diagnosis, prevention and treatment of infectious diseases. A very promising application of functional genomic data is the identification of molecular 'signatures' or biomarkers exhibiting the molecular composition or physiological 'status' of a biological sample. In collaboration with the Dutch Dairy Board we recently started biomarker research focusing on gestation and health. In the future molecular biomarkers, probably in combination with sensor technologies, will be applied to measure or predict health and/or quality parameters of livestock species and the products derived thereof. In this paper, prospects, challenges, and requirements for the application of genomic markers in precision livestock farming will be discussed. Several examples of research efforts into this direction will be presented.

Measuring of vacuum near teatend inside the teatcup with piezoresistive pressure transmitters and usage of output signal for steering milking machines

Ströbel, U., Rose, S. and Brunsch, R., Leibniz-Institut für Agrartechnik Potsdam-Bornim e.V., Verfahrenstechnik in der Tierproduktion, Max-Eyth-Allee 100, 14469 Potsdam, Germany

Many milking systems are equiped with new-fashioned sensor technology for diary management. But milking systems at the moment have no mechanism to inhibit temporary vacuum maxima, but vacuum maxima can abet udder inflammation. The aim of the present study is, to investigate, if piezoresistive pressure transmitters are appropriate to get integrated in the four teatcups direct and close to the teatend, for measuring and steering the vacuum permanent in the milking system. Furthermore the transmitters have to be resistant against chemicals, which are used for cleaning in the milking system. If it is possible by means of transmitters to measure and to control the vacuum at the teatend very exact, then the number of udder inflammations could be decreased and with this effect, the average endurance of milk cows can be increased. Possibly this could bring a positiv effect to the image of dairy farming, so in last consequence, it could lead to a better income of many farmers. Before the transmitters can be integrated permanently in a milking system, much different probations have to be conducted. For this study it is necessary to investigate transmitters of three different producers, which are known for high precession. It is important to investigate, whether the vacuum data, from transmitter measured are exact enough and if the transmitters can get integrated in a carrier, which is special constructed for this case and is integrated between milktube and teatcup. The connection between both elements can easy be achieved, by plugging a short milktube on the pipe of the carrier and on the teatcup. For measuring the vacuum it is necessary to use the 'wet methode' according to DIN ISO 6690 for milking systems. At this, one water tank, four water flow meters and the DIN ISO artificial teats are used for simulating the cow. The water tank is connected to the water supply and has a level controlled water tap, so that the liquid level stays on a permanent level during the milking process all the time. With the flow meters the flowing water amount can become adjusted on a constant level. There is a device in use called 'BoviPress' as vacuum measurement device, which is produced by the company A&R Trading GmbH. Herewith it is possible to measure the pressure parallel at ten measuring points. So four measuring points, direct in opposite to the transmitters can be allocated. Furthermore there are four measuring points direct at the teatend for measuring the vacuum curve. The artificial teats have a measuring point for teatend vacuum. The other two measuring points are used for measuring the machine vacuum and the vacuum at the milkmeter. This method will be used for each of the three transmitter brands. Then with statistics, the error rate between the reference (BoviPress) and the used transmitter will be accounted. At least the minima and the maxima measured during each trial, become compared. With this method, the transmitter, which is most suitable for installation in the milking system can be found. If it is possible actually, to use transmitters for steering vacuum in milking systems, this study will show the result and which one of the tested, is the most suitable. So more developments can be done to make the transmitters consistent for a long term use in the the milking system. When the transmitters fullfill these requests, the steering of the vaccuum in milking systems is possible.

Diurnal feeding patterns of individual dairy cows fed with an automatic precision feeding system
Abbink, N.[1], André, G.[2], Bleumer, E.J.B.[2] and Pompe, J.C.A.M.[1], [1]Wageningen University, Farm Technology Group, P.O. Box 17, 6700 AA Wageningen, Netherlands, [2]Wageningen UR, Animal Sciences Group, Animal Production Systems, Edelhertweg 15, 8219 PH Lelystad, Netherlands

Automatic precision feeding systems can supply fresh feedstuffs to dairy cows continuously, while conventional feeding systems deliver fresh feedstuffs once or twice a day. Previous research showed that a group of 10 cows that was fed with an automatic precision feeding system responded to the continuous supply of feedstuffs with a diurnal feeding pattern with less pronounced peaks compared to the situation of conventional feeding. This could lead to the suggestion that automatic precision feeding systems can feed a given number of cows with fewer feeding places compared to conventional feeding systems. Research on conventional feeding systems has shown that various conditions - such as a high pressure per feeding place - may stimulate individual cows to exhibit a diurnal feeding pattern that differs from the rest of the herd. For an automatic precision feeding system it is not clear whether the more even diurnal feeding pattern is caused by the phenomenon that all cows spread their feeding visits more evenly over the day or that individual cows feed at different times. The objective of this study was to determine whether automatic precision feeding systems affect the diurnal feeding patterns of individual cows and whether lactation stage and season have an influence on these patterns. We collected information on the feeding visits of 94 Friesian Holstein cows over a period of 6 months from the log files of the automatic precision feeding system at the High-tech Experimental Dairy Farm of Wageningen UR. For each feeding visit we determined the begin and end time, the duration of the visit, dry matter intake and feeding rate. We applied exploratory data analysis to reveal relations between these variables and differences in patterns between cows and within cows, both during lactation and season. We tested apparent relations and patterns statistically by means of longitudinal data analysis. The data analysis showed that diurnal feeding patterns differed between individual cows during lactation and season, but not within cows - different patterns between cows were found for the duration of the visits, dry matter intake per visit, feeding rate, total feeding time per day and number of visits per day. Some cows combined longer visits to the automatic precision feeding system with a lower dry matter intake than other cows, which resulted in a lower feeding rate. This implies that some cows utilize the automatic precision feeding system more effectively than others.

Estimation of feed intake of dairy cows by means of feeding behaviour caracteristics
Azizi, O.[1], Kaufmann, O.[2] and Hasselmann, L.[2], [1]University of Kurdistan, Animal Science, P.O. Box 416, Sanandaj, Iran, [2]Humboldt - Universität zu Berlin, Faculty of Agriculture and Horticulture, Invalidenstrasse 42, 10115 Berlin, Germany

The transition period and early lactation is critical important to the health and productivity of dairy cows. In this period more than 80% of the total health disorders occur. In the most cases this disorders are metabolic related and caused by a negative energy balance. To get information about arising problems in an early stage, sensor based animal monitoring systems are used. This systems analyse among others characteristics of animal behaviour. Therefore it is necessary to analyse the relations between behaviour patterns and physiological parameters. In this context the research was directed on the relationships between feeding behaviour and feed intake. Data (feed intake and time, spend on feeder) from 70 lactating dairy cows were collected from the 2nd to the 15th week of lactation. The monitoring was conducted by using an electronic feeding system, which was equipped with an electronic identification of each individual cow. The visits at feeder were clustered in meals based on a 'meal criterion'. The effects of parity, stage of lactation and milk yield level on feeding behaviour and feed intake were investigated. A second study determined the effects of metabolic - related production diseases on feeding behaviour and feed intake. The high correlation between feeding behaviour characteristics such as meal duration suggests that measuring the time spend eating could be used to estimate the feed intake. Moreover the monitoring of feeding behaviour might be helpful to detect the cows` risk for metabolic – related production diseases at an early stage.

The effect of different geometrical wind tunnels in aerodynamic characteristics and ammonia mass transfer process in aqueous solution

Saha, C.K.[1], Zhang, G.[1], Ye, Z.[2], Rong, L.[3] and Strøm, J.S.[1], [1]Aarhus University, Agricultural Engineering, Blichers Allé 20, P.O. Box 50, DK-8830 Tjele, Denmark, [2]China Agricultural University, Agricultural Structure and Bio-environmental Engineering, College of Water Conservancy & Civil Engineering, Mail Box. 67, 100083 Beijing, China, [3]Aalborg University, Civil Engineering, Sohngårdsholmsvej 57, DK-9000 Aalborg, Denmark

The fundamental studies on emission factors are still required for modelling ammonia and odour emissions from manure surface in livestock buildings. The objectives of this paper are to determine the effects of geometrical sizes of wind tunnels on ammonia emission. Three different sizes of wind tunnels were used for the studies with different air velocities and turbulence intensities. Different perforated plates were installed at the inlet to generate three turbulence levels over the liquid surface. The dimensions of the wind tunnels used were (1) 6m×0.5m×0.5m, (2) 4.2m×0.35m×0.35m and (3) 2m×0.15m×0.15m. The aerodynamic characteristics were examined to further assess the performance of wind tunnels. Four inlet air velocities with three turbulence intensities were studied to estimate the changes of ammonia mass transfer process and aerodynamics characteristics. The mass transfer coefficient increased with increasing turbulence intensities. The emission rates measured using the three wind tunnels are shown to be strongly correlated. The relationship between the emission rate with air velocity, turbulence intensity and geometrical size are established. The effects of wind tunnel dimensions on the emission coefficients were addressed.

Monitoring pig's activity using manual visual labelling

Borgonovo, F.[1], Leroy, T.[2], Costa, A.[1], Aerts, J.-M.[2], Berckmans, D.[2] and Guarino, M.[1], [1]Faculty of Veterinary Medicine, Department of Veterinary and Technological Science for Food Safety, Via Celoria, 10, 20133, Milano, Italy, [2]Faculty of Bio-Science Eng., Katholieke Universiteit, M3-BIORES: Measure, Model & Manage Bioresponses, Kasteelpark Arenberg 30, B3001 Heverlee, Leuven, Belgium

Animal welfare is defined as the state of an animal's health and physics. One of the best ways to evaluate the well-being of the animals is to watch the animal's behaviour this can be also a good feedback of environment quality. The aim of the study was to develop a labelling tool to classify and predict aggressiveness of pigs without disturbing the animals. In our trial 33 pigs housed in two pens were monitored using an infrared sensitive CCD camera placed at the height of 5 metres above the pen floor. The camera was connected to a PC with built-in frame grabber using a coaxial cable. Video recordings of the two pens were made throughout 24 hours a day. The lens was focused on the pigs and images were captured with a resolution 768x586 and a frame rate of 1 frame per second (fps). After downloading data, in the laboratory, the files were visually checked and labelled and each image was analysed and manually labelled. In total were 326,936 observations (correspond to 24 hours). In the manual visual labelling the recorded images were observed and was given a score in according to the behaviour shown from the animals; from '0' for no activity, '1' for aggressive episodes as fighting or struggling, '2' in case of biting one another, '3' for abnormal behaviour as nuzzling or suckling one another to '4' for the feed assumption when the pigs were lined up at the trough. The behaviour scores obtained from the labelling were reported in Excel files, together with automatically measured activity and occupation indices of the pigs. A second analysis was done in the file images with behaviour score 1 (aggressive episodes), these were ranked in the four following categories: to the pig's neck; grount to grount; to the pig's lateral side; to the pig's back side. From this analysis it was found that the fighting episodes were shown by pigs in low percentage (less than 1%) of the observation time and were prevalently grount to grount (30%) and pig's neck aggressions (26%). While the pig's lateral side and the pig's back side were shown in low percentage (22% each). The results shown that the aggressive episodes, occurred mainly in the time slot between 12.00 to 19.00, had a specific mean value of the activity index (0.019 with the mean value of the Std Dev: 0.014) and a mean value of the occupation index (0.571 zones 1 and 4, 0.217 zones 2 and 3 with the mean values of the Std Dev were of 0.103 in zones 1 and 4 and 0.036 in zones 2 and 3).

Online quantification of the excitement of a stallion exposed to an oestrous mare

Avezzù, V.[1], Jansen, F.[2], Guarino, M.[1], Pecile, A.M.[3], Quanten, S.[2] and Berckmans, D.[2], [1]University of Milan, Department of Veterinary and Technological Sciences and Food Safety, Via Celoria 10, 20133 Milan, Italy, [2]Catholic University of Leuven, Division Monitor, Model & Manage Bioresponses, Kasteelpark Arenberg 30, 3001 Leuven, Belgium, [3]University of Milan, Department of Veterinary Clinical Sciences, Via Celoria 10, 20133 Milan, Italy

Exposure of a stallion to an oestrous mare causes a state of excitement, which is reflected in behavioural and physiological (heart rate, cortisol,...) changes. The objective of this research was to quantify in a non-invasive and online way this excitement of a stallion when exposed to an oestrous mare, based on the online measurement of heart rate and physical activity. The performed experiment can be divided into two parts. In the first part the stallion was taken outside for a short walk. In the second part the stallion was locked up again in his box and exposed to an oestrous mare. During the experiment the stallion's heart rate and physical activity were continuously measured using the Polar Equine RS800, which measures heart rate as interbeat intervals, and the ActiGraph GT1M, which gives a measure for physical activity. The relationship between the stallion's physical activity and its heart rate was described by a mathematical model. Since a stallion can be considered as a complex, individual, time varying, dynamic system, this model has to be estimated in real time for each stallion individually in each experiment. The time varying character of the system requires that the model is validated in real-time and re-estimated if necessary. This real-time input-output modelling allows determination of the component of the stallion's heart rate which is related to physical activity. The results of the experiment show that the measured heart rate can be related almost completely to the physical activity of the stallion during the first part of the experiment (walking outside without exposure to oestrous mare). However, when the stallion is exposed to the oestrous mare, the component of the stallion's heart rate related to the physical activity accounts for a significantly smaller part of the total heart rate. This shows that the component of the stallion's heart rate which is not explained by the physical activity can be used to quantify in an online and non-invasive way the excitement of a stallion exposed to an oestrous mare.

Water saving through smarter irrigation in Australian dairy farming: use of intelligent irrigation controller and wireless sensor network

Dassanayake, D.[1], Malano, H.[1], Dassanayake, K.B.[2], Langford, J.[3], Douglas, P.[1] and Dunn, G.M.[4], [1]The University of Melbourne, Department of Civil and Environmental Engineering, Swanston Street, 3010, Melbourne, Victoria, Australia, [2]The University of Melbourne, Department of Land and Food Resources, Swanston Street, 3010, Melbourne, Victoria, Australia, [3]The University of Melbourne, UNIWATER, Swanston Street, 3010, Melbourne, Victoria, Australia, [4]The University of Melbourne, Department of Land and Food Resources, 1 College Cresent, Dookie Campus, 3647, Dookie, Victoria, Australia

As a classical practice of border-check irrigation, time to cut-off of the water supply is determined when the waterfront reaches two third of the irrigation bay. This would results waste of water, higher labour cost and lower productivity of pasture. In order to address these issues, a sensor based border-check irrigation system including a real time feedback control has been developed. The system consists of a wireless sensor and actuation network, a central host/user interface, which collects stores and displays real time information, and central control system software. This paper describes a new Intelligent Irrigation Controller (IIC), which can estimate event based optimal irrigation time whilst keeping better pasture growth in dairy farms. Furthermore, it discusses how to cut off the labour cost furthering current investigation. IIC automatically monitors real-time inflow and advance wetting front times at two locations via the wireless sensor network and estimates the infiltration properties and optimal irrigation times using an inverse solution. Passing the time to cut off back to the sensor network or using another suitable method, irrigation inflow is terminated at the end of the optimal time to cut-off. The new system is currently under evaluation at a trial farm in Dookie, Australia, and initial results indicate up to 43% saving in water use and substantial savings in labour maintaining better pasture growth.

Mathematical modelling of infectious diseases: transmission of foot-and-mouth disease in Pantanal Matogrossense region, Brazil

Ternes, S.[1], Aguilar, R.[2], Maidana, N.[3] and Yang, H.[4], [1]Embrapa Agriculture Informatics, Av. André Tosello, 209, 13083-886 - Campinas, SP, Brazil, [2]Embrapa Pantanal, R. 21 de setembro, 1880, 79320-900 - Corumbá, MS, Brazil, [3]National University of Mar del Plata, Diagonal J. B. Alberdi 2695, Mar del Plata, Argentina, [4]State University of Campinas, R. Sergio Buarque de Holanda, 651, 13083-859 - Campinas, SP, Brazil

Outbreaks of foot-and-mouth disease (FMD) have resulted in the slaughter of millions of animals, despite this being a frequently non-fatal disease for adult animals, though young animals can have a high mortality. FMD is probably the most important livestock disease in terms of economic impact. FMD primarily affects cloven-hoofed domestic and wild animals, including cattle, pigs, sheep, goats, and water buffalo. FMD status is an important determinant of international trade in livestock products, and the existence of FMD is an effective barrier from the markets with the highest prices for these products. Therefore, many resources have been and still are dedicated to surveillance, control and eradication of this disease. Humans are very rarely affected. Foot-and-mouth disease virus (FMDV) is the prototypic member of the Aphthovirus genus in the Picornaviridae family. This picornavirus is the etiological agent of the acute systemic vesicular disease that affects cattle and other animals worldwide. It is a highly variable and transmissible virus. There are 7 immunologically distinct serotypes and over 60 subtypes. The emergence of new subtypes in a region leads to failures of immunity of vaccines used and as a result, may appear outbreaks. These genetic differences between the agents of disease are the reason for imposing sanitary barriers or buffer zones. Experience in South America shows that sooner or later, the integrity of such buffer zones is breached. Consequently, a scheme was devised which is based on ecosystems and which takes into account the dynamics of FMD, the farming systems and cattle movements to identify primary endemic areas (i.e. virus maintenance areas), secondary endemic areas (i.e. areas of virus propagation) and epidemic areas (i.e. areas of explosive outbreaks). Epidemiological studies clearly determined the existence of different ecosystems that provide the conditions necessary for the maintenance of the virus. Thus, extensive areas have all the elements so that the agent remains in activity with emergence of epidemics in certain seasons, leading to the false conclusion that the disease has characteristics of cyclical presentation. We are proposing an approach that integrates quantitative methods of analysis with observations of the occurrence of outbreaks and epidemics in the field. The multidisciplinary study, which involves collecting samples of field, processing laboratory and mathematical modelling, is essential to understand the dynamics of transmission of FMD and analyse the impact of control. The Pantanal of South America is one of the most immense and biologically rich environments on the planet. Often referred to as the world's largest freshwater wetland system, is situated in South America, mostly within the Brazilian states of Mato Grosso and Mato Grosso do Sul. To analyse the dissemination of FMD in Pantanal Matogressense Region we use deterministic dynamic systems. These mathematical models are represented by ordinary and partial differential equations, which allow us to evaluate the average behaviour and the spatial dissemination of the disease, respectively.

Feeding soy hulls to robotically milked high-yielding dairy cows increased milk production

Halachmi, I.[1], Shoshani, E.[2], Solomon, R.[2], Maltz, E.[1] and Miron, J.[1], [1]Agricultural Research Organization (A.R.O), P.O. Box 6 The Volcani Centre, Bet Dagan 50250, Israel, [2]Ministry of Agriculture & Rural Development, Extension Service, P.O. Box 28, P.O. Box 28, Bet Dagan 50250, Israel; halachmi@volcani. agri.gov.il

If the milking frequency in an automatic milking system (AMS) is increased, the intake of concentrate pellets in the robot may be raised accordingly. Consumption of a large quantity of starchy grains within a short time can impair the appetite, decrease voluntary visits to the milking stall, and lower the intake of dry matter (DM) and neutral detergent fiber (NDF). Therefore, the hypothesis to be tested in this study was that conventional starchy pellets fed in the AMS could be replaced with pellets rich in digestible NDF without impairing the cows' motivation to visit a milking stall voluntarily. Sixty cows were paired according to age, milk yield, and days in milk, and were fed a basic mixture (BM) along the feeding lane (16.2 kg (DM)/d per cow), plus a pelleted additive (6-14 kg (DM)/d per cow) that was consumed in the milking stall and in the concentrates self feeder (CSF), which they could enter only after passing through the milking stall. The 2 feeding regimens differed only in the composition of the pelleted additives: that for the control group contained 52.9% starchy grain, whereas that for the treatment group contained 25% starchy grain, plus soy hulls and gluten feed as replacement for part of the grain and part of the low digestible NDF rich feeds. During the first 2 months, i.e., 0-60 days in milk (DIM) a cow received 10-12 kg of concentrate pellets. After 60 d, concentrate feed was allocated according to milk production: 25 kg/d or less entitled a cow to 2 kg of concentrate feed; over 25 kg/d received 1 kg/d additional concentrate feed per 5 kg/d additional milk production, and 60 kg/d or more received 9 kg of concentrate. The concentrate feed is split between the robot and a feeder, the actual feeding levels are in table 1. The 2 diets resulted in similar frequencies of voluntary milkings (3.12 vs. 3.16 visits/d per cow in early stages of lactation). Average milk yields were higher in the treatment group (42.7 vs. 39.7 kg/d per cow in early stages of lactation), and percentages of milk protein and milk fat were similar in the 2 groups. The results suggest that the proposed pellets rich in digestible NDF can be allocated via the AMS, in conjunction with the BM in the feeding lane. Thus, it appears to be possible to use the AMS to automatically encourage selected high-yielding cows to maintain a high frequency of voluntary milkings, by increasing the amount of pelleted concentrates allocated in the robot, without any negative effects on appetite, milk yield, and milk composition.

Analysis of capacity reserves in automatic milking systems

Harms, J. and Wendl, G., Bavarian State Research Center for Agriculture, Institute for Agricultural Engineering and Animal Husbandry, Prof.-Dürrwaechter-Platz 2, 85586, Poing, Germany

In automatic milking systems, the exploitation of the technical capacity of the system is important for economic reasons. Nevertheless the farmer often doesn't know if there are capacity reserves and how to mobilize them. One of the reasons for this is probably a lack of possibilities to compare the own farm to other farms. Furthermore it is difficult to estimate how changes in system management or animal parameters will influence capacity. The aim of the project was to develop a tool for detection of weak points of an existing system in order to improve system utilization and to reduce mechanisation costs. Based on a theoretical approach combined with results of practical farms and time measurement on experimental farms a mathematical model was created to predict system capacity expressed as yearly milk yield, number of daily milkings / visits or possible herd size. The model parameters were divided into animal related parame-ters like average milk flow or milk yield, farm management related parameters like utilization of the system, additional visits or milking frequency and technical parameters like cleaning duration or duration of attaching the teat cups. It could be shown, that regarding a single technical parameter, improvements have only small effects on system capacity compared to animal or management related parameters. Reducing the time for attaching the teat cups by 10 sec. for example, increased yearly milk yield by approx. 1.8%. A better milk flow of 2.3 compared to 2.0 l/min led to an improvement of approx. 8.3%. Reaching a better utilisation of the sys-tem of 5% resulted in 5.5% more milk per year. Nevertheless, when summing up improvements of all technical parameters (e.g. attaching teat cups, cleaning teats, system cleaning, …), the effect on capacity was comparable to the effects of animal or management related parameters. As the effects of changes in one parameter are not independent of the other settings, improving a certain parameter showed different efficiency concerning the capacity of the system depending on the initial scenario. The developed model can help to detect weak points of a certain system or a certain farm, compared to other systems or farms. Furthermore it offers a possibility to pre-estimate the effects of improvements on system capacity. Both is essential for the farmer's decisions and might help to realize capacity reserves.

Deterministic grass model

Huzsvai, L., University of Debrecen, Böszörményi 138., H-4032, Hungary

With this model, we aimed to assess the production and capacity of animal support of grass land reserves on sand soil in Hungary. Our goals were: minimizing the number of measurements and required input parameters needed for running a model based on 1 measurement/day/square meter. Previous botanical measurements revealed that the grass in question is composed mainly of C_4 species. So we used a hypothetical C_4 plant to model the grass reserve with. The model language is Visual Basic, and we used the widely known program; MS Excel as input-output platform. Two methods were defined: (1) growth and development of the grass; (2) impacts of grazing on the grass. In the model, development is determined by temperature, while determinants of grass growth are biological age of the plant, photosynthetically active radiation, temperature, available water and nutrients. Volumetric expansion (GAI) is split into two periods; the first is unlimited by photosynthesis, while the second is a source limited period. Simulation of grazing was based on grazing characteristics of sheep, assuming a stubble-field with the height of 3 cm. Because of the remarkable fluctuation of rain water in Hungary, rainfall is the main determinant of grass production in the country. Hereby, we composed a simple capacitive, water circulation model that calculates the amount of water available for the grass, in the top 30 cm layer of the soil. Components of the model are: calculations for the potential and actual evaporation and transpiration (mm day^{-1}); amount of rainfall (mm day^{-1}) accounting for interception. Additionally, we assess the extent of water stress that depends on potential evapotranspiration (PET), GAI and actual humidity content of the soil. The input weather file contains global radiation (MJ m^{-2}); daily average of air temperature (°C); relative humidity (%) and wind speed (m s^{-1}). To keep the model simple, only 3 soil characteristics are included, namely: saturated soil water content; drained upper limit (DUL) and lower limit (LL) (m^3 m^{-3}). To start the simulation, one needs to set the starting date (using Julian date format, it is represented as the number of days passed from 1st January); actual GAI value; plant height (cm); number of days between two grazing and actual humidity content of the soil (m^3 m^{-3}). After running the simulation, the results show the grass yield in each period, expressed in dry matter, hay value and green mass (g m^{-2}). Furthermore, an output file is generated as well, containing the following parameters on a daily basis: date; canopy dry matter (g m^{-2}); plant height (cm); GAI (m^2 m^{-2}); biological age; PET; actual evaporation and transpiration (mm day^{-1}); soil water content (m^3 m^{-3}) and water stress (0.00-1.00). Model validification occurred in the years 2005-2007, by using the volume of yields mowed from sample areas on a regular basis. Difference between the simulated and observed yield data did not exceed 10% in any of the 3 years. The model concludes that highest grass yield over years occurs when implementing a grazing technique with 35 days intervals between two grazing period.

Urine sensors for sheep and cattle

Betteridge, K.[1], Carter, M.L.[2] and Costall, D.A.[1], [1]AgResearch Grasslands, Climate, Land and Environment, Private Bag 11008, Palmerston North 4442, New Zealand, [2]Enertrol Ltd, 240 Mangaone Rd RD9, Feilding 4779, New Zealand; keith.betteridge@agresearch.co.nz

Dynamic nitrogen (N) models are widely used to estimate N emissions to the environment. Models need to be parameterised to the local conditions and need to adequately reflect the biological system being evaluated. N leaching and nitrous oxide emissions are major concerns in most agricultural systems and both are driven by the load of N in urine patches in the field. These patches are increasingly being incorporated into N models but, through lack of data relating to their distribution in the field, random distribution is assumed with a Poisson probability distribution of overlapping patches being assumed. As animal grazing and resting behaviour is often not random, especially in hill pastures, stock camps are common and may receive a disproportionately large amount of excreta on only a small proportion of the paddock. This paper will describe urine sensors and trial results from their use on both ewes and cows. The equipment logs the time of each urination event and a GPS worn by the animal logs the position of the urine patch in the pasture. Results will demonstrate the extent of urination frequency differences within groups of sheep and cows; how well these data relate to published data; and the spatial distribution patterns showing the existence of stock camps for both species in hill pastures. The implications of camping on the creation of critical source areas of N emissions will be discussed, as will the problems of scaling small plot trial results on flat land to large paddocks in hill country. As urinary N concentration of individual urination events for a given animal is highly variable during and between days, knowledge of the urinary N concentration of each event is highly desirable. We will present early results from the use on grazing sheep of a modified urine sensor that estimates urinary N concentration. Data gathered with our urine sensors will enable N models to be upgraded to more closely represent urinary excreta returns to pasture and, therefore, to better predict N emissions. This can include evaluating N emission mitigation strategies manipulation of urination frequency and urinary N concentration. Using GIS tools farmers may also be able to target mitigation strategies (e.g. nitrification inhibitor application) to areas identified as critical source areas.

Effects of different ratios of nonfiber carbohydrate to rumen degradable protein on the performance, blood and rumen samples of Holstein cows

Rafiee, H., Aboureihan campus, Tehran university, Tehran, Iran

Ruminants are provided with tow major sources of protein, microbial protein and rumen undegradable protein. Microbial protein is a good quality protein because of its amino acids content and post ruminal digestibility. Microbial yield in the rumen depends largely on the availability of carbohydrates and nitrogen (N) in the rumen. Currently, nonfiber carbohydrates (NFC) are used as a source of energy in dairy diets. In research trials, dietary protein was changed by replacing high-protein supplements with cereal grains, sources of NFC. Therefore, it remains unclear whether the reported effects on cow performance were achieved independently of alterations in ruminal metabolism and the postruminal supply of N and carbohydrate. Our hypothesis was that finding an optimum NFC: RDP ratios in diets of mid lactating dairy cows would be valuable. Nine multiparous midlactation Holstein cows averaging 171 days in milk and 24 Kg of milk/d were assigned into a replicated 3×3 Latin square design to study the effects of altering NFC to rumen degradable protein (RDP) ratio on performance and blood samples. NFC: RDP ratios were 4.07 (diet 1), 3.71 (diet 2) and 3.34 (diet 3). Ratios were achieved through altering RDP content of diets while NFC was held constant at 40% DM. Urea was supplemented as a source of RDP to decrease the ratio. RDP contents were 9.8, 10.8 and 11.8% DM respectively. Results were analyzed by MIXED models with effect of cow(square) as a random and days of sampling in each period as repeated measures. The effect of treatments on N-NH3 rumen liquor was significant and linearly increased ($P<0.01$). Treatments had no effect on milk yield and fat yield. Protein and solid non fat (SNF) percentage of milk increased linearly when cows were fed diets with greater RDP ($P<0.01$). Milk urea nitrogen (MUN) increased linearly when cows were fed increasing amounts of RDP ($P<0.01$). The effect of treatments on plasma urea nitrogen (PUN) was significant and linearly increased ($P<0.01$). Results confirming that low producing cows are less likely to respond to altering NFC: RDP ratio, but the ratio equal to 3.71 was the best performance in midlactation dairy cow.

Recording of water intake of suckler cows to detect forthcoming calving
Mačuhová, J., Jais, C. and Oppermann, P., Intitute for Agricultural Engineering and Animal Husbandry, Prof.-Dürrwaechter-Platz 2, 85586, Germany

Automatic recording of different parameters on individual basis in livestock animals can help to detect some deviation from normal production and/or behaviour sooner than by direct animal observation. The aim of this study was to test if it is possible by recording of water intake to detect forthcoming calving and start of lactation resp. to detect problems with lactation after calving in suckler cows. Therefore, an amount of daily water intake, daily duration of water intake and daily number of watering place visits were recorded during period since two weeks before calving (day −14 to −1), at calving day (day 0) and until two weeks after calving (day 1 to 14) in German Simental and German Yellow suckler cows (n=22). They were kept together with other animals of suckler cow herd in the free stable with deep bedding expect few days around the calving, where they were kept in calving box. The experimental measurements were performed during one winter period. All animals were fitted with electronic ear tags for individual animal identification. To record the water intake the individual drinking cups (Texas Trading®) fitted with in-line flow meters were used. The data were not normal distributed and therefore nonparametric test as Friedman test and Wilcoxon test for paired data were used for statistical data evaluation. High variation of tested parameters was observed during the tested period in individual animals. The day had not any significant effect ($P>0.05$) on all tested parameters during two weeks before calving. At calving day only the duration of drinking was significantly shorter than before calving ($P<0.05$). Daily water intake and also daily number of watering place visit did not differ between period before and on calving day ($P>0.05$). Water intake was significantly ($P<0.05$) higher after calving than before or on calving day. The number of visit was lowest on calving day, however, differed significantly only between calving day and period after calving ($P<0.05$). Duration of drinking was significantly ($P<0.05$) longer after calving than before or calving day. In conclusion, it was unfortunately not possible to recognise forthcoming calving by drinking behaviour in suckler cows. The changes in drinking behaviour (expect duration of drinking) could be observed only after calving day. Problems with lactation after calving could be possible to identify, because water intake increases significantly after calving in normal lactating animals.

Near body temperature measurements in broilers with a wireless sensor

Bleumer, E.J.B., Hogewerf, P.H. and Ipema, A.H., Wageningen UR, Animal Sciences Group, Edelhertweg 15, 8219 PH Lelystad, Netherlands

Growth of broilers is highly depending on feed intake, body temperature and environmental temperature. In practice temperature in a pen is measured by use of a few temperature sensors hanging from the ceiling. This method gives an indication of the temperature of the surroundings of the animal. Measuring near body temperature of an individual animal provides more information about the temperature in the direct surroundings of the animal. In a pilot study carried out in 2007 a wireless sensor network was applied for recording near body temperature profiles of individual broilers. The network consisted of wireless sensors, repeaters, a gateway and a pc. The sensors with a sensirion SHT10 humanity sensor, a processor, memory, radio and a battery were attached to the wings of four 5-6 weeks old broilers. The housing of the wireless sensor was in direct contact with the skin of the animals. Data was transmitted to a gateway that was connected to a PC for data recording. Eight repeaters hanging from the ceiling and equally spread over the compartment of a larger pen in which the broilers were kept were used to relay the signal of the wireless sensors to the gateway. The sampling frequency of the sensors varied between 6 and 60 measurements per hour. The data consisted of temperature, humidity, voltage of the battery and RSSI (Received Signal Strength Indication). RSSI information can be used for estimating the location of an animal in the pen compartment. In total there were about 2600 chickens in the 125 m2 compartment. It is possible to register a temperature that is related to the body temperature of an individual broiler with a wireless sensor. During the testing period the near body temperature showed large fluctuations. In dark periods (2x 4-h periods per day) the temperature was higher compared to the light periods (2x 8-h periods per day), probably because the animals are more grouping during dark periods. The ambient temperature was the same during light and dark periods. Broiler behavior was not affected by the sensor nodes. In younger chickens further miniaturization of the technology will be necessary.

Postharvest food participatory technology development among rural farmers in Nigeria
Abiodun, A.A. and Ogundele, B.A., Nigerian Stored Products Research Institute, Postharvest, NSPRI, KM. 3 ASA-DAM Road, PMB 1489, Ilorin, Kwara State 234, Nigeria; bukkyma2001@yahoo.com

Some rural communities in three selelected states, viz: Kwara, Oyo and Osun states of Nigeria were analysed for participatory food storage technologies development. Predominant occupation of rural dwellers among the communities studied were farming with little or no knowledge of food storage and processing. Participatory Technology Approaches(P. T.A.) of livelihood analysis, crop prioritization, problem tree, venn diagram, daily activity profile as well as seasonal calendar were employed based on the training experience received from CBDD in the year 2005. The findings reveal that there is necessity to influence the rural communities in term of food storage technologies, organise adequate post-harvest training and empowering the rural households. In conclusion, intervention of relevant stakeholders in food handlings, storage, processing and other post-harvest areas is necessary.

Adoption of automation and information technologies in relation to milking and animal monitoring on New Zealand dairy farms
Dela Rue, B. and Jago, J., DairyNZ, Private Bag 3221, Hamilton 3240, New Zealand

The New Zealand dairy industry is the world's largest exporter of milk product. Its international competitive advantage has been built on the low-cost production systems of year-round pastoral farming and efficient resource management. Key trends impacting on animal management over the decade to the 2006/07 seasonhave been: fewer herds (11,630, ↓21%) with increasing herd size (337 cows, ↑62%), more intensive farming (2.8 cows/ha, ↑12%) and increased milk solids (fat + protein) production per cow (330 kg milk solids (MS), ↑10%) and per hectare (934 kgMS, ↑26%). Total employee numbers have increased as farms amalgamate to gain the benefits of scale, and unpaid labour has decreased due to fewer family members working on farms (total labour units ↑39%, unpaid labour ↓27%, paid labour ↑82%). In addition to the challenges of growth and associated environmental issues, farm labour has emerged as a key constraint facing the dairy industry and has potential to limit productivity gains in the long term. The NZ dairy industry recognises that automation and information technologies can reduce the reliance on a highly skilled and experienced workforce. Industry reports indicate an increasing rate of technology adoption; however, there is no industry-wide data to support this. A survey was conducted in 2008 to provide objective data that describes dairy farmers' technology use and milking practices in relation to herd and milking management. This information will be used to track technology adoption rates and changes in milking practices over time. This will direct research focus and extension needs for industry good activities aimed at labour efficiency and productivity on dairy farms. The research was undertaken on a quantitative basis by surveying farmers drawn from a national database of approximately 9000 dairy farmers. Two populations of dairy farmers with herds of 100 cows or greater were sampled; those that milk in Herringbone dairies (79% of all dairies) and those that milk in Rotary dairies (21% of all dairies). The 12 minute telephone survey documented responses for 252 Rotary operators and 280 Herringbone operators (532 in total). The survey achieved the target Confidence Level of 95% and a Confidence Interval of 6% for both populations. The median age of the Herringbone (HB) dairy was 20 years, and Rotary (R) dairy was 10yrs. Rotaries milked larger herds: (R=567, HB=314), had more sets of cups (R=41, HB=26) and had a higher ratio of sets of cups to people required to milk (1 person: R=29, HB=17, 2 people: R=37, HB=27, 3 people: R=49, HB=37). 83% of farmers had a computer with herd management software in their home and 13% at the dairy. The use of automation to reduce manual tasks is low: Cup removers 18% (HB=9%, R=54%), Teat spraying 18% (HB=10%, R=49%), Drafting 4% (HB=2%, R=11%). The use of information technologies is extremely low in all dairy types: Electronic ID 5%, Electronic milk meters 2%, In-line mastitis detection 2%, Automatic weighing 4%, Automatic heat detection <1%. The use of automation technologies located at the dairy is increasing as farmers trade labour for capital, but still remains very low. Information technology use in relation to daily monitoring of cow health and reproduction is minimal. The value of information technology and automation will need to be demonstrated in pasture-based dairy systems if uptake is to be increased.

Chain-wide communication exchange in the German pork industry: an empirical analysis

Plumeyer, C.H. and Theuvsen, L., University of Goettingen, Department of Agricultural Economics and Rural Development, Platz der Goettinger Sieben 5, 37073 Goettingen, Germany

In the mid-1990s agri-food supply chains faced several food crises. Therefore, legislation on food production was changed, and now supports exchange processes of chain-wide information. Legal obligations in order to exchange information throughout food supply chains (for instance, food chain information, salmonella monitoring) are complemented by chain-wide certification systems. Nevertheless, empirical studies show that despite these legal and private obligations to transfer information remarkable communication deficits can still be observed. The German pork production is a good example for a food supply chain which lack of communication between supply chain partners. In order to improve inter-firm information exchange in food supply chains, IT-based system nowadays often complement traditional communication media (such as letter, fax). Such an IT-based solution was just recently introduced in the German pork chain. It is part of the mandatory salmonella monitoring system and is managed as part of the IT-based salmonella database 'Qualiproof' owned by the German certification system Q&S GmbH (Quality and Safety GmbH). The Qualiproof database allows to inform pig fattening farms – online as well as offline – about their salmonella affection and status. Hence, this information can serve as basis for farmers' animal health measures. It is considered as an essential part of a systematic salmonella management on pig farms. Nevertheless, the increasing salmonella affections and worsening of the salmonella status of many pig plants show that farmers are obviously not able or not willing to apply this information. Therefore, other scientific studies still consider the interface between farmers and slaughterhouses as an obstacle for information transfers along the pork production chain. Against this background, this paper aims at analyzing the status quo and the determinants of information use and information exchange in the German pork production. To do so, it uses the Q&S operated salmonella monitoring system as an example. Between April and May 2008 a national survey of pig fattening farms was conducted in order to identify obstacles to information exchange as well as processing and suggest solutions to this problem. The sample size (N=873) allows in-depth statistical analyses. Preliminary results indicate that intrinsic motivation is paramount for successful information processing and salmonella management. The identification of determinants on information exchange and processing in food supply chains provides insights into successful IT architectures. This includes hardware and software solutions that support chain-wide communication.

Utilization and comparison of three ground based sensors for nitrogen managment in irrigated corn
Shaver, T.M., Westfall, D.G. and Khosla, R., Colorado State University, Soil and Crop Sciences, 1170 Campus Delivery, 80523, USA

Precision farming has been a major research focus of agronomists for over a decade. Much of this research has been directed towards enhancing the efficiency of farm inputs without negatively impacting profitability or the environment. Studies have shown that remotely sensed imagery (particularly normalized difference vegetation indices, or NDVI) can provide valuable information about in-field variability in corn. Currently there are several devices commercially available that determine NDVI and numerous studies specific to ground-based NDVI sensors have shown that NDVI readings adequately quantify corn variability and correlate well with many variables that affect corn yield. However, little work has been done to determine which of the commercially available sensors performs best or how variables such as site specific management zone (SSMZ), corn row spacing, wind and sensor movement speed over the canopy affect NDVI readings. Also, little has been done using ancillary soil and crop variables such as leaf and soil N content and plant height in conjunction with NDVI to increase grain yield correlation that could ultimately aid in N requirement determinations. Previous studies have demonstrated a correlation of NDVI with the N content of corn leaves suggesting that NDVI sensors could be useful in verifying previously delineated SSMZ in irrigated corn. Four site years were examined and sensor readings were collected across three SSMZ at the V8, V12, and V16 corn growth stages. While NDVI readings did show a relationship with delineated SSMZ it was only consistent at corn growth stage V12. There are several sensors that determine NDVI, however utilization, climatic and management variables may affect NDVI readings. Three sensors (Crop Circle™, green and red GreenSeeker™) were examined across different crop variables, environmental variables, and management factors under greenhouse conditions to determine how these variables affect NDVI. The Crop Circle™ sensor performed best at distinguishing N differences across a wide range of corn growth stages. The red GreenSeeker™ sensor performed well but was more affected by corn row spacing and sensor movement than the Crop Circle™ sensor. The green GreenSeeker™ displayed very high NDVI variability and the results bared little resemblance to N application rates except at the V11 and V12 corn growth stages. Past studies have also illustrated a positive relationship between NDVI and variables that affect yield. Three available NDVI sensors were again compared (Crop Circle™, green and red GreenSeeker™) to see which best determines N variability and if ancillary soil and crop variables used in conjunction with this NDVI can increase the correlation of NDVI with grain yield. Sensor readings were collected across four N application rates at the V8, V10, V12 and V14 corn growth stages. Stepwise regression was conducted to determine the best possible correlation using all variables. Overall the Crop Circle™ and red GreenSeeker™ sensors adequately distinguish N variability. The Crop Circle™ had the highest correlation with yield at the V14 corn growth stage when used with leaf N content and plant height in a multiple regression that resulted in an R^2 of 0.95.

Implementation of recording of good agricultural practices in an agro-entrepreneurial information system
Caceres, V., AGRODIVERSIDAD S.A.C., Management and Production Software, Jr. Eulogio del Rio N° 1063 – Huaraz - Ancash, 51, Peru

The agro-exportation companies of the Callejón of Huaylas do not have an agro-enterprise information system that has implement the registries of the agricultural practices, that are now an important condition, for certification and commercialization in international markets. The objectives of the present project were to analyzes design and implement an agro-entrepreneurial information system that contains the recording of good agricultural practices for the best handling of agricultural production, as established in the EUREPGAP protocol. The computer science project SIA-EUREPGAP (V.1.0.0) has been elaborated development with the methodology of software called Rational Unified Process (RUP), an object-oriented software, by using the life cycle of the evolutionary prototype, using as tools for its programming and development to VisualBasic. Net 2005 and the SQL Server Express 2005. The purpose of this computer science project was to develop a software that serves as s tool for management, auditing, control, and documentation of the recording of good agricultural practices of the EUREPGAP norm, that allows it to be used by any organization or enterprise attempting to meet the standards of international quality for its agricultural products.

Discrimination of a soil-borne disease complex due to simultaneous infection of *Heterodera schachtii* and *Rhizoctonia solani* in sugar beet

Hillnhütter, C., Sikora, R.A. and Oerke, E.-C., University Bonn, INRES-Phytomedicine, Nußallee 9, 53115 Bonn, Germany; chillnhu@uni-bonn.de

In sugar beet fields, damage due to soil-borne nematode and fungal root pathogens often appear in clusters. In addition, these soil-borne disease agents can occur either alone or simultaneously on one plant making detection and optimum treatment difficult. In greenhouse experiments susceptible, tolerant or resistant sugar beet varieties were inoculated with the fungus Rhizoctonia solani and/or the nematode Heterodera schachtii alone or concomitantly. Attempts were made to discriminate between the occurrence of each disease alone or in combination prior to visual diction of disease symptoms. Concomitant infections of these two pathogens in sugar beet plants can lead to synergistic based symptoms that cause accelerated symptom development which over time can yield to increased mortality of the plants. The use of site-specific preventative plant protection measures requires precise detection techniques that can discriminate between the occurrence of single or complex disease as well as the distribution of the pathogens. To determine variation in symptoms of different sugar beet varieties by the pathogens R. solani and H. schachtii a hyperspectral sensor was used at different times after infection. Hyperspectral vegetation indices were calculated from the leaf reflectance data to determine vitality changes and accordingly the injury level of the plants. The potential of this method was investigated in greenhouse trials to devolve it in the future to field trials.

Mobile internet in e-government: the case study of Hungary

Szilagyi, R.[1] and Szilagyi, E.[2], [1]UD CASE FAERD, Boszormenyi str. 138., 4032 Debrecen, Hungary, [2]UD Faculty of Business Administration, Kassai str. 26, 4028 Debrecen, Hungary

Due to a notable change of the regulation of public administration electronic governmental services have become an actual issue on an operational level in Hungary since 2005. According to the Hungarian public opinion the current situation is considered rather unsatisfactory: there is much room for improvement in the field of E-governance. This paper seeks to examine the current Hungarian social and technical conditions of a more important part of e-governance: m-governance. So in the first part of the paper we try to give a brief overview (diagnose) of the Hungarian m-governance situation, then we would like to represent our empirical results. The paper tries to sketch a case study of Hungary and tries to give a suitable answer for the question: are more problems with these electronic (mobile) services than positive experiences or advantages in Hungary.

Spatial variability in irrigated cotton in relation to soil electroconductivity
Bauer, P., Stone, K. and Busscher, W., USDA-ARS, 2611 West Lucas Street, 29501 Florence, SC, USA

Variable rate irrigation technology has been installed on a number of center pivot irrigation machines in the Southeast USA. Although the technology exists for being able to irrigate on a site-specific basis, knowledge is lacking on how to optimize spatial application of water to crops. We conducted this research for two years in producer fields with center pivot irrigation systems. Our objective was to evaluate the relationship between cotton (*Gossypium hirsutum*) growth and plant water status and soil electroconductivity. Data were collected from two fields, one field in 2006 and one field in 2007. Soil electroconductivity was determined for both fields using a Veris 3100 Soil EC system. In each year, 10 areas were selected and measures of plant height, leaf stomatal resistance, and soil water content were collected throughout the growing seasons. In addition, whole field NDVI was measured from an aerial platform twice each season. The results indicate that there appears to be potential for using soil EC to delineate irrigation management zones as a first step in optimizing spatial water application.

Environmental benefits of ecologically friendly paddy fields: a choice experiment approach

Aizaki, H., National Institute for Rural Engineering, National Agriculture and and Food Research Organization, Department of Rural Planning, 2-1-6 Kannondai, Tsukuba, Ibaraki, 305-8609, Japan

In Japan, traditional agricultural ecosystems have provided a variety of habitats for wild animals and plants. Changes in agriculture, especially paddy fields, have affected the number of species as well as the wildlife population. Over the past decade, concern with regard to the role of paddy fields as a wildlife habitat has been growing. A considerable number of paddy fields, however, do not meet the requirements for serving as a wildlife habitat. In order to promote ecologically friendly paddy fields that meet these requirements, it is necessary to garner public support. In addition, implementing such a plan would result in environmental benefits for the public. In this study, the choice experiment, which is one of the stated preference methods, was used to evaluate the environmental benefits obtained by implementing plans involving spreading ecologically friendly paddy fields. In a choice experiment, each respondent is presented with questions and is asked to select the best plan from among two or more plans that are characterized by two or more attributes. In our study, the plans for promoting ecologically friendly paddy fields had four attributes: the first was the number of intermediate egrets in each 10-hectare paddy field, which would serve to indicate the condition of paddy fields as a wildlife habitat. Birds play the important role of acting as indicators of the ecological effects of changing agricultural practice. The second attribute was the creation of a paddy field for bird watching (bird-watching field). The third attribute was the creation of a paddy field where children can catch animals such as loaches, crucian carps, and aquatic insects living in the field (eco-field). The fourth attribute was the contributions for the plan. It was assumed that the cost for implementing the plan would be met through donations from each household. Four hundred and twenty-two households living in Tsuchiura city, Ibaraki, served as the respondents. On the basis of a simple model, the amount of money the respondents were willing to pay for the implementation of a plan that increased the number of intermediate egrets by one for every 10 hectares was 3,715 yen per year, and the amount they were willing to pay for constructing a bird-watching field and eco-field was 642 and 1,630 yen per year, respectively.

Usage of information technology over the combine harvester users: a case study for Antalya region
Yilmaz, D., Canakci, M. and Akinci, I., Akdeniz University, Faculty of Agriculture, Department of Agricultural Machinery, Faculty of Agriculture, Department of Agricultural Machinery, 07070, Turkey

Combine harvesters have important role for farming. Information Technology (IT) in combine harvesters has become to a key factor for the use of these machines. The complex process technology inside of the harvesters has to be managed by drivers to maximize the performance of the machines (ha/h, t/h) by taking into account the necessities of the plant production (minimal grain losses, grain damage). IT has been supported the drivers to reach these goals. The objective of the study is to usage of IT over the combine harvester users: A case study for Antalya Region. IT devices, which are used in combine harvester in Antalya-Turkey, have been explained. For this purpose, a questionnaire has been carried out over the combine harvester users in Antalya region. Easiness usage and necessity of IT in combine harvester were investigated for combine harvester users. Topical problems were determined. Some suggestions were proposed to be increased the IT usage.

The economic impact of contract farming: a case study in Maharashtra state of India

Kulkarni, S.[1], Grethe, H.[2] and Van Huylenbroeck, G.[3], [1]KULeuven, Namsestraat 69, 3000 Leuven, Belgium, [2]University of Hohenheim, Faculty of Agriculture and Food Policy, Schloss, Osthof-Süd, 70593 Stuttgart, Germany, [3]University of Ghent, Department of Bioscience Engineering, Copure Links 653, 9000 Ghent, Belgium

The present investigation aimed at understanding the functioning and economic effects of contract farming in a potato growing region of India from a farmers' perspective. It compared contract and non-contract farmers in terms of net returns, and access to inputs, credit and extension services in order to contemplate the factors motivating farmers' participation in contract farming. Personal interviews using a standard questionnaire were conducted with contract farmers (n=53) and non-contract farmers (n=41) in the Pune region of Maharashtra state of India using an ex-post facto survey research design. The farmers were engaged in contracting with Frito Lays. Ltd. for chip quality potato production. A logit model was used to analyze the factors influencing the choice of farmers to participate or not in contract farming and costs and returns from contract and non-contract situations were compared. The Mann Whitney test was used to compare the satisfaction level of farmers regarding the provision of extension services, inputs and credit. The results indicated that factors including education, distance to credit source, age and total land holding had a positive influence on participation in contract farming whereas off-farm income and membership of agricultural cooperative groups had negative influence on participation. The comparison of costs and returns between the two groups indicated higher net returns for contract farmers which is attributed to higher yield, higher selling prices and less complicated marketing channels. Predetermined prices, lack of middlemen in the supply chain reducing the marketing costs and an organized marketing channel are the advantages of contract farming compared to conventional farming which suffers from fluctuating prices. Contract and non-contract farmers differed significantly in terms of their satisfaction for access to quality inputs (seeds, fertilizers, chemicals), provision of market information and credit availability with contract farmers having better access to inputs and facilities.

Determination of supporting and repressive elements of the integrated animal health system
Fick, J. and Doluschitz, R., University of Hohenheim, Institut of Farm Management, Computer Application and Business Management in Agriculture, Schloß - Osthof - Süd, 70599 Stuttgart, Germany

Food scandals, animal diseases and a stronger concern of consumer in food production cause an increasing attention of politics and society (e.g. EU animal health strategy). Livestock farmers and veterinarians are urged to assure and optimise animal health amongst others through data exchange and documentation. Currently available data (systems) from these process participants are often decentralized, deficient and redundant in Germany. Up until now the approaches of the livestock farmers and the veterinarians regarding to the applications of the animal health systems are analysed. Via cluster analysis different groups of supporting, undecided and not supporting the system could be identified. For instance the characteristic of the group supporting/undecided veterinarians is younger than the veterinarians who decline the application of the system. To this group belongs a higher part of mix-animal-surgery. Further this group owns a higher percentage of mobile IT-equipment and has a superior willingness to invest in ICT. The group keeps ¾ of the interviewed veterinarians and also includes undecided veterinarians. These results lead to the next step – a demonstration of the system to a focus group of veterinarians. A first version of the integrated animal health system was developed. Now the aims are to identify of the system in order to that the system will meet the requirements of the target groups. Possible elements are innovators, aspects of the novelty, information sources and communication relations. To find out supporting and repressive elements there will be used a selection of qualitative methods like participant observation, interviews and demo workshop. Aspects of innovation and diffusion research and as well of usability research will be the basis of this research set. Results of other projects have shown that veterinarians are hard to get involved into technology changes. Therefore we assume that this group will have a central influence on success or failure of this project and the major focus will be on this group. Expected results of this set of methods are an intensive exchange of the participating veterinarians, changes in the approaches of the veterinarians regarding to the applications of the animal health system, identification of supporting and repressive elements of the system, observation of learning effects and improvement of the usability. Further it is anticipated that the diffusion process will start at this stage of the project. All this will lead to an improvement of the system and will raise the chances of adoption of the system by the process participants.

Cost and benefit aspects of it-based quality assurance and traceability systems: results of a Delphi-survey
Roth, M. and Doluschitz, R., Institute of Farm Management 410c, Computer Applications and Business Management in Agriculture, Schloss Hohenheim, 70599 Stuttgart, Germany

Food scandals have led to the establishment of several approaches to guarantee quality and traceability of food along the supply chain of animal products. But most of these approaches are isolated applications which are not compatible among each other. Facing these challenges, the IT FoodTrace project was established to create an IT solution that ensures traceability and quality along the supply chain of meat and meat products. Although an integrated traceability solution will generate benefits for the stakeholders, it might be possible that the acceptance of such a solution is low, due to costs and labour input. Other studies not only showed this general finding, they also claimed that statements concerning the profitability of a traceability solution are a major issue for acceptance. Therefore the aim of this study is to identify and to quantify relevant cost- and benefit-aspects of stakeholders. Another aim is to detect the acceptance of IT-based traceability and quality assurance systems among potential stakeholders. Case studies will be used to show detailed information about costs and benefits of IT-based traceability and quality assurance systems throughout the meat supply chain. The study will be carried out by means of a Delphi-survey. In two successive surveys about 50 experts will be asked about their opinion concerning cost and benefit aspects as well as acceptance issues of IT-based traceability and quality assurance systems. Case studies with the focus on cost and benefit aspects on all stages of the meat supply chain will complete the project. The first of two Delphi-surveys showed a great participation rate. So far, around 50 Experts (mostly experts from the meat supply chain, but also from authorities and universities) completed the questionnaire and were giving additional comments. First preliminary results show, that e.g. the willingness to pay an extra for meat products with guaranteed traceability is estimated low or even not existent in the eyes of experts. This result differs from the findings of a case study, where the interviewed persons stated a positive willingness to pay. More than 86% of the experts estimate license costs as well as general running cost as the most relevant cost factors of an IT-based traceability solution, whereas 56% don't think, that IT-based traceability solutions will lead to an increase in internal controls costs within processing companies. Overall, 92% of the experts state acceptance of an IT-based traceability solution, if economical benefits will be certain.

Scientific based expert models, a method to downsizing information to many farmers
Mayer, W., Progis Software GmbH, Postgasse 6, 9500 Villach, Austria

'Agriculture employs more people and uses more land and water than any other human activity', is expected to feed a world population that will increase from six to ten billion by 2050 and is faced with the challenges that global warming, the reduction of natural resources and the higher demand for bio-energy as of environmental care taking, will bring along. Further, there are many national and transnational regulations which are additional barriers for this sector. Well tried cultivation methods will be insufficient for the future. The agri-forest- environment must look for new models enabling fast and precise decisions based on facts. Rural areas need better management. New systems must be introduced fast and worldwide. The implementation of advanced ICT-software technologies, capacity building and a sophisticated know-how transfer are the prerequisites for a successful future. Rural area management has to be done within a network of scientists, IT-experts, qualified advisers and a range of other stakeholders of the food/feed production chain. PROGIS has developed a GIS (Geographical Information Systems) platform called WINGIS with applications for rural area management and is the only provider worldwide of an integrated ICT-based software technology for rural area management. The core of the product is an expert data base containing all data necessary for modern land management. It is equipped with data coming from local agri-forest scientists, who integrate the latest findings on cultivation into this data-base. Users of the software have therewith access to latest expert know-how. With GIS (orthoimages or field or forest polygons or even Google or Virtual Earth) there can be planned, simulated and documented, calculated details, developed business plans or work out data for a necessary insurance covering environmental risks, for several years. Advisors can optimise group farmers needs, as logistics, compare farm business anonymously, or optimize farm management by profiting from group-synergies. The PROGIS software provides also modern land management tools which cover biodiversity, sustainability, multipurpose land use, evaluation of land use potentials and more, but in any case, the close collaboration with scientists, who are feeding the system ongoing is necessary in order to face successfully the ever changing demands.

Prototype of an academic ISO11783 compatible task controller

Ojanne, A.[1], Kaivosoja, J.[1], Suomi, P.[1], Nikkilä, R.[2], Kalmari, J.[2] and Oksanen, T.[2], [1]MTT Agrifood Research Finland, Vakolantie 55, FI-03400 VIHTI, Finland, [2]TKK Helsinki University of Technology, Automation and Systems Technology, Otaniementie 17, 02015 TKK, Finland; ttoksane@cc.hut.fi

Research of tractor-implement automation and mobile communication between work units and Farm Management Information System (FMIS) has been important research topics at MTT Agrifood Research Finland. Practical research and development require technology platform to be adjustable by researcher themselves. To fulfil this demand, an academic version of Task Controller (TC) was designed and constructed. The ISO 11783-10.2 standard describing task controller and management information system data interchange was the basis of the development work. The standard specifies most properties very exactly, but in some parts freedom is given to the designer. The prototype documented in this paper fulfils most of the specifications of the standard. Panasonic CF-P1 Pocket PC with GPRS connection was used as the TC device. The programming environment was MS Visual Studio and programming language C#. The TC was tested in field in ISOBUS sprayer environment (http://autsys.tkk.fi/en/Agrix/Basic). The ISO 11783-10.2 was mostly followed, but there are some properties, data formats etc. which are not fully completed in all details, and are still under development. The development consisted creating TC interfaces to Implement ECU and Tractor ECU via ISOBUS and FMIS server. The designed TC realises the basic needs for farm field operations: (1) Transferring task files from FMIS to TC in XML format. (2) Initializing the TC and IECU according to needs. (3) Adjusting the treatment zones, described in polygons, according to task file declarations. (4) Collecting sensor data in IECU and storing it with time and spatial data in a logfile. (5) Sending the logfile to FMIS via wireless internet. The TC following the ISO 11783-10.2 standard was proven to work in field tests. The used programming environment and language was proven appropriate. Further research and development work is needed to define content of the information gathered from the implement and structure of task file XML and logfiles, especially when connecting the system with different implements.

Assessing whole-farm risk positions by means of a risk barometer: an internet application for arable and poultry farmers

Van Asseldonk, M.[1], Baltussen, W.[2], Hennen, W.[2] and Van Horne, P.[2], [1]Institute for Risk Management in Agriculture, Hollandseweg 1, 6705 CN Wageningen, Netherlands, [2]Agricultural Economics Research Institute, Wageningen University and Reserach Centre, Hollandseweg 1, 6705 CN Wageningen, Netherlands

Farming is a risky occupation. Farmers are confronted with a continuously changing landscape of possible price, yield, and other outcomes that affect their financial returns and overall welfare. Risk-management involves the selection of methods for coping with all types of risks in order to meet the decision-maker's goal while also taking their risk attitude into account. This means that calculating the risk-return trade-off in designing risk management strategies is an important target in agricultural business. A considerable amount of general work had been done on issues such as on-farm risk management and the use of some well-established risk-sharing instruments such as commodity futures and conventional insurance. However, a large proportion of the past work on financial instruments for risk management has been done using partial analysis, i.e., assessing the merits of using each particular risky asset or risk instrument alone. Moreover, many previous analyses were very general in nature and not tailored to individual (farm) production circumstances. There is insufficient recognition of the fact that an individual farm manager can influence and improve his or her own risk situation through a range of on-farm and off-farm risk management strategies. Besides that, the individual risk situation also depends on farm specific characteristics such as farm size and financial situation. Therefore the goal of this study is to develop an instrument that measures the (relative) riskiness of a certain farm. This so-called risk barometer assesses the risk position of the farmer on basis of both the riskiness of the farm system and the risk aversion of the farmer. The risk barometer captures the following main risk sources: yield risk, price risk and interest risk. The expected values and variances of yields are elicited on the basis of relative subjective estimates of the farmers themselves (self-rating approach). Next, the opportunities for modeling risk positions are elaborated. In the presentation we will elaborate on the concept of a risk barometer and the merits of such internet tools for farmers. This concept is illustrated for a typical Dutch arable farm and poultry farm.

Variable rate seeding for French wheat production: profitability and production risk management potential

Dillon, C.R.[1], Gandonou, J.[2] and Shockley, J.[1], [1]University of Kentucky, Agricultural Economics, 403 C.E. Barnhart Building, Lexington, KY 40546-0276, USA, [2]Texas State University - San Marcos, Agriculture, 601 University Drive, San Marcos, TX 78666, USA

Variable rate seeding has met mixed results regarding economic feasibility with success in few locations but lack of success for many regions. Discussion with French producers reveals that some of these farmers are practicing variable rate seeding as a means of managing the existence of rocky soils. The relevance of this practice for French farmers is due in part to the small size of the production unit and their subsequent relatively good knowledge of differences within fields. This permits and encourages producers to manage by subsections of fields rather than treating a field uniformly. Additionally, French farmers often use highly intensive production practices with substantial input levels resulting in high yields; these conditions also lend themselves to consideration of variable rate application. The purpose of this study is to partially fill the gap of precision agriculture research for French agriculture, especially regarding economic assessment of variable rate application. Specifically, planting date and variable rate seeding production practices are considered regarding their profitability and ability to reduce temporal fluctuations in wheat yield thereby lowering production risk faced by the farmer. The inclusion of both of these production management decisions will allow consideration of the interactive effects of planting date, which does not vary by soil type, and plant population which potentially varies by soil type. Biophysical simulation using the Decision Support System for Agrotechnology Transfer model (DSSAT v4), provides the underlying yield response data based on 20 years of historical weather. This data is incorporated into an economic optimization, mathematical programming model considering resource constraints and the risk attitude of the producer. Sensitivity analysis with respect to alternative soils, seed price, wheat price and cost of variable rate seeding technology is conducted. Preliminary results show merit of variable rate seeding for the base scenario and the potential of using variable rate seeding to reduce production risk borne by the producer, depending upon technology cost. Thus, evidence of the economic feasibility of the practice of variable rate seeding by early adopting French producers is found, demonstrating support for their decision.

The potential benefit of site-specific phosphorus and potassium management for small fields

Pena-Yewtukhiw, E.M.[1], Rayburn, E.B.[2], Yohn, C.W.[2] and Fullen, J.T.[3], [1]West Virginia University, Plant and Soil Sciences, P.O. Box 6108, Morgantown, WV 26506, USA, [2]West Virginia University, West Virginia University Extension Service, P.O. Box 6108, Morgantown, WV 26506, USA, [3]Fullen Fertilizer, Rout 3E, Union, WV 24983, USA

Precision nutrient management begins with soil testing, based on a greater volume of information regarding the soil fertility status of a given field area. That information comes at greater cost. These costs are balanced by the economic benefit of finding areas of the field with greater nutrition stress and secondarily by finding areas of the field with reduced input needs. University fertilizer recommendations were designed with the vagaries of field-average soil sampling in mind. The soil test result was a composite of many values across the sampled area, whose variation was unknown. In the USA, there is a belief that benefits of precision nutrient management accrue to large grain production fields. In the Appalachian region, topography and landholding patterns result in smaller fields, many under mixed forage management (haying and grazing). Such fields are not usually considered for precision nutrient management. Our general objective was to evaluate that premise, assuming that grid soil sampling 'captures' more of the within-field variation in bioavailable phosphorus (P) and potassium (K). Existing data were inadequate to answer this question because of an insufficient number of 'research' fields within a common soil/landscape physiography. Our specific objectives were, for a large number of grower fields, to determine: a) the magnitude of variation in grid-sample bioavailable P and K, relative to a single composite sample; and b) the potential benefit of precision nutrient management, relative to the 'field-average', for fertilizer P and K recommendations. In 2005 and 2006, 33 fields in Jefferson County, West Virginia, USA, lying in the western Chesapeake Bay watershed, ranging in size from 2 to 14 ha (total of 250 ha) and under mixed forage management, were divided into grids ranging in size from 0.4 to 1.5 ha (average of 10 grids per field at 0.8 ha per grid) and sampled for bioavailable P and K. At the same time, each field was randomly sampled to give a single composite sample per field, typical of current sampling protocols. Fertilizer P and K recommendations for alfalfa-grass production, for each field under both grid and composite sampling protocols, were compared. Fertilizer P and K valued at \$2.20/kg P_2O_5 and \$1.65/kg K_2O, respectively. Grid sampling and soil analysis were valued at \$15/ha. Variable rate application of fertilizer(s) was valued at \$17/ha. Among the 33 fields, the means and coefficients of variation (CV's) were unrelated to each other for bioavailable P or K. Means and CV's for these soil test parameters were unrelated to field size. There was a positive relationship, across the 33 fields, between field-average bioavailable P and K. Across the entire 250 ha, grid sampling, combined with current university recommendations, resulted in an average redistribution of fertilizer P and K at rates of 23 kg P_2O_5/ha and 44 kg K_2O/ha, worth \$124/ha, relative to field-average recommended uniform amendment applications. Nearly \$60/ha was saved by not amending already fertile field areas, while \$64/ha was attributable to greater amendment of less fertile grids. Economic benefits ranged from \$0 to \$255/ha for an individual field, were greater than the \$32/ha 'breakeven' for 70% of the fields, and were dependent on soil test parameter mean and variation.

ICT mass customization in flowers & food: feasibility of service-oriented BPM platforms

Verdouw, C.N.[1], Verloop, C.M.[1] and Ten Voorde, H.[2], [1]LEI Wageningen UR, P.O. Box 29703, 2502 LS The Hague, Netherlands, [2]Cordys, P.O. Box 118, 3880 AC Putten, Netherlands

Firms in Flowers & Food industry are characterized by high uncertainty of both supply and demand. At supply side, they have to cope with great variability in their business processes because of dependency on living materials. They may reduce uncertainty by improving process control, but remain vulnerable to weather conditions, pests, decay and other incontrollable factors. At demand side, this type of firms face a high degree of demand uncertainty among others because of weather-dependent sales and changing consumer preferences, especially if it is upstream in the supply chain. Furthermore, there are many environmental legislations and high consumer requirements on food safety. Information is of crucial importance to manage this uncertainty by enabling timely and network-wide information sharing and flexible response. This makes great demands on especially the interoperability and agility of supporting information systems. However, information systems in Flowers & Food are often patchy (island automation), characterized by poor integration and a lot of manual data re-entering. In order to achieve the required interoperability and agility, this paper proposes an ICT mass customization approach enabled by Business Process Management (BPM) platforms based on Service Oriented Architecture (SOA). In business literature, mass customization is broadly advocated as core approach to combine flexibility and customization with efficiency and standardization. It relates the ability to provide customized products or services through flexible processes with the ability to produce in high volumes at reasonable costs. Recently, application of mass customization principles to ICT has become possible by the progress at different lines of developments, especially the evolvement of modular software, shift from data-oriented to process-driven software, evolvement of Enterprise Application Integration (EAI), focus on architecture (model-driven software) and isolation of business logic in formalized rules. ICT mass customization combines the advantages of standard and customized software. It enables on-demand configuration of information systems from standard components with standardized interfaces. BPM platforms combine these developments in integrated and application-independent toolsets for business process definition, modelling, simulation, deployment, execution, monitoring, analysis, and optimization. In the paper, first a conceptual architecture of ICT Mass Customization using BPM platforms is developed. Next, the implementation of this architecture in the Cordys tool is described. Cordys is a single toolset that offers comprehensive BPM and SOA capabilities, including service bus technology, composite application development, business process modelling, business rules definition, data definition mapping, workflow execution, business activity monitoring and business intelligence. Finally, the applicability to agriculture is illustrated by means of a case-study in arable farming, in which integrated functionality for a geo-fertilizer advice process is composed from web services of different involved organisations for legislation, soil analysis, LAI (Leave Area Index) map and Advice Calculation. The EDITeelt+ standard is used for the content of the communication between the components. It is concluded that BPM platforms such as Cordys are promising toolsets to meet the high requirements on ICT interoperability and agility in Flowers & Food industry.

Agro living lab: a new platform for user-centered design in agriculture

Haapala, H.E.S., Seinäjoki University, Research and Development, Keskuskatu 34, 60100 Seinäjoki, Finland

Agro Living Lab in Seinäjoki Finland is a new platform for user-centered R&D. According to current research, usability can be the limiting factor for the use of new technologies in agriculture. Agro Living Lab is an attempt to produce more usable products and services through the involvement of end-users in R&D process. According to IST 2006 Conference in Helsinki Finland: 'Living labs move research out of laboratories into real-life contexts to stimulate innovation. This allows citizens to influence research, design and product development. Users are encouraged to co-operate closely with researchers, developers and designers to test ideas and prototypes.' Currently Living Lab concepts are being used in approximately 70 locations in the world for different needs in R&D. Our scope in Seinäjoki is to facilitate the R&D process of agricultural machinery and IT manufacturers and service providers. The Living Lab concept was chosen as the methodology to bring research and development nearer to the end-user and students. Involvement of students is a central resource for the Agro Living Lab so that a special Living Lab pedagogy is currently being developed. Agro Living Lab Seinäjoki is a joint concept of Seinäjoki University of Applied Sciences and Seinäjoki Technology Centre Ltd.

SARI®, computer software for sectioning and assessment remote images for precision agriculture uses

Garcia-Torres, L.[1], Gómez-Candón, D.[1], Caballero-Novella, J.J.[1], Peña-Barragán, J.M.[1], Jurado-Exposito, M.[1], Castillejo-Gonzalez, I.[2], García-Ferrer, A.[2] and López-Granados, F.[1], [1]Institute for Sustainable Agriculture, Crop Protection, Apartado 4084, 14080 Cordoba, Spain, [2] University of Cordoba, School of Agronomy, Campus of Rabanales, 14080 Cordoba, Spain; luisgarciatorres@uco.es

Software named Sectioning and Assessment of Remote Images® (SARI®) has been developed to implement precision agriculture strategies through remote sensing imagery. SARI® is designed to divide remotely sensed imagery up into 'micro-images' of 'micro-plots' and to determine the quantitative agronomic and environmental indicators of each micro-plot. To validate the uses of SARI® studies on weed patches aggregation, crop-weed competition and herbicide application prescription maps were achieved using ground-taken data and remote images of weed infested plot in southern Spain. The work herein developed has shown that remotely sensed imagery managed with SARI® can play an important role in developing cost-effective precision agriculture.

Hybrid wireless networks for advanced communication services in agriculture

Boffety, D.[1], Chanet, J.P.[1], Hou, K.M.[2], André, G.[1], Amamra, A.[2], De Sousa, G.[1] and Jacquot, A.[1], [1]Cemagref, UMR TETIS, 24 Avenue des Landais, BP 50085, 63172 Aubière Cedex, France, [2]Université Blaise Pascal, LIMOS, Complexe Scientifique des Cézeaux, 63177 Aubière Cedex, France

Nowadays, broadband connectivity plays a key role for economic prosperity and quality of life as descent road and public transport network. Furthermore, wireless communications enable to extend broadband networks (xDSL or satellite) to rural and isolated areas. The FP6 NET-ADDED project aims at developing and validating technical methods to improve the deployment and operation of hybrid wireless technologies, in coherence with the growing demand of international broadband communications. The project themes and validation sites of the project are the following: (1) broadband internet access for rural schools in Morocco, in support to the remote continuing education programme of the Ministry of Education; (2) access to e-Government and e-Commerce services for rural tourism user communities in Greece and Turkey; (3) access to educational contents for Non-Governmental Organisations in Burkina Faso and Benin; (4) remote connection of Teacher Training Centers and High School in Cambodia; (5) use of satellite as part of National Telemedicine Network for continuous medical education on laparoscopic surgery in Turkey; (6) wireless Sensor Networks (WSNs) dedicated to precision agriculture for farming user communities in France. The objectives of this paper are to describe this last theme. Farmers need two types of communication services: classical broadband access services (mail, professional web site, etc.), and more dedicated communication services such as data transfer, environment and animal monitoring using wireless sensor network (intelligent irrigation etc.). The objectives of our work are focussed on the development of a hybrid wireless network platform for advanced communication services in agriculture, which enable to provide these two kinds of service everywhere in the farm (building, fields…). The proposed hybrid wireless network is composed of different wireless mediums: WiFi and Zigbee. The idea is to use mesh WiFi network to implement the network backbone and to use and ZigBee/WiFi Gateway to reach the WSN (ZigBee wireless sensor node) dedicated to data acquisition. The core of our devices is the LiveNode platform. The LiveNode platform has the following hardware and software component: (1) a hardware node build around the component based-concept, named 'LIMOS Versatile Node' (LiveNode), which enables to build sensor, gateway, access point by using one or several basic component; (2) a real time Operating System, named 'LIght weighted Multi-threading Operating System' (LIMOS); (3) a wireless routing protocol with an integrated Quality of Service (QoS) support, named 'Communication Inter Vehicle Intelligent and Cooperative' (CIVIC). In France, an experimental site including 3 farms has been deployed and several services based on the LiveNode Platform are offered to farmers: soil moisture and irrigation management, animals' survey, mobile internet access… Different algorithms of QoS have been compared to estimate the bandwidth on such hybrid wireless networks. The network design will be described in this paper and the obtained results will be presented and discussed.

Multi scale analysis of the factors influencing wheat quality

Har Gil, D.[1], Svoray, T.[1] and Bonfil, D.J.[2], [1]Ben-Gurion University of the Negev, Department of Geography and Environmental Development, Beer-Sheva, 84105, Israel, [2] Agricultural Research Organization, Field Crops and Natural Resources, Gilat Res. Center, M.P.Negev, 85280, Israel

In Israel the economical importance of wheat is great, as it constitute the main growth in areas with low precipitation. The baking industry requires production of high wheat grains quality and Gluten Index (GI) is a fast and prevalent method for estimation wheat gluten quality. During the last years it was found that a significant part of Israeli fields produced wheat grain characterized by pure GI, unsuitable for baking. Gluten Index can be influenced by a combination of biotic and abiotic factors such as availability of water and nutrition, climate and whether conditions prevail during wheat ripening, wheat cultivar and others. Although GI represents wheat quality, the factors influencing it not yet characterized quantitatively. This study aim to (a) understand which factors influence GI (b) grade there influences (c) determine if there is a difference in there influence at different scales – national, regional, farm and within field scales. The study was based on testing the relationship between wheat quality from fields across Israel to growing season weather conditions and management factors as fallow, emergence date and wheat cultivar. The spatial and temporal variability in soil type and growing conditions creates a wide range of wheat quality. GIS was used to interpolate climate factor as: total rain; spring rain; number of days where maximum temperature was above 30 °C during March/April/the whole season; number of heat spells days (above 35 °C and below 25% relative humidity); and number of days were minimum temperature was above 22 °C, that were spatially joined to the wheat fields layer. Trials were organized with grower cooperators across Israel and over 450 commercial fields were sampled for GI at each season for three growing seasons. A quantitative investigation on the effects of growing season weather and the management factors on wheat quality are in progress, using multiple regression, cluster and data mining analysis. Preliminary results for one season show that the factors that most determine GI at the national scale are the growing region, wheat cultivar, ripening type and number of days with maximum temperature above 30 °C during the grain filling stage. Considering these results during planning sowing probably would decrease low GI problem and will increase high quality wheat grains production.

CAPRICORN: a windows program for formulating and evaluating rations for goats
Ahmadi, A. and Robinson, P.H., University of California Davis, Department of Animal Science, One Shields Ave, 95616 Davis California, USA; abahmadi@ucdavis.edu

CAPRICORN is a package of least cost computer rations formulation programs for goats. The programs have been developed from animal nutrient requirements and feed nutrient analyses in the US National Research Council Bulletin entitled 'Nutrient Requirements Of Small Ruminants: Sheep, Goats, Cervids, And New World Camelids (2006)'. CAPRICORN is designed to run under Windows 95, 98, 2000, XP and Vista operating systems. This paper describes some of the programs in the CAPRICORN package. Examples of data entry screens and ration printout screens will be shown.

Computer model use in decision making system for planning and management of agrotechnologies
Yakushev, V.V., Agrophysical Research Institute of the Russian academy of Agricultural Sciences, 14 Grazhdansky prosp., 195220 Saint-Petersburg, Russian Federation

The decision support system (DSS) which enables generating planned and operational decisions in agronomy and realizing the information support of farm practices implementation in precision farming (off-line and on-line regimes) has been developed in Agrophysical Research Institute. DSS contains a database, a knowledge base (declarative and procedural), and also GIS tools. Procedural knowledge - the mathematical models completed in the form of individual program modules. The integration of mathematical models to DSS is examined for differential fertilizer application rate calculation, forecasting of phenological stage onsets, assessment of agrotechnologies and farm practices by economical and environmental criteria. The procedure of program modules connection to DSS in the form of dynamic linked libraries (DLL) comleted on certain rules is considered. It is noted, that the main requirement of integration DLL in DSS is correct description of export procedures and functions, and also an operation with them. Practical realization of the interface between DSS and mathematical model remains for system users, who will have to determine a method of library loading (statistical or dynamic), the order of a data transfer to export procedures and functions, and also the order of representation and storage of the gained results.

Methodological approaches to estimating of an optimal time instant of agrotechnical practices
Bure, V.M.[1] and Yakushev, V.P.[2], [1]Saint-Petersburg State University, Applied mathematics and Control Processes, Universitetsky prosp., 35, 198504 Saint-Petersburg, Russian Federation, [2]Agrophysical Research Institute Russian Academy of Agricultural Sciences, 14 Grazhdansky prosp., 195220 Saint-Petersburg, Russian Federation

The estimation problem of an optimal time instant of agrotechnical practices realization arises at planning stage. Expected losses per the time unit, related to overestimating or, on the contrary, underestimating of a time instant of required agrotechnical practices it is possible to estimate, as a rule. The losses can be directly expressed in terms of money (for example, downtime for machinery, the losses connected with necessity of additional man power etc.). In certain cases the magnitude of losses can be resulted from expert estimation of the relative undesirability of the error connected with overestimating of a time instant of required agrotechnical practices in comparison with its underestimating. Frequently it is possible to determine time limits for agrotechnical practices realization beforehand. The problem comes to estimation of a time instant of required technological operation realization within some time interval. The mathematical model based on probability distribution is proposed. To estimate probability distribution it may be used the statistical information. Optimal solutions are determined and the procedure of their practical implementation under statistical data is proposed.

Investigation on relations between organic matter, Zn and Cu content in some soils of Guilan

Norouzi, M., Naraghi, M., Alidoust, E., Honarmand, M. and Forghani, A., Guilan university, Soil science, Rasht, 0098, Iran; mehdi_uni2000@yahoo.com

Soil is a precious wealth that is gained with difficultly but easily will be lost. Lake of adequate recognition and knowledge about importance of soil and easy access to it by everyone, have resulted in society lake of care and attention toward this vital article. In dry and semidry parts of world, such as Iran, not only organic matter returns to soil rarely but also because of overpowering function of micro-organism, has a quick dissolution. In more than 60 percent of Iran's cultivated soil, the amount of organic matter is less than 1% and in considerable parts of it less than 0.5%. But surveys, specified that, amount of organic matter in soils of Guilan, because of its special climate, is more than 1% and in some cases even reaches 6% to 8%. Chemically micro-elements affect each other's attraction capacity in soil. As organic matter is also one of the active, dissolved elements in soil, we decided to measure organic matter, Zn and Cu relation in some parts of Guilan country's soil. Consequently, we did sampling in 40 areas of guilan. Amount of organic matter by Walkey-Black method and complete amount of Cu and Zn was determined by digestion way and atomic absorption spectrometry. Investigation of relation between data showed that, there is no significant relation between amount of organic matter, Cu and Zn. Only in 1% level of confidence, a significant relation between Cu and Zn was observed. Also Cluster analysis of forty soil samples by Ward's method, based on scattering of organic matter, Cu and Zn, was performed. Considering organic matter distribution, Cu and Zn were on lowest Cluster analysis level, and samples were classified in five, four and four categories

Optimal cropping patterns in Saudi Arabia using mathematical sector model

Al-Abdulkader, A.M.[1], Al-Amoud, A.I.[2], Awad, F.S.[2], Al-Tokhais, A.S.[3], Basahi, J.M.[4], Al-Moshailih, A.M.[2], Al-Dakheel, Y.Y.[5], Alazba, A.A.[2] and Al-Hamed, S.A.[2], [1]King Abdulaziz City for Science and Technology, Natural Resources and Environmental Research Institute, Riyadh 11442, Saudi Arabia, [2]King Saud University, Riyadh 11451, Saudi Arabia, [3]Ministry of Water and Electricity, Riyadh, 22111, Saudi Arabia, [4]King Abdulaziz University, Jeddah, 21589, Saudi Arabia, [5]King Fisal Universty, Alhasa, 28111, Saudi Arabia

Saudi Arabia is an arid country with limited natural resources and harsh climatic conditions. Thus, efficient allocation of the limited natural resources among development sectors such as agriculture is vital to Saudi Arabia. The main objective of this paper is to highlight some of the preliminary findings of a 5-year research project conducted in Saudi Arabia to determine the optimal cropping patterns that maximize the national net returns subject to the limited natural resources, mainly, available water supply, irrigation water and arable land. A mathematical sector model is used to achieve the ultimate goal of the research project. The economic structure of the sector model consists of three main activities: crop production, marketing and international trade activities. Three main constraints are applied in the sector model: resource constraints, crop balances and trade balances. This paper will answer what, why and where to plant major crops in Saudi Arabia that maximizes the national net returns given the limited resources.

Agronix, neural network for fertilisation management

Megna, A., Spada, V., Campisi, R. and Manzini, V., A.D.M. srl - Agricultural Data Management, software development, Via Cairoli, 71, 97100 Ragusa (RG), Italy

With the increasing diffusion of ICT, the use of decisional systems for better management of agricultural practices becomes more increasingly significant. The need to improve the quality of yields that takes into account not only the nutritional requirements of crops but also the optimization of agricultural resources with a view to reducing the environmental impact of agricultural practice was the stimulus to developing AGRONIX, a multilingual web oriented application, offering a sophisticated architecture to manage and to support decisions in the field of fertilization and irrigation. With its core based on neural network, AGRONIX is able to adapt to different pedo-climatic situations and to the variability of crop needs and technical yield conditions, to solve some unpredictable factors or purely local connotation that are not easily quantifiable and therefore could never weight in the planning of a plan of fertilization. The scientific knowledge work together with the neural network that mimics the structure of the human brain imitating the mechanisms of learning. According to data that is entered, neural networks correct the parameters of the model to find the relationship between the data. The neural network is able to simulate a perfectly natural phenomenon also very complex, can seize and reveal the influence of particular combinations of elements not known and therefore not used by any model. Compared to a general model based on scientific knowledge, that can be so far from the correct forecast for local characteristic and exceptional facts not taken into consideration, the network, however, can 'incorporate' these anomalies in their knowledge. From knowledge comes new knowledge and the model is more suit to the farm situation taking into account all relevant factors that affect that specific production. The result is a plan fertilization ad hoc in terms of quality and production as well as the optimization of resources. The ability to include and to acquire new knowledge from specific and local information, the internet accessibility and the multilingual support makes this application usable to world level

Geo-fertilizer advice modeled using BPM and SOA

Verloop, C.M.[1], Wolfert, J.[1] and Beulens, A.J.M.[2], [1]LEI Wageningen UR, Plant Systems Division, P.O. Box 29703, 2502 LS The Hague, Netherlands, [2]Wageningen University, Information Technology Group, P.O. Box 8130, 6700 EW Wageningen, Netherlands

On a farm a lot of processes happen, in the cultivation of the products the customers want. These products have to cope with more and more stringent constraints on food quality and food safety. Because of this, the processes as well have to be of high quality, meeting food and environment standards, and highly effective and efficient. Moreover, information about the products and the processes must be made available to and shared with the partners in the agricultural supply chain network, and finally to the customer. Processes are increasingly automated to realize this. In the food economy information sharing more and more becomes a competitive factor. Agri-food companies increasingly participate as networked enterprises in multi-dimensional, dynamic and knowledge-based networks. Standardization in information integration of processes, applications, data and physical infrastructure are important to realize this. For setting-up and changing integrations quickly, a rapid (re-)configuration approach is needed. This requires component-based information systems, independent standard components, standardized interfaces and communication between components, a central repository of published components and standardized procedures for selection and implementation of components. The business processes in real life must be leading in designing and using information systems. A combined approach of Business Process Management (BPM) and Service-Oriented Architecture (SOA) is considered to be very suitable to meet these requirements. In this paper the process of geographic-specific fertilization of a parcel is described. Instead of uniformly distributing the fertilizer over the parcel, it is more effective and efficient to supply on each smaller area the amount that is really needed. Currently the information needed is supplied on hardcopy. A pilot for a fully automated information system to provide a geographic-specific fertilization advice is created, based on BPM and SOA principles. The basic principles of BPM and SOA are explained and applied to the geo-fertilizer advice case. The process model describes the business process in more detail using the formal business process modeling notation (BPMN). The process is modeled using the aspects legislation, soil analysis, LAI (Leave Area Index) map and Advice Calculation. The EDITeelt+ standard is used for the content of the communication between the components. Modeling is done in an Enterprise Application Integration (EAI) software suite, named Cordys. This application can be used to automatically generate business process modeling language (BPML) from the BPMN models that form the basis for standard web services. In this way, a service-oriented architecture is generated from the business processes. This architecture type is the basis for actual information systems that are used by the various actors in the value chain. It is concluded that using BPM and SOA, it is possible to make a model of the geo-fertilizer advice. The model illustrates the use of standardized, re-usable components and the use of a standard for communication content between the components. In this way, the model forms a basis for setting up a complete architecture for total farm management and, broader, for the agri-food supply chain. Implementation and execution of the model in the Cordys tool really automates and optimizes the process, and thus helps meeting the constraints of the modern food economy.

A new algorithm to use oblique view images as a low cost remote sensing system
Hartmann, K., Lilienthal, H. and Schnug, E.,

Remote sensing (RS) and its usage for agriculture is well documented and RS imagery products for agricultural purposes are offered worldwide. Most of the available imageries are produced from airborne and satellite data. As these systems operate above the clouds they are dependent on weather conditions and are not very flexible in the acquisition and data delivering time. In contrast ground based oblique view imagery can be used very flexible and almost weather independently. A new algorithm was generated to calculate pseudo-nadir imagery from oblique viewing. This method needs minimum input information only and integrates elevation information if available. The algorithm was validated on a small scale using artificial texture and on a field scale. Imagery was taken with a commercial digital camera. The oblique view algorithm could generate pseudo-nadir images which fit to measured data in the field. It could further be used for classification purposes of different growing zones. The results are promising and offer a possibility for a flexible low cost and multitemporal RS system.

Fostering quality of argumentative computer-supported collaborative learning within academic education in the agri-food sciences

Noroozi, O.[1], Biemans, H.[2], Mulder, M.[3] and Chizari, M.[4], [1]Fellow member of the academic board of TMU, Iran and PhD student at Wageningen University, Wageningen, 6700 EW P.O. Box 8130, Netherlands, [2]Wageningen University, ECS Department, Wageningen, 6700 EW P.O. Box 8130, Netherlands, [3]Wageningen University, ECS Department, Wageningen, 6700 EW P.O. Box 8130, Netherlands, [4]Tarbiat Modares University, Agricultural extension and education, Tehran 14115-111, Iran

In the information and communication era, Computer Supported Collaborative Learning (CSCL) is becoming increasingly prominent in higher education, and universities are progressively bringing argumentation in CSCL environments into the mainstream of their educational programmes. These environments allow students to enhance and/or support learning in general and to solve authentic problems collaboratively in particular. The study proposed here is designed to promote the quality of the learning process and results in using argumentation in CSCL environments in Agri-food sciences (human nutrition, agricultural education, food, environmental, and rural studies). As part of their regular study program, students in Agri-food sciences have to solve authentic problems with strong value and interest connotations. To be able to do so, they should combine both concepts and theories from different disciplines and perspectives, while this is supposed to be difficult for students. Argumentation in CSCL environments could be a possible instructional strategy to foster these learning processes. Moreover, different factors like student, peer, task, tutor, medium and instruction can affect the effectiveness of CSCL argumentation-based environments. This study will focus on fostering quality of argumentation in CSCL environments by identifying the influential factors which influence the effectiveness of these strategies.

Analysis the role of Information and Communication Technologies (ICTs) in improvement of food security Iran's rural households

Lashgarara, F., Mirdamadi, S.M. and Farajjolah Hosseini, S.J., Islamic Azad University (IAU), Science & Research Branch, Agricultural Extension & Education, Ashrafi Esfahani Blvd., 1477893855 Tehran, Iran

Access to sufficient and desirable food is one of the principles of any developing and healthy society. Therefore, a priority among development goals in any country is access to food security. One of the important solutions for attainment to food security is Information and Communication Technologies (ICT's). The purpose of the research was to identify influencing potentials of information and communication technologies on improvement of food security Iran's rural households. This research has been done in eight provinces of Iran during 2006-2007. This is an applied research with descriptive- survey methodology. The main tool to collecting data was questionnaire. The population for this research was 253 extension experts from eight provinces (Qum, Ilam, Kordestan, Qazvin, Semnan, Kerman, Tehran, and Lorestan) from this population 170 were selected through a stratified sampling. The results showed that in the view point of experts, the situation of food security available is unfavorable. However, information and communication technologies can have an important role to improve food security of rural households. The results of multiple-regression analysis showed that eight variables of: providing information, help to increase food production, help to agricultural production marketing, considering to address needs, improvement of interactions and communication, appropriate technologies, access to and content of old technologies determined 78% of variance of food security of rural households and improvement variable.

E-government services for famers and experience in the North-East Hungary region

Szenas, S.Z.[1] and Herdon, M.[2], [1]Central Agriculture Office, Kossuth street 12-14, 4024 Debrecen, Hungary, [2]University of Debrecen, Böszörményi street 138, 4032 Debrecen, Hungary

After the Hungarian accession to the EU the area payment system was separated into two parts namely SAPS and TopUp. The aim of the government is the fast and punctual administration of the payments and to give quick access to the available money for the agricultural farmers. The e-administration was introduced in the year of 2007 for those farms which owned more than 200 ha area. This e-administration was accessibility for every farm in 2008. After a pre-registration the administration was available with the government portal. Hajdu-Bihar county is the third biggest county in Hungary as far as agriculture in concerned. Most of the customers asked for help from the Central Agriculture Office (MGSZH). Out of the 21,000 entitled farmers 19,400 received payments with the help of the officers in 2008. In the year of 2009 the area payment system is facing a significant change. The so-called SPS system is to be introduced and the farmers will be allowed to claim their payments through the governmental portal only. In the SPS system the production was separated from the payments. The customer is entitled to the payments on the base of basic numbers, which is calculated by the previously claimed the payments. The condition of the SPS is the cross compliance and its detailed workout is being underway.

Water delivery optimization program, of Jiroft Dam irrigation networks, by using genetic algorithm

Rahnama, M.B. and Jahanshae, P., Shahid Bahonar Univerity of Kerman, Irrigation, 22 Bahman Blv., 7616914111, Iran

Water is the most important critical liquid for living. With the population increases and human's need ever-increasing to water supply. The optimum use of water sources at the irrigation networks is needed. One of the effective factors in operation from irrigation networks, to achieve suitable efficiency at the project level, is water delivery program. In addition to time of exploitation from channels, water delivery scheduling is considers too, because while designing it the type of water delivery plan effects on construction size, channel capacity, and finally construction cost. For the reason that traditional optimization methods for water delivery had many limitations, therefore, genetic algorithm method was used for optimization. That has a numerical optimization method with search ability of specific maximum or minimum parameters index to remove these limitations. In this research, Jiroft irrigation network has been considered. The distribution channels embranchment located random in different irrigation blocks, and obtained the best discharge of each embranchment and complete time of irrigation plan by genetic algorithm parameters. The results shows the best irrigation plan with different objectives supply, are the reduction of water distribution channel capacity to embranchments, the reduction of necessary time for watering plan fulfillment and the reduction of time waste in each irrigation period.

Designing, implementation and institutionalizing the national agricultural innovation system
Mortazavi, M. and Ranaie, H., Institute of Technical and Vocational Higher Education of Agriculture, P.O. Box 13145-1757, Tehran, Iran

In this article we aim at reviewing the institutional developments and paradigms of agricultural research systems in different historical eras. After delineating the basic features of agricultural researches, this article is to point out that these institutional developments demonstrate a move towards the use of systems approach in the process of designing agricultural research systems as well as facilitating the more participation of stakeholders in designing, implementation and evaluation of agricultural research goals, policies and plans. Then we aim at studying the developments of agricultural research system in Iran and point out that many of Iranian public institutes of research has been established during past eight decades. We claim that Iranian agricultural research system is in the primary phase of institutionalization and this is contrary to considerable institutional capacities existent in the country. Iranian agricultural research systems has not been developed with the aim of integrating different capacities and actors around a common focal point and has not changed itself to create a synergic virtue among them. It has been recommended to consider such issues in overall policy making process and establishing a National Agricultural Research System (NARS) in Iran to be able to institutionalize Agricultural Knowledge & Information System (AKIS) and ultimately achieving National Agricultural Innovation System (NAIS).

The analysis of modeling methods applied to forest growth research
Komasilovs, V. and Arhipova, I., Latvian University of Agriculture, Faculty of Information Technologies, Liela street 2, LV-3001, Jelgava, Latvia

The application of information technologies in the forest sector enables to increase their competitiveness and economic efficiency. The efficiency of the forest sector depends on the application of modern information technologies in private and state forest management planning process. In order to increase forest management planning efficiency it is necessary to investigate methodological resources for problem solution preparation and funding. It is necessary to work out methods and models as well as their computer realization in connection to system modeling. Nowadays there is an actual problem connected with capital value detection and future management planning of growing forest not only in Latvia but also in other European Union countries. Forest is such a type of resource that grows for long period, and therefore it needs to plan future use of wood resources carefully. Errors and experiments are unacceptable in the real world due to time and resource limits, and in this condition the simulation and modeling approach is appropriate. The efficiency increase in forest management related to the development of new data processing programs and data transmission programs. The research purpose is to describe architecture of virtual forest system. The main idea of such system is to simulate growth of a single tree in competitive conditions among other trees, in other words to create growing virtual forest in which users can improve forestry related skills, such as forest taxation and territorial division. In addition, researchers could use this system as virtual testing ground. By changing various growing conditions, it is possible to forecast future state of forest and analyze impact of changes. Such system would be useful for student education, experimenting and testing various methods and algorithms. Intended architecture of the system is client-server with various modules types. One of the main core modules is data generation subsystem where tree growth simulation takes place. Another important module is visualization subsystem used on client's side. In addition, it planned to have interface to other forestry systems used in Latvia for data comparison and reporting facility. Base platform selected for this system is Java EE 5; it means that architecture design is Java oriented. Finally, the use of simulation and modeling methods makes it possible to save money and time for experiments and at the same time get results of research close to the real world situation.

The role of e-signature implementation in regional development

Zacepins, A. and Arhipova, I., Latvian University of Agriculture, Faculty of Information Technologies, Liela street 2, Jelgava, LV-3001, Latvia

Nowadays implementation of e-signature is one of the tasks of Latvian government. The introduction of e-signature is considered by the government not only at the state level, but also by the management level of enterprises for meeting their needs. Today in Latvia there are not so many services where inhabitants can use e-signature and therefore it is needed to give people the possibility to use e-signature in more areas. In the paper the role of e-signature implementation is described how use e-signature for signing online banking transactions. E-signature means inclusion and application of Information Technologies for more efficient and modern ensuring of the functioning of the state, self-governments and the enterprises related to them, as well as for the establishment of mutual links between population and organizations. E-signature is a form, how state and self-government can use the new technologies for their advantage in order to ensure more comfortable availability of information and services for population and enterprises, to improve the quality of rendered services and provide more opportunities for regional development. Implementation of using e-signing feature in banking area will be projected and developed using for an example one of the privates banks in Latvia. Different aspects of this process for the regional development are introduced and analyzed. Foreign countries and other banks experience is described and used as well. As a result of the research e-signature service is described and the guidelines of the e-signature implementation for online banking are produced. E-signature is useful service, which can ease the process of signing all types of contracts and documents, tracking and filling them in one easy step. Implementation of e-signature for online banking will help to populate and advertise this service in our country and will boost the process of e-signature implementation in other business and government areas too.

Get and hold clients on your web project

Razgale, V., Latvian University of Agriculture, Faculty of Information Technologies, Liela iela 2, Jelgava LV 3001, Latvia; vitarazgale@tvnet.lv

When we move our business operations to virtual environment, we face with conception of e-business. On this article I'm going to talk about question How to get and hold clients on almost any Web Project. Under Web Project I assume e-shops, e-magazines, blogs, e-bank and other e-business directions. There are no problems to create Web page and to place it on Internet, but when we want to earn money with it, we need to think about getting and attracting clients. And when we have them, we want to keep and make them to return again and again. For explore this topic, I found a lot of interesting theoretical materials about good Web Projects, about colors and information layouts. The main value for developing the paper, I made a questionnaire for students about e-business. What is important for them and what is the main reason why are they dedicated to some Web Project. From theory, to attract customers to Web Project, first goes the advertisement and good chosen keywords for search programs like Google. As the second come the information on your Web Project and how it all looks and is it easy to use the information. If the Web page is tawdry customers will not choose it and will close it as soon as opened. After analyzing the questionnaires I found important aspects for developing e-business when the clients are young people. They will open Web address if their friends will tell about Web Project rather than clicking on banners on other Web Projects. They want something that attracts them and keeps in touch. Information has to be up to date and about the topic. Layout is important, but they can get used to it, if the project is what they need.

A data warehouse for decision suport in the Brazilian beef industry

Massruhá, S.[1], Barioni, L.[2], Lacerda, T.[1], Lima, H.[1], Da Silva, O.[2] and Narciso, M.[1], [1]Embrapa Agricultural Informatics, Av. Andre Toselo,209 Barao Geraldo, 13083-886, Brazil, [2]Embrapa Cerrados, Parque Estação Biológica - PqEB s/n, 70770-901, Brazil

Livestock production systems are very diverse, with their diferent types (e.g. extensive grazing, feedlots, crop-livestock and silvopasture systems) heterogeneously distributed in space. The diferent systems vary on the use of inputs (fertilizers, industry byproducts, grain, etc), magnitide of response to environmental factors (e.g. climate), intensity of natural resources exploitation and their impacts on soil, water, atmosphere, biodiversity, farmers income, regional economy and society. Nowadays beef production is intensifying with extensive grazing being replaced by more intensive systems. Different outpus and impacts are expected for a given association of a production system and a regional agroeconomic and agroecologic context. Also a interaction system vs. region may be expected, which would imply in the existence of superior spatial arrangements of the different production systems and better diagnostic of opportunities for public and private policies. Such analysis obviously require the use of mathematical models, scenario studies and optimization procedure which demand large quantity of socio-economic and agro-ecological analysis from several sources and themes integrated in space and in time. Therefore, the construction of a data warehouse may be the first step in order to achieve data integration. It involves migration of data in one direction: from the data sources to the warehouse and from there to the data analysis tools. In this process, data need to be understood and organized in an uniform representation. A model of a data warehouse, the procedures and methodologies to manage data in this domain and the challenges on communication and information technology are discussed. Finally, the integration model of the datawarehouse with analytical tools and maps for data analysis and reporting is presented.

Studying the importance of web portals in an agriculture educational Institution through the evaluation of the web portal of ITVHE

Mokhtari Aski, H., Rajabbeigi, M. and Mortazavi, M., Institute of Technical & Vocational Higher Education of Agriculture, P.O.Box 13145-1757, Tehran, Iran; mokhtari21@hotmail.com

The Institute of Technical and Vocational Higher Education of Agriculture is the major planner and organizer for educating agriculture work force in Iran in higher education. ITHVE was among the first organizations harnessing a web site in the family of research and education organizations of agriculture sector of Iran that recently migrated to a web portal. Although, by design and implementation of an information system such as the aforementioned web portal, the organization aims at improving the efficiency, the result in most cases is not true. One important cause of such misbehaviour is that the information system is not matched with the working environment of the organization. The web portal as an information system expected to be effective in improving the efficiency of the organization through coordinating other applications such as automated secretariat, finance, budget, public affairs and advertising educational packages and courses etc. The effective use of the existing web portal can make ITVHE more competitive in educational environment of Iran and neighbouring countries. The first step in improving an information system, i.e. web portal, is to understand it. In this paper, researchers hypothesized that the existing web portal is not working properly according to the major missions of ITVHE. In this regard and for proving the hypothesis a group of experts were consulted for determining the principal criteria for evaluation of the web portal and its audience. In the mean time the opinions of the target audience were examined using a questionnaire. In the end, the hypothesis proved to be true based on the findings of the research and some improvements like new interface, alternative organization schema of the content etc. were recommended.

Assessing factors effecting on application of internet and website in educational and research of graduate students in college of agriculture

Mirdamadi, M. and Alimoradian, P., Science and Research Branch Islamic Azad University, Agriculture, Hesarak, 1477893855 and Tehran, Iran

As network computing and tools for learning, teaching, and administration gain more power and accessibility, integrating technology into the educational process is becoming a major initiative for most colleges and universities. This study was to assess factors effecting on application of internet and website in educational and research of graduate students in college of agriculture at Islamic Azad University, Science and Research Branch. This was a descriptive and correlation method research. Population for this study were all graduate students in college of agriculture studying at 2005-2006 academic year (n=671). From this population (n=225) of them were selected through random sampling. After collecting questionnaires, data was analyzed by using SPSSwin software. The finding shows that there are a significant correlation in using internet in school and during undergraduate levels, English language skill, familiar with internet programs, be able to use search engineer, computer skill, age, grade point average (GPA), student encouragement by instructor, having personal computer at home, educational level, field of study, and job status. Result from multiple-regression analysis indicate that about 42% of the differences in the amount of using the internet are predicted by independent variables of Familiar with internet, English language skill, familiar with internet programs, and age.

Precise agriculture and mezo- and microclimate production agroecology system
Nasonov, D.V. and Uskov, I.B., Agrophysical Research Institute, Agroclimatology, 14 Grazhdansky prosp., 195220 Saint-Petersburg, Russian Federation

Diversity of a soil cover and a landscape variety of agroecological facies make specific demands to differentiation of agrotechnologies and processing methods in systems of precise agriculture. Microclimatic heterogeneity takes place and, as consequence there is a necessity of an estimation of productivity taking into account the factors influencing it. According to structure of influencing factors three categories of productivity are allocated: potential productivity, climatic productivity and really possible productivity. Last two categories of productivity define a choice of system of processing methods and the agrotechnologies which are a basis for systems of precise agriculture. Potential productivity is defined by intensity of a stream of sun radiation, genetic possibilities of a kind and a grade of culture and architectonics of crops. Climatic productivity is limited, besides, by provision of crops with heat and a moisture. Really possible productivity decreases regarding climatic productivity depending on an agrophysical and agrochemical condition of soil. Development of technological programs on management of process of cultivation of a crop includes an estimation of each factor of efficiency and ways of management of these factors' components. It is necessary to notice that a background estimation for differentiation under the specified factors of industrial fields are mesoestimations of productivity of region and a mesoclimate of a soil-climatic zone. Transition from a background estimation to differentiated on fields and contours is carried out on the basis of special measurements and maping of microclimatic and soil factors of productivity. At the heart of such estimations it is necessary to use balance methods of power- and mass- exchange in the «soil – crops – atmosphere» system. The choice of the processing methods included in technological programs of precise agriculture, is carried out on the basis of possibilities of component's management on heat, moisture and mineral food. Maping the listed factors in systems of precise agriculture should be made with GPS binding, providing realisation of technological operations differentiated on each of factors of productivity. Results: maps of meso- and a microclimate, maps of really possible productivity as integrated indicator for listed above factors in frameworks of meso- and microlandscape soil-climatic heterogeneity on an example of Leningrad region, Gatchina area, farm of Menkovo and Rozhdestvenno.

Web platform for simulation of agricultural mechanization technologies

Muraru, V.M., Pirna, I., Cardei, P., Muraru Ionel, C. and Sfaru, R., Inma Bucharest, IT, 6, Ion Ionescu de la Brad, sector 1, Bucharest, 031813, Romania; itcc@zappmobile.ro

Research and technology development have been the foundation of impressive productivity gains in the agricultural sector. The ability of the sector to conserve natural resources and protect the environment depends, in part, on the technologies used. Agricultural research is the source of new technologies, and important new technologies have emerged that may benefit the environment if adopted. On other hand, on the market is a large offer of machines and installations designed to agriculture. However, many new and current agricultural technologies are complex and require a much higher level of human capital and managerial skills than in the past, increasing the costs of their adoption. Certain technologies may be economically desirable over time, but require substantial capital investment (for example, certain precision farming technologies). Every technology is based on a set of farm machinery and implements. The paper proposes a web platform which includes o series of data referring to the cultivation technologies, mechanization technologies and farm machinery and implements. Using the information from the databases the end user will be able to simulate varied mechanization technologies for varied crops using the farm machinery and implements with a variety of characteristics from different manufacturers. The databases will contain also, connection information between equipments (power source – tractors, farm machinery and implements) for technologies that will be visualized as natural are possible.

A fine-tuned phosphorus strategy for sustainable production

Söderström, M.[1], Ulén, B.[2] and Stenberg, M.[1], [1]Swedish University of Agricultural Sciences, Dept. of Soil and Environment, P.O. Box 234, SE-532 23 Skara, Sweden, [2]Swedish University of Agricultural Sciences, Dept. of Soil and Environment, P.O. Box 7014, 750 07 Uppsala, Sweden

Phosphorus (P) is a limited resource necessary for food production. The use of P in agriculture is also an environmental problem. The discharge of P to the Baltic Sea in 2000 was estimated at 34500 tonnes, of which 14.5% came from Sweden. The variability of P within farms as well as within fields is substantial and a large amount of P is applied on farm land without P-requirement even if current P-recommendations are used. An optimization of this problem, i.e. risk for P loss and profitable and sustainable agricultural production, necessitate an integrated approach. In this on-going project (2007-2009), the intention is to develop and apply an integrated environmental and production index (EPI) for estimation of P requirement within fields. The basis is the current P recommendations which should be expanded to include factors important for assessment of risk for P loss. The latter should preferably be easy to measure to facilitate practical implementation and it should be possible to apply it site-specifically. Field work is carried out at two farms in Sweden, Logården in the southwest and Hacksta in the east. At both these farms, P in drainage water from single fields or field parts is measured continuously. Extensive sampling of the soil as well as data from GPS-based yield mapping is also available. The Swedish P-index (PI-S) developed by Djodjic & Bergström was used for assessing risk for P-loss. In 2007 and 2008, two fields was fertilized according to current P recommendations, a few fields were not P fertilized at all and two fields per year were fertilized according to a strategy developed in this project. This strategy is to some extent building upon the PI-S, but simplified considerably. Factors for assessing risk for P-loss was: DPS (degree of P saturation in the soil extract), local slope inclination and plant-available P in the subsoil. These were combined with plant-available P in the topsoil and expected local yield, which are used for P requirement estimation from a production point of view. A linear reduction percentage from each of the risk factors was estimated. The total reduction percentage was fairly well correlated with the more complex PI-S if estimated for different field parts. Variable-rate application files were produced and applied using the Yara N-Sensor terminal as control unit. In this manner, highest fertilization rate will be applied on parts of the field with low P in the topsoil, a high yield potential and a low risk for P loss. We hope to highlight how this type of optimised within-field fertilization affects both yield and P loss in the drainage water, both on short and on long term. In addition to the continuous monitoring of P concentration in the drainage water, data from the project will be used for the calibration of P models which can be used for scenario modelling. If successful, project results could be used to fine-tune current P fertilization recommendations. Even if the most advanced available precision agricultural equipment is used, precise requirement estimations are needed to find and manage hot-spots. This is one measure that will reduce the environmental impact of agriculture.

A system to increase safety and security of agricultural products by image information

Tanaka, K.[1,2] and Hirafuji, M.[1,2], [1]National Agriculture and Food Research Organization, Field Monitoring Research Team, 3-1-1 Kannondai Tsukuba Ibaraki, 305-8666, Japan, [2]University of Tsukuba, Graduate School of Life and Environmental Sciences, 1-1-1 Tennodai Tsukuba Ibaraki, 305-8577, Japan; tanaka.kei@affrc.go.jp

The trouble about agricultural products, such as camouflage of food's indication label and use of an unregistered agricultural chemical, is increasing in Japan. Therefore, a consumer is sensitive to safety and security of agricultural products. A farmer can give security to a consumer and increase value of agricultural products by appealing the safety of them. Disclosure of farmer's information and an agricultural work history has already been performed by some farmers. In addition, showing images that were taken at the field periodically gives reliability further to the existing information. Already we have developed Field Server which has wireless LAN, a weather sensor, and a camera, and installed in some fields. Field Server stores meteorological data and images into a server periodically. Several hundreds of images are stored every day per one Field Server. Since a vast quantity of images are continuation of the image which hardly changes, images have no information value by just showing. By extracting only images which are meaningful for a user, images acquire value. We developed the system to show images which a user wants to see from vast quantity of images by image change detection and clustering. The developed system enabled it to detect images of the appearance of a worker or a maintenance vehicle. The detection image can raise reliability of agricultural work history by linking them.

An analysis of ICT adoption trends on Irish farms
Murphy, D.J., Teagasc, FMTS, Kildalton College, Piltown, Co. Kilkenny, Ireland; dervlamurphy@hotmail.com

A set of questions in the 2004 National Farm Survey (NFS) identified the overall level of ICT usage on Irish farms at 40% compared to 46% in Central Statistics Office (CSO) for all households in the Republic of Ireland in 2004. The analysis showed that ICT adoption on Irish farms was slower than internationally reported. The existence of a digital divide influenced by systems of farming and the growth of part-time farming was a major finding. The NFS gathers information on almost 1,200 farm households on an annual basis. Teagasc has conducted the NFS and its predecessor the Farm Management Survey (FMS) annually since 1972. The primary function of the NFS is to collect and analyse information relating to farming activities. The NFS is part of the FADN (Farm Accountancy Data Network) system, which is administered by the Institute of Agricultural Economics (VUZE). The ability to supplement the main NFS survey with additional questions on new farm practices, is an invaluable source of data on trends in technology adoption. In 2008 the same questions establishing the availability and utilisation of ICT on Irish farms were asked in order to establish trends and factors influencing adoption and usage. The influences of variables on adoption rate such as farm system, scale, age, education level and farm efficiency will be explored. The study will also attempt to quantify the benefits and efficiency gains of farmers who use ICT compared to similar farms who do not use ICT.

Options for precision horticulture in Gala orchards based on site-specific relationships between environmental factors and harvest production parameters

Manfrini, L.[1], Corelli Grappadelli, L.[1] and Taylor, J.A.[2], [1]University of Bologna, Dipartimento di Colture Arboree, Via G. Fanin, 46, 40127, Bologna, Italy, [2]The University of Sydney, Australian Centre for Precision Agriculture, McMillan Bld, A05, NSW, Australia, 2006, Australia

Management zones (or classes) are often used in cropping and viticulture systems but they have not been widely reported in horticulture. A management zone is a spatially contiguous area where a particular treatment may be applied. There are several data sources, such as soil sensor, elevation, canopy sensors and production data, available to help derive and interpret management zones. In many cases, horticulture producers already have some of these but historically they have not applied them spatially to differential crop management strategies. Monitoring crop production (crop yield and fruit quality parameters) was undertaken in a 2.2 ha block in an apple orchard (cv. Gala) near Sydney, NSW, Australia. The block was managed as a single unit but consisted of two sections, North and South, cut by an internal road. Environmental information in the form of an elevation and apparent electrical conductivity (EC_a) survey were already available from the orchardist. The elevation data was used to model micro-climatic variations, particularly net radiation interception, using the SRAD model. The environmental information was then classified to form zones and crop monitoring undertaken to quantify how much of the crop variation was driven by these environmental zones (EC_a, elevation, micro-climate). The paper's null hypothesis is that production variation in an apple orchard is not influenced by environmental variation (and thus management zones are not a useful tool for differential management). The environmental data showed two distinct soil types in the apple orchard which were classified into two contiguous zones. The crop monitoring showed there was also considerable spatial variation in apple yield and quality which provides an opportunity to differentially manage production. The first interpretation of the results indicated that the environmental data poorly explained the variation observed in the production data. An inverse model approach was subsequently undertaken (effectively analyzing whether crop production explains environmental variation). This was done with data on the number of fruit/tree, mean fruit weight and percentage of fruit harvested in the first day of picking (an indication of fruit maturation). The inverse modeling identified patterns in the data that were clearly due to management effects, despite the orchard being managed as a single entity. When the north and south block effects are taken into account the influence of soil type on production becomes more evident. This indicates that understanding, recording and modelling management in orchards is very important for precision horticulture. The paper concludes with a short discussion on some of the ecophysiological reasons why environmental differences causes production variation in this area and how management may overcome this.

A fuzzy inference segmentation algorithm for the delineation of management zones and classes

Pedroso, M.[1], Guilleume, S.[2], Charnomordic, B.[3], Taylor, J.[4] and Tisseyre, B.[5], [1]Embrapa, SGE, Av W3 Norte (final), 70770-901 Brasilia, DF, Brazil, [2]Cemagref, UR TEMO, 361, rue JF Breton, 34033 Montpellier Cedex 1, France, [3]INRA, UMR ASB, 2 Place Pierre Viala, Bât 29, 34060 Montpellier, France, [4]INRA, UMR LISAH, 2 Place Pierre Viala, Bât 21, 34060 Montpellier, France, [5]Cemagref - SupAgro Montpellier, UMR ITAP, 2 Place Pierre Viala, Bât 21, 34060 Montpellier, France

Management zones and classes are a common method of simplifying multiple data sources into a map that is usable and practical for management. They form an interim step between uniform (average) and continuous (true site-specific) management structures. The creation of management classes can be done manually, through observation of maps and expert knowledge, or statistically. Statistical approaches have mainly focussed on classification algorithms, particularly the fuzzy or hard k-means algorithm. These approaches have two main problems, firstly there is usually no spatial domain in the classification thus small, unmanageable zones may be generated which must be reclassed and secondly there is no clear indication of the optimum number of classes to generate without a posterior expert interpretation. An alternative approach is to use segmentation-based algorithms. However, there has been little work published on the use of segmentation algorithms for the delineation of management classes or zones. This paper examines the application of an existing fuzzy inference system segmentation algorithm to generate management classes/ zones for precision agriculture. The fuzzy inference algorithm is encoded in the software package FisPro. The software was initially designed for food quality applications but its general nature makes it applicable to other domains. The program utilises a tessellation algorithm to convert point data into polygon data. The polygons are then aggregated based on a simultaneous consideration of the attribute values and of the neighbourhood. The algorithm takes into account the transitions between zones by defining fuzzy kernels and supports in the attribute space and in the 2-D space. The systems is halted at either a user-defined number of zones or when certain criterion is reached. The ingenuity of this approach is two-fold. Firstly, it is a segmentation algorithm which generates zones (not classes). This avoids the issue of small, irregular shaped classes observed from current recommended classification techniques. Secondly, the fuzzy system inference allows expert knowledge to be incorporate in the zoning process (unlike the purely statistical k-means approach). Thus, this approach should correct the problems associated with current classification algorithms. However the output from the software still need to make agronomic as well as statistical sense and this is the aim of the paper. FisPro is currently a univariate system. It is undergoing reprogramming into a multivariate system. The first stage of this paper will analyse the efficacy of management classes derived from multiple data sources (yield, soil, canopy sensing etc) using either a) univariate analysis on multiple data followed by map aggregation (compression) or b) multivariate FisPro analysis which generates one final map. In the second part of the paper these outcomes will be compared with the results from a classification algorithm (k-means clustering) run on the same data set.

An integrative approach of using satellite based information for precision farming: Talking.Fields

Bach, H., Migdall, S. and Ohl, N., Vista Remote Sensing in Geosciences, Gabelsbergerstr. 51, 80333 Munich, Germany

Sustainable food production and food security is one of the central challenges of this century. Limited acreage and high demand for resources drive up prices. This situation is aggravated with the current growth in bio-energy production. In order to achieve higher efficiency in agricultural production as a solution to these problems, precision farming is one option. Its application in farming practice is growing from year to year. In order to apply site specific measures, a geo-information service is needed to communicate the required spatial information to the farmers. Spatial inputs to the geo-information service can consist of satellite data. At present the following shortages in the use of satellites for agricultural applications can be observed however: the spatial information about crop development is not yet fully operationally available; the communication between geo-information service, farmer and field is missing; and satellite-based, integrative applications are missing. The goal of the approach in Talking.Fields is to overcome these shortages and increase the efficiency of agricultural production via precision farming, using a geo-information service that applies satellite-based data sources and techniques in an integrative way. Three components – satellite navigation, satellite communication and Earth Observation are central elements in Talking.Fields. Satellite sensors from Earth Observation deliver spatial information about the crop development on the fields. In-field sensors measure weather conditions and provide the current environmental situation. The farmer communicates the applied and planned farming measures to the geo-data service. An agricultural geo-information service integrates all input data and translates it into advices for farming measures. The recommended agricultural measures are delivered to the farmer. Satellite navigation is used for steering of tractors and site specific farming activities. The effectiveness of the proposed measure can be validated using a satellite observation acquired x-days after the conducted measure. These components of Talking.Fields will be demonstrated for fertilizer applications using information on biomass and nitrogen status from satellite images as one central input. The demonstration case will also provide an economical estimate of the benefit of precision farming measures. Thus the farmer can better decide whether starting with site-specific fertilization is economically sound and if yes, which fields should be targeted first. For the provision of spatial crop information, satellite data might offer the only economically meaningful way. Presently, mostly tractor based sensors are used for precision farming measures. It will be demonstrated, that the same information can also be obtained from satellite providing an economically interesting alternative to in-field sensors. As a summary, cost-saving and eco-friendly technologies are applied to cultivate fields precisely to account for the changing soil conditions and to optimize yield formation. This concept allows implementing precision farming via Talking.Fields.

Advanced web technologies for environment-related mashups

Maria Koukouli, M.K. and Alexander Sideridis, A.S., Agricultural University of Athens, Informatics Laboratory, 75 Iera Odos Str., 11855 Athens, Greece

As the new technologies and services become popular through the use of Web 2.0, new opportunities emerge for the development and use of efficient tools for collecting information. Advances in Web technologies make it a powerful channel for sharing the news and information about environmental issues. The latest trend is the development of environment-related mashups. These applications cover a broad spectrum of topics such as pollution, status of hurricanes, climate change, industrial impact, information about endangered destinations in the world, forest fires and natural disasters. In the current study the technology used in the development of mashups will be presented. Furthermore, the existing environment-related mashups will be analysed in order to examine the quality and quantity of information in this field that is already published on the internet. The analysis of the existing applications is driving us towards designing an environment-related system which improves capabilities and performance of widely applied similar systems. In particular, conclusions drawn are used in developing a system for the citizens' health data in combination with special environmental conditions for a specific major rural area of Greece. The analysis shows that the flexibility of mashups enable these applications to be expanded and used for a variety of causes. Special conditions in rural or urban areas, information related to food scares as well as any other data can be collected through the use of mashups in order to extract reliable conclusions for the prevention and protection of the citizens' health, the preservation of nature and contributing in the efforts for sustainable development. A successful design for an environmental mashup will offer not only a tool of knowledge for the internet users but also a dynamic tool for the researchers in Earth sciences.

Towards a data model for the Integration of LADM and LPIS

Inan, H.I.[1], Milenov, P.[2], Van Oosterom, P.[3], Sagris, V.[2], Zevenbergen, J.[3] and Yomralioglu, T.[1], [1]Karadeniz Teknik University, Department of Geodesy and Photogrammetry, 61080 Trabzon, Turkey, [2]DG JRC of the European Commission, Institute for the Protection and Security of the Citizen, GeoCAP Unit, Via E. Fermi 2749, 21027 Ispra, Italy, [3]Delft University of Technology, OTB Research Institute, Jaffalaan 9, 2628 BX Delft, Netherlands

One of the aspects of the Common Agricultural Policy (CAP) of the European Union is to focus on the management of subsidies to the farmers. For this purpose, member states (MS) have established Integrated Administration and Control Systems (IACSs), including Land Parcel Identification System (LPIS) as the spatial component of IACS. The declared agricultural parcel is a subject of the payment calculation as well as for administrative control. Declared agricultural parcel can be unstable over time and space. Therefore, the reference parcel (RP) is used as basic unit of LPIS. RP can be either cadastral parcel or production block. Some of the MS used their cadastral system as a starting point for the creation of their LPIS; others made use of dedicated production block systems. The initiatives of Land Administration Domain Model (LADM) which concentrates on the standardization of different LA systems and LPIS Core Model (LCM) which is focussed on standardization of different LPIS applications are two main building blocks of this study. In this study, for a more effective use of LA data (cadastre parcels) as RPs, a LADM – LCM collaboration/ integration data model is developed and some issues are discussed. Within the model, sub parcels as a spatial refinement of cadastral parcels are proposed for the classification of basic land cover types in cadastral parcels. This classification also provides a good functionality for the management of declared agricultural parcels. For the administrative data, farmers are associated with payment entitlements as a proof of their rights to be paid for agricultural activities. Yearly aid applications which are main registries for subsidies are associated with farmers. They are composed of declared agricultural parcels and farmer sketches. Declared agricultural parcels as the main component of aid applications are associated with cadastral parcels through sub parcels. Several issues have been discussed concerning the type of collaboration between LADM and LCM. Two main ones are included in this paper. Representation of farmers is one of them. In the model, farmers are designed as specialization of persons in the LADM because LADM person classes provides the functionality of representing natural, non-natural and also group of them. This type of persons may also be farmers. Association with farming rights and restrictions is the other issue. It is a fact that the only right IACS/LPIS is about is the right to be paid (entitlement). However, in the collaboration model, the functionality of tracing related rights and restrictions through cadastral parcels is also presented. The collaboration model included in this paper may be regarded as an initial step towards LADM – LCM collaboration. The model covers only cadastral parcels as RPs. Cadastral data is not used in an integrated manner even in the MS where cadastre data is used for LPIS set-ups. Instead, cadastre data sets are extracted from original data sets. Therefore, with the collaboration model, integrated use/communication of data sets is proposed. Currently, it seems the model may only be useful for a few countries which already use cadastral data for their LPIS. However, it may be a good solution for the future.

The use of video segmentation technique to build a lightweight tool for remote monitoring of agricultural process through Internet

Hirakawa, A.R., Saraiva, A.M. and Amancio, S.M., Escola Politecnica da Universidade de Sao Paulo, Dept. of Computer & Digital System Engineering, Av. Prof. Luciano Gualberto, 13, 158, 05508-010, Brazil

The use of video resources in web-based experiments over computer networks, known as weblabs, is very useful for remote monitoring and controlling of agricultural in field processes. On this context, and considering that bees are important for agriculture as natural pollinators, a research consortium with: LAA (Laboratório de Automação Agrícola da Escola Politécnica) and the Laboratório de Abelhas of the Ecology Department of the Biosciences Institute (IB) is conducting the ViNCES (Virtual Network Center of Ecosystem Services) project. ViNCES aims researching on web-based information system and advanced Internet to provide better tools for polinization and photosynthesis studies. The bee weblab, called BBBee, allows remote research and observation of mandassaia specie bees (Melípona quadrifasciata athidioides), including climate, audio and video data acquisition from hives. Real time audio and video are broadcasted over the Internet and through a local area network using streaming format. The bandwidth requirements for video transmission are usually high and the video quality, after compression and decompression using conventional standards, is, in general, low and with loss of information. On this scenario, this work presents a video coding and decoding (CODEC) algorithm to enhance the quality of the transmitted images, and, at the same time, reduce the transmission and storage requirements. The algorithm makes use of segmentation and tracking techniques of video objects in such videos. In addition, the algorithm allows selection of tracking object (in the ViNCES, the bees) to enhance its information quality and make the stream coding for sending through Internet. A complete CODEC was implemented and tested with actual video of bees. The data rate and quality of the resulting videos were evaluated and measured using objective quality metrics. The experimental results shown that a video coded using the proposed algorithm has 50% of size compared to the same video coded using MPEG4, also reaching better image quality.

Wine traceability and wireless sensor networks

Gogliano Sobrinho, O. and Cugnasca, C.E., Escola Politécnica - USP, Agricultural Automation Laboratory, Av. Prof. Luciano Gualberto, travessa 3, n° 158, sala C2-56, 05508-900 - São Paulo - SP, Brazil

The purpose is to present the continuation of a research about traceability in the Brazilian wine supply chain. The first phase of the research was conducted during two years at the Agriculture Automation Laboratory from Escola Politécnica - USP, in São Paulo, Brazil, which led to the modeling of an information system intended for collective use, based on a Service Oriented Architecture. The project aimed at the development of an information system for maintenance of traceability data for the Brazilian wine industry. A functional prototype of the system, which provides more than 70 web services to be used by local clients at the participant wineries and its associations was also built. The project started with an extensive bibliographic review and evolved with several contacts and reunions with Brazilian wine producers and researchers at Embrapa Grape and Wine at Bento Gonçalves, a branch of The Brazilian Agricultural Research Corporation, where the main wine producers in Brazil are located. Since 2005, traceability data maintenance is mandatory for all food producers intending to export their products to any European Union country. Besides that, final consumers have more and more been demanding information about food products. In the second phase of the research, in conjunction with a series of tests of the functional prototype to be conducted at the experimental winery from Embrapa Grape and Wine, the installation, configuration and use of wireless sensor networks in vineyards are being evaluated. Through the use of such networks, it is possible to collect data directly related to the quality of the grapes as well as climate conditions in real time. A low cost, web based, local client for interacting with the proposed system is under development. Due to the use of the Internet for remote data transmission, several issues involving data security are also being considered. The development of a web-based information system capable of surpass those difficulties is one of the goals of the research.

Electronic claiming of SAPS: the Hungarian case

Csótó, M.[1] and Székely, L.[2], [1]Budapest University of Technology and Economics, Information Society Research Institute, Stoczek street 2-4., 1111 Budapest, Hungary, [2]Corvinus University of Budapest, Institute of sociology and social policy, Közraktár utca 4-6., 1093 Budapest, Hungary

In 2007 the Hungarian ARDA (Agricultural and Rural Development Agency) – which is responsible for paying out supports financed by the European Agricultural Guarantee Fund (EAGF) and the European Agricultural Fund for Rural Development (EAFRD) and implementing market measures – introduced an electronic claiming system for European funds, especially the Single Area Payment Scheme (SAPS). After a successful pilot period (with the participation of the country's biggest farms) the system was made available for all of the 200 000 registered farmers in Hungary. Surprisingly enough, in its very first year 95% of the claims were sent to the ARDA electronically. Why is it so fascinating? The rate of the Internet-connected households (35%) and Internet users (45%) are still low in Hungary. ARDA also conducted a readiness survey among farmers, and the findings of the survey are mirroring the Hungarian situation: only every third farmer has an access to the internet and the skills needed for using the e-claiming systems are also lacking. So, the basic question is: what was the key factor of this success? The answer is simple: the advisor network of the ARDA. The Agency laid special emphasis on the communication of the system to the farmers and also worked out a method - using Hungary's governmental gateway (the Client Gate) – which let advisors submit the electronic forms with the previous agreement and involvement of the farmers. After all, the electronic claiming system is a success story, but are there any long term benefits for the farmers? Do they see the advantages of ICT and will they be encouraged to use the technology in everyday farming? Could this system be the killer-application?

Efficient knowledge transfer for advisers and farmers
Quendler, E. and Boxberger, J., Insitut fuer Landtechnik, Nachhaltige Agrarsysteme, Peter Jordan Strasse 82, 1190 Wien, Austria

Farmers need proven and new knowledge of engineering matters to solve technical problems and manage investments in their agricultural business. This claim will be complicated through public budget consolidation, the current information flood, the extension of working life, which increases the knowledge requirements of each farmer and their disorientation as well as the importance of lifelong learning through further education and advice. According to recent budget restriction in the Bavarian government, in the future the state aims only to be involved in providing those goods and services which the private sector is not willing to provide. If disadvantages for these matters, especially due to delayed take up of new technology - technical progress in farming -, the related national economic concerns and knowledge transfer should be avoided, compensating means are necessary. For determination of effective measures, it is important to differentiate between knowledge transfer activities due to public concerns and commercial, cost-effective operation interests. In the field of agricultural engineering it is necessary to include further criteria to guarantee an effective and undisturbed transfer. The reason is the missing availability of advice in engineering for farmers in the private advice sector. The relevant criteria are the limited number of public consultants, the farm related amount of investments for sustainable existence of farm businesses, key-competences of Bavarian farms, the advice requirements of the majority of farms and the possibilities of a work sharing cooperation between public and private advice services. The public consultants have to act as supra-regional multiplier, as knowledge engineers, who support public and private consultants of other disciplines in operating with complementary knowledge in engineering matters on farms. Other instruments for efficiency increases in knowledge transfer within an organisation are the shorten of knowledge transfer ways, application of new information and communication technologies, reorganisation according communication channels, improved networks between actors in knowledge transfer and timely consultant profiles next to the cooperation possibilities with private advisory organisations and the building up of demand oriented core capabilities.

I3S: integrated solution support system for water stress mitigation

Kassahun, A.[1], Krause, A.[2] and Roosenschoon, O.[2], [1]Wageningen UR, Department of Social Sciences, Information Technology Group, Hollandseweg 1, 6706 KN Wageningen, Netherlands, [2]Alterra, Wageningen UR, Centre for Geoinformation, Droevendaalsesteeg 3, 6708 PB Wageningen, Netherlands

Water stress is a global problem with far-reaching economic and social implications. The mitigation of water stress at regional scale not only depends on technological innovations, but also on the development of new integrated water management tools and decision-making practices. Water stress from an integrated multi-sectoral perspective combines the wide variety of existing and new type of analysis and mitigation options to deliver optimal and adaptable solutions to water stress. AquaStress is an EU funded integrated project delivering interdisciplinary methodologies enabling actors at different levels of involvement and at different stages of the planning process to mitigate water stress problems. The purpose of Work block 4 (WB4) as part of the AquaStress project is the development of supporting methods and tools to evaluate different mitigation options and their potential interactions. The main task of WB4 is the integration of knowledge, information, data and methods within the context of IT systems. Information produced during phase 1 (by WB's 1 & 2) will be assembled in an Integrated Solutions Support System (i3S). The i3S will be enriched with the options considered in phase 2 (WB3). According to the mission of work block four (WB 4) of the AquaStress Project, key science and knowledge outputs have to be brought together and integrated in a computer based infrastructure. This system puts the outputs at the disposal of the user community, finally assisting stakeholders to resolve problems arising from water stress and to profit from 'lessons-learned' in other, similar cases. In that regard the support system's objective was 'to enhance the selection process of water stress mitigation options by providing a suite of tools that can effectively support the participatory development of a water stress mitigation plan.' At the current state the i3S is fully operational with a growing amount of knowledge and data. The i3S provides public access not only to knowledge and data but also to water management tools and an extended glossary of terms from various management domains. The tools interact with each other through a common knowledge base. Some tools also support OpenMI as a means of exchanging data between models. The i3S can be used in a number of ways, for many different purposes, by different types of people and organizations. It allows project teams to define traceable and executable processes they should follow to solve water management problems. The query and presentation tool of i3S allows stakeholders to browse the knowledge base and evaluate options thereby enabling them to actively participate in the selection of water stress mitigation options. Questionnaire and multi-criteria assessment tools of the i3S make it possible to bring together different perspectives of stakeholders and reach a common understanding. A dedicated knowledge base editor allows experts to add data, information, knowledge and experiences and by doing so, share it with stakeholders and other interested parties.

Technological advances in developing Web applications for the agri-food sector

Weres, J., Kozłowski, R.J., Kluza, T. and Mueller, W., Poznan University of Life Sciences, Institute of Agricultural Engineering, Department of Applied Informatics, Wojska Polskiego 28, 60-637 Poznań, Poland; weres@up.poznan.pl

A need for prompt and reliable information dedicated to distributed clients: farmers, institutions and food processing engineers involved in the agri-food sector results in development of professional Web-based information systems. Selection of effective tools for a software developer is of critical importance to support software quality standards. In the paper the most advanced technologies and environments related to the recent version of Microsoft .NET Framework were analyzed and described, and applied to develop several Web applications for the agri-food sector. The methods were based on ASP.NET 3.5, C++/CLI and C# 3.0 standards available in Microsoft Visual Studio 2008, enhanced with AJAX to make Web applications more interactive, with LINQ to work with data, with CSS to build attractive and consistent web sites, and with Silverlight to develop and distribute rich Web applications. The advanced technologies described in the paper were exemplified with the following Web-based systems developed by the authors: 1. A system for analysis, design and management of cereal grain drying and storage, composed of a module containing a set of cereal grain drying databases, computational modules for determining quantities characteristic for drying air properties, simulation of the moisture content changes in technological processes of drying and performance of various drying systems and driers, and a decision support module for selecting appropriate equipment and conditions for drying cereal grains. 2. A system supporting protection of rapeseed plantations. 3. A system supporting management of agricultural engineering faculty with respect to research, teaching and verification of knowledge.

An Internet-based decision-support system for winter oilseed rape protection

Kozłowski, R.J., Poznan University of Life Sciences, Institute of Agricultural Engineering, Department of Applied Informatics, Wojska Polskiego 28, 60-637 Poznań, Poland; rkozlowski@up.poznan.pl

Winter oilseed rape is a very popular agricultural crop in many countries including Poland. Information about the most dangerous pests and diseases in oilseed rape cultivations are very important. Delivery of this information is the main task of institutions offering expert agricultural advising. The information demand by farmers is steadily increasing and can effectively be met by employing modern communication technologies. Computer aided decision support systems helping the farmer in undertaking essential for his plantation decisions are the alternative for traditional ways of the access to the information on agricultural production. The computer-based methods proposed in this paper support the farmers in undertaking the right decisions and improve the flow of data between experts and farmers. In this paper an Internet-based decision-support system was presented to assist farmers in making the best decisions concerning the protection of their winter oilseed rape plantations. The client-server decision-support system designed delivers information about pests and diseases of winter oilseed rape and the methods of combating them. The decision-support algorithms make it possible to identify the pests and diseases on the basis of morphology or characteristic damage to plants, and select the best methods of eliminating these threats.

Neural image analysis for agricultural products

Nowakowski, K., Poznan University of Life Sciences, Institute of Agricultural Engineering, Department of Applied Informatics, Wojska Polskiego 50, 60-637 Poznań, Poland; k.nowo@up.poznan.pl

The computer image analysis form perspective of acquiring information about investigated objects is a very important technology. Thanks to advanced techniques we can precisely analyze images. Complex optic devices allow us to see what is hidden for human eyes. However the character and the quantity of received information requires man's participation in unambiguous interpretation. Artificial neural networks (ANN) with consideration on their similarity to human brain can replace man in this task. Very significant stage of neural modeling is a choice of characteristic tags. They are indispensable to correct inference. The process of preparation of learning files is also important. Finding an effective method of transformation of images to information acceptable by an artificial neural network is fundamental for correct operation of ANN. Well planned and conducted learning process guarantees correct operating model. The conjugation computer image analysis and artificial neural networks permit us to enlarge the range of uses these technologies. From practical point of view such connection permits full automation of relevant processes of identification or classification and eliminates necessity of human's participation. The work presents the neural model to identify selected kinds of corn kernels damages. It also presents original computer system designed to convert a picture on the basis of a color and shape criterion to learning files.

Neural identification of selected kinds of insects based on computer technology for the image analysis

Boniecki, P., Piekarska-Boniecka, H. and Nowakowski, K., Poznan University of Life Sciences, Institute of Agricultural Engineering, Department of Applied Informatics, Wojska Polskiego 28, 60-637 Poznań, Poland; bonie@up.poznan.pl

There have been noticed growing explorers' interest in drawing conclusions based on information of data coded in a graphic form. The neuronal identification of pictorial data, with special emphasis on both quantitative and qualitative analysis, is more frequently utilized to gain and deepen the empirical data knowledge. Extraction and then classification of selected picture features, such as color or surface structure, enables one to create computer tools in order to identify these objects presented as, for example, digital pictures. The work presents original computer system designed to digitalize pictures on the basis of color criterion. The system has been applied to generate a reference 'learning' file for the neural system to identify selected kinds of insects.

Utilization of the computer image analysis and artificial neural network models for measuring lamb's intramuscular fat level content

Przybylak, A. and Boniecki, P., Poznan University of Life Sciences, Institute of Agricultural Engineering, Department of Applied Informatics, Wojska Polskiego 28, 60-637 Poznań, Poland; bonie@up.poznan.pl

Solution of the problem of identification of quantity of the intramuscular fat, on the basis of information from ultrasonographic images of living animal, has the essential utilitarian meaning. Currently, the estimation of fat content level is made after the slaughter, while the running tests attempt to answer the question whether it is possible to perform analysis in real time on live animal using an ultrasonograph combined with a computer equipped with a suitable software based on generated neural models. First stage of studies concerned finding a feature allowing to identify and predict fat in muscle tissue by artificial neural networks based on ultrasound images. The quality of network's response was checked, both on the basis of dataset containing information about the brightness of each pixel from the grey-tint data table and on the basis of analysis of the distribution of pixels. At the moment, both methods allow for the classification in the dataset, but accuracy of the results should be increased. This suggests a need to find new representative features for the dataset, which may be combined with currently existing data.

In situ sensors for agriculture

Charvat, K.[1], Gnip, P.[1], Jezek, J.[2] and Musil, M.[2], [1]WirelessInfo, Cholínská 1048/19, 784 01 Litovel, Czech Republic, [2]Czech Centre for Science and Sociely, Radlicka 28, 150 00 Praha 5, Czech Republic; charvat@wirelessinfo.cz

Wireless Sensor Networks open new possibility for on line information access in precision farming. The presented paper describe new possibilities, how new sensors technology could increase of quality of agriculture decision. The paper is focused on description of integration of new sensors measurement with existing Web based system for precision farming. Paper conclude work of previous research projects, butt also experiences from current commercial solutions. The Field decision making system (FDMS) running by farmers and service organizations in crop production must be supported by Complex data collection system (CDCS) in Geographic information system (GIS). CDCS can be improved by integrating sensor web applications. Effectiveness of any information system can be evaluated on the basis of its ability to deliver relevant, accurate, and timely generally weather, soil and crop condition at time. Conventional model of agricultural practices is data collection in GIS being added by web based preliminary decision support systems for crop feeding (include soil and crop fertilizer) and crop protection(against weeds, pests, fungi, etc). The Prefarm system was combined with weather control model, which bring initial on-line data to above mentioned models. From agriculture point of view the focus is on data collection, detail monitoring of local weather during whole year on-line in the setting grid of monitoring area – air temperature, rain fall, air humidity, air pressure, wind. Detail monitoring of soil and crop condition in vegetable season – 'data on time' – soil moisture, soil temperature, temperature in crop to extend collected data from system PREFARM, BACCHUS, DIVINO to more detail analyze advantages and disadvantages for crop protection on application and fertilizer. Data processing, weather prediction model – basic information for above mentioned models, short time (one –two days) and middle time (7 days) weather prediction. This prediction follow global weather forecast in region, but main aim of that is focus on specific crop. For example: data from weather forecast are presented for farmers as graphic presentation of weather conditions and demand on diseases and pests: the weather forecast in Agro meteorological models is different for each crops, because each crop has a specific condition of growing and different focus in food change. Also genetic of each crop is different. Some varieties are more resistant against to specific diseases, some less. Monitoring of statistical volumes for cost evaluations and final result of application as follow: time and date of application; rate of application; area of application; method of application; weather during application; objective of application; person, who applied; person in charge.

PCR detection for mapping in-field variation of soil-borne pathogens

Jonsson, A.[1], Almquist, C.[1,2] and Wallenhammar, A.-C.[3], [1]Division of Precision Agriculture, Department of Soil and Environment, SLU, P.O. Box 234, SE 532 23 Skara, Sweden, [2]Eurofins Food/Agro AB, P.O. Box 905, SE 531 19 Lidköping, Sweden, [3]HS Konsult AB, Boställsv. 4, SE 70227, Sweden

Soil-borne diseases are responsible for annual yield losses in many agricultural crops. A prerequisite for successful development of a sustainable plant production is the availability of efficient analysis of soil-borne the diseases such as *Aphanomyces eutheiches* causing root rot in peas, *Plasmodiophora brassicae* causing club root in oil seed rape. Spatial variability within fields and variations between fields in the occurrence of *Plasmodiophora brassicae* and *Aphanomyces euteiches* were determined on farms in south and central Sweden using quantitative PCR-assays. The molecular methods developed are validated by traditional bioassay techniques. Soil was sampled using GPS from fields where the disease occurred and the results are presented as an interpolated disease map. Species-specific primers and Taqman fluorogenic probes were designed to amplify small regions of *P. brassicae* ribosomal DNA. Total genomic DNA was extracted and purified from soil samples using commercial kits. the amount of pathogen DNA was quantified using a standard curve generated by including reactions containing different amounts of a plasmid carrying the *P. brassicae* target sequence. Regression analysis showed that the assays were linear over at least 6-7 orders of magnitude ($r^2>0.99$) and that the amplification efficiency was >95%. A considerable (100-1000 times) variation in DNA–content was observed in the fields sampled for *P. brassicae* and *Aphanomyces euteiches*. Relations between the occurrence of pathogens and soil parameters such as pH-value, soil type, clay content, plant available macro- and micro nutrients were evaluated. The amount of pathogens were also correlated to electromagnetic conductivity (EM38). Molecular methods for routine diagnosis will enable producers to respond to market opportunities by securing a more intensive crop rotation. The results will constitute basic data for an evaluation of the benefit of a systematic detection of soil borne pathogens by a quantitative PCR-method in combination with chemical soil mapping aiming at an efficient application in Precision Agriculture.

Expert system for crop selection in Punjab region

Jassar, S.[1] and Sawhney, B.[2], [1]Ryerson Univeristy, 350 Victoria Street, Toronto, On, M5B2K3, Canada, [2]Punjab Agricultural University, Ludhiana (Punjab), 141004, India

'Expert Systems' is one of the important application oriented branches of Artificial intelligence. In the past decade, a great deal of expert systems had been developed and applied to many fields such as office automation, science, and medicine including agriculture. Crop selection is a crucial and decisive task under the dynamic environment of agricultural systems created by differences in climate, soils, cultivation practices, topography and available resources. For farmers and farm production managers, a decision regarding crop selection is a difficult task. Numerical methods have failed because understanding about crop systems are qualitative based on experience and cannot be mathematically represented. Expert System are Computer programs that are different from conventional computer programs as they solve problems by mimicking human reasoning processes, relying on logic, belief, rules of thumb opinion and experience. Since a number of factors have to be considered for crop selection, a solution to this predicament is an expert system that uses all available information to select the best suitable crops. In this project, the development of an expert system for selecting crops in a region in Punjab is presented. The system acting as an intelligent consultant asks a set of questions and then suggests an appropriate crop. Front end is developed using java and it is user friendly. It recommends crops to a farmer at an early stage of crop planning based on location, climate and farm level information pertaining to soils and available resources. The expert system was evaluated by specialists using farm data from a selection of farmers to assess its performance. It was found to be a valuable tool for crop selection.

Application of multispectral and hyperspectral vegetation indices for prediction of yield and grain quality of spring barley in Hungary

Milics, G.[1], Burai, P.[2], Lénárt, C.S.[2], Tamás, J.[2], Papp, Z.[3], Deákvári, J.[3], Kovács, L.[3], Fenyvesi, L.[3] and Neményi, M.[1], [1]University of West Hungary, Biosystems Engineering, Vár 2., 9200 Mosonmagyaróvár, Hungary, [2]University of Debrecen, Centre for Agricultural Sciences and Engineering, Dept. of Water and Environmental Management, Böszörményi út 138., 4032 Debrecen, Hungary, [3]Hungaria Institute of Agricultural Engineering, Department for Mechanization of Plant Production, Tessedik Sámuel u. 4., 2100 Gödöllő, Hungary

The objective of this paper is to determine the role of multispectral and hyperspectral based vegetation indices for predicting yield and grain quality of spring barley in Hungary. Spring barley is commonly used in Hungary as a raw material for beer production. In order to fulfil the quality expectations of the beer industry, grain protein content prediction and measurement can play important role on spring barley production and marketability. Multispectral vegetation indices were based on Landsat satellite pictures (25 m geometric resolution, 7 bands spectral resolution), while hyperspectral indices were based on AISA DUAL Airborne Hyperspectral Imaging Systems images 1 m geometric resolution, 359 bands spectral resolution between 400 and 2,450 nm). In order to be able to compare data with the calculated indices, reference yield and quality data was collected prior to and during harvest. For yield data collection AgroCom Terminal and Yield Mapping System was applied. Quality (protein content) data was collected by two methods: (1) by hand in a systematic grid prior to harvest; (2) by Zeltex On-Combine Grain Analyzer during harvest. Hand collected samples were analyzed in laboratory. All collected data was converted to 25 m and 1 m pixel size maps by means of interpolation techniques (ArcGIS) in order to be comparable with multispectral (25 m resolution) and hyperspectral (1 m resolution) images and vegetation indices. Results showed that prediction of grain quality compared to quantity was achievable with higher confidence in both cases. Correlation between multispectral vegetation indices and a yield was in the best case r=-0.5854/n=206/, while between hyperspectral vegetation indices and yield correlation showed much lower results. At the same time correlation between multispectral vegetation indices and grain protein content was r=-0.8118/ n=206/, while between hyperspectral vegetation indices and protein content/hand collected samples/the best result was r=-0.5033. Due to calibration failures data collected by Zeltex Analyzer was not correct, therefore comparison of such data was not carried out.

Analysis of the intra-field variability in vineyards for the definition of management zones by means multi-variant analysis

Díez-Galera, Y.[1], Martínez-Casasnovas, J.A.[1], Rosell, J.R.[2] and Arnó, J.[2], [1]University of Lleida, Department of Environment and Soil Science, Av. Rovira Roure 191, E25198 Lleida, Spain, [2]University of Lleida, Department of Agro-Forestry Engineering, Av. Rovira Roure 191, E25198 Lleida, Spain

One of the fields in which precision agriculture principles are being applied to gain control over the production system by recognizing and managing variability is viticulture. This is because winegrapes are considered as a high value crop. In managing vineyards, previous experiences have been mainly addressed to identify and to map management zones according to yield and or plant vigour, since viticulturist work under the hypothesis that different yield or vigour zones should result in different quality grapes. The present research addresses the analysis of intra-field variability in vineyards by means of multi-variant techniques (principal component analysis - PCA) by considering not only yield and vigour variability but reference soil properties and grape quality parameters. The research was carried out in four vineyard fields located in Raimat (Lleida, NE Spain): The fields present different areas, varieties and irrigation system: Cabernet Sauvignon (plot 12, 5 ha and sprinkle irrigation), Pinot Noir (plot 15, 11.1 ha, drip partial root drying; plot 30, 5.3 ha, sprinkle irrigation) and Syrah (plot 44, 17.7 ha, drip partial root drying). In all the fields, three types of variables were analyzed: soil properties (pH, electric conductivity, organic matter content, calcium carbonate content, water retention availability for plants, texture), vegetation and yield parameters (normalized difference vegetation index – NDVI, yield), and grape parameters (pH, probable alcoholic degree, total acidity, weight of 100 berries, absorbance at different wavelengths, anthocyans and total polyphenols contents). The sampling density varied between 4.3-25.6 samples per hectare. The PCA reveals logical inverse relations between vigour/yield and grape quality parameters in all cases. The variance explained by the two PCA varies between 33-51%. If the grape quality parameters are grouped in a quality index, the explained variance slightly increases to 38-52%, and to 33-70% in the case of considering the quality index together with NDVI and yield. It could be indicating that the considered soil variables have low influence in grape yield and quality variability. The study was completed with a cluster analysis combined with a multiple rang test, in order to confirm the potential variables to define management zones. The results show that clusters (two) differentiated from yield, NDVI and soil properties do not separate grape quality parameters except for the Pinot Noir under drip irrigation field. However, if clusters are differentiated on the basis of NDVI and yield, there are significant differences in the clusters with respect grape quality parameters. These results confirm that NDVI and yield are the best variables for management zoning to differentiate grape quality.

The use of OpenMI in model based integrated assessments

Knapen, M.J.R., Alterra, Wageningen University and Research Centre, CGI, Droevendaalsesteeg 3, 6708 PB Wageningen, Netherlands

Integrated policy assessments are tools to find out whether, why and how policies are developed and what the possible options for these policies are. Performing an integrated assessment involves determining the economic, social and environmental impacts of these options. In model based integrated policy assessments, there is a need for linking models and data from different domains. The OpenMI (Open Modelling Interface and Environment) provides a blue print captured in a standardized set of interfaces to describe, link and transfer numerical data between models on a time step basis. It is an interfaced based open standard that enables simulation models of environmental and socio-economic processes to be linked. Although the original focus of the OpenMI was on hydrological models, it could (and wishes to) grow into a standard for the mentioned environmental and socio-economical domain. Having a single standard for model integration, instead of inventing new 'standards' all the time would be beneficial. Using OpenMI in some of the EU 6th framework program projects for integrated assessment can serve as catalyst for the further development of OpenMI into these other domains. To successfully use the OpenMI standard for these integrated assessment projects, requirements from the applicable other domains need to be included. This paper will describe some of those requirements both from a modeller's and a policy evaluator perspective. Based on an internal working copy of the Java version of OpenMI the required conceptual and technical changes have already been implemented and tested. At the moment an organizational process is taking place to see how the OpenMI Association, that governs the development of the OpenMI standard, can support these modifications. This is ongoing and beyond the scope of this paper.

ICT adoption constraints in horticulture: comparison of the ISHS and ILVO questionnaire results to the EFITA baseline data sets

Taragola, N.[1], Van Lierde, D.[1] and Gelb, E.[2], [1]Institute for Agricultural and Fisheries Research (ILVO), Social Sciences Unit, Burg. Van Gansberghelaan 115 B. 2, 9820 Merelbeke, Belgium, [2]Center for Research on Agricultural Economics, 9 Hagalil St., 76601 Rehovot, Israel; nicole.taragola@ilvo.vlaanderen.be

Sustainable agricultural and rural development are currently issues of strategic importance worldwide. Information and Communication Technologies (ICT) have the potential to accomplish significant economic, social and environmental benefits. The objective of this paper is to determine the constraints limiting adoption of ICT in horticulture, and compare them to the EFITA baseline data sets. EFITA is conducting since 1999 a review survey of ICT adoption constraints in agriculture. Comparable data sets have been added from additional sources – the Agrocomputerage fairs in Germany, AFITA conferences with the most recent being the AFITA 2008 conference to be held in Tokyo, Japan, and more. These surveys provide a baseline for two comparative surveys in horticulture. The first survey in horticulture was conducted in Berlin, Germany at the 2004 ISHS symposium on horticultural economics and management, and will be updated at the 2008 ISHS symposium, to be held in Chiang Mai, Thailand this December. A second survey was organised in 2005 by ILVO on a sample of 208 horticultural businesses in Flanders, Belgium. Comparing the results of these questionnaires identifies technology innovation adoption commonalities and insights which suggest remedial steps to expedite ICT adoption and the necessary research priorities. The results of the ISHS questionnaires till now have revealed the following comparable adoption constraints: 'end user (ICT) proficiency', 'lack of training', 'ICT benefit awareness', 'time', 'cost of technology', 'system integration' and 'software availability'. In further detail - participants from 'developed' countries stressed as constraints: 'no perceived economic benefits', 'do not understand the value of ICT', 'not enough time to spend on technology' and 'how to get a benefit from the use of ICT'. Respondents from 'developing countries' stressed the importance of the 'cost of technology' and 'lack of technological infrastructure', suggesting that these are threshold constraints for ICT adoption. The results of the ILVO questionnaire are in line with the ISHS survey and the EFITA surveys over time, indicating a shift from ICT technical proficiency as a primary limiting factor towards the lack of understanding 'how to get a benefit from the various ICT options' – this being a challenge to research, extension and the ICT market services. In the final paper these results will be completed with the outcomes of the new surveys organised in the second half of 2008.

Microsprayer accuracy for application of glyphosate on weed potato plants between sugar

Nieuwenhuizen, A.T.[1], Hofstee, J.W.[1], Van De Zande, J.C.[2] and Van Henten, E.J.[1], [1]Wageningen University, Farm Technology Group, P.O. Box 17, 6700 EW Wageningen, Netherlands, [2]Plant Reserach International, Field Technology Innovations, P.O. Box 16, 6700 AA Wageningen, Netherlands

Arable farming is faced with increasing pressure on the environment and shortage of labor. Therefore manual removal of volunteer potato plants is becoming more costly and some full-field spraying methods are not environmentally friendly and not effective enough. Volunteer potatoes have become an increasing problem in arable farming because they are not controlled good enough. However, accurate removal is needed because cropping problems arise from the unprotected weed potato plants. Problems in the crop rotation are: spread of *Phytophthora infestans*, host to several soil nematodes, and competition to the crops. The volunteer potato weed problem urged for a better control strategy during the growth season where farmers now manually apply glyphosate. A detection and microspraying system were developed.,The microspraying system was evaluated for its performance –drift and precision– for application of glyphosate to volunteer potato plants. In the laboratory the microsprayer moved with velocities of 0.5, 1.0, and 2.0 m/s at a height of 25 cm above the ground surface. In detail, the microsprayer consists of needles with an orifice of 1 mm^2 and five fast acting control valves. The sprayer sprayed dark colored fluid on paper sheets with a predefined droplet pattern of 50 droplets, the visible deposition of the droplets on the sheets is called a footprint. The fluid was more viscous than water to prevent a splashing effect on the plant leaves. For different fluid pressures and the three different travel velocities, footprint measurements were taken. The droplet sizes varied between 3.30 µg and 14.8 µg. From the sheets of paper with footprints we analyzed the number of droplets that were released and the number of unwanted small droplets – drift – that were formed during the droplet formation in the air beneath the needles. This was done using an image processing programme. An Anova was used to identify differences between the treatments and subsequently we found the travel velocity and fluid pressure with the lowest amount of unwanted microdroplets and the highest accuracy of positioning of the droplets. As a result the system showed the best ratio of 48 large droplets and 63 microdroplets when the fluid pressure was 2.1 bar and the mass of the single droplets was 4.60 µg corresponding with an accuracy in X-direction of 1.63 ± 8.75mm and in Y-direction of 0.03 ± 0.94mm. However, when we looked at positioning accuracy in X and Y-direction from the targeted position, then the fluid pressure of 2.0 bar and the mass of the single droplet of 14.8 µg performed best with a deviation of 1.02 ± 6.78 mm in the X direction and 0.00 ± 0.90 mm perpendicular to the driving direction, this last setting corresponded with 49 large droplets and 110 microdroplets. The results show that it is feasible to target predefined positions using the microsprayer. However, the number of microdroplets should be reduced as their effect on crop plants can be lethal due to small amounts of glyphosate that is in the microdroplets. Ongoing research investigates the dose effect relation between a more viscous fluid containing glyphosate and volunteer potato plants.

Authors index

A

Aarnink, A.J.A.	215, 218
Abbink, N.	320
Abd El Lateef, E.M.	299
Abiodun, A.A.	334
Acevedo, E.	184
Achten, V.T.J.M.	91, 105
Acosta, L.E.	162
Adamek, R.	22
Aerts, J.-M.	242, 323
Agam, N.	279
Agelet-Fernández, J.	186
Aggelopoulou, A.D.	29
Aguilar, R.	326
Ahmad, H.	144
Ahmadi, A.	141, 357
Aimrun, W.	167
Aivazov, G.S.	300
Aizaki, H.	342
Akinci, I.	343
Al-Abdulkader, A.M.	361
Al-Amoud, A.I.	361
Al-Dakheel, Y.Y.	361
Al-Hamed, S.A.	361
Al-Moshailih, A.M.	361
Al-Tokhais, A.S.	361
Alazba, A.A.	361
Alchanatis, V.	10, 30, 35, 39, 279
Alexander Sideridis, A.S.	384
Alferova, D.V.	291
Alidoust, E.	360
Alimoradian, P.	375
Almquist, C.	397
Alves, M.C.	274
Amamra, A.	355
Amancio, S.M.	386
Ameseder, C.	137, 140
Amin, M.	167
Ananyev, I.P.	292
Andersen, J.C.	99
Andersen, N.A.	99
Andonovic, I.	227
André, G.	148, 211, 212, 216, 320, 355
Andujar, D.	42, 303
Ang, S.	298
Antler, A.	210
Aparecido Alves, L.R.	182
Arhipova, I.	370, 371
Arkhipov, M.V.	291
Arnó, J.	20, 21, 186, 400
Aronsson, P.	58
Arshad, M.	6
Åstrand, B.	75
Avezzù, V.	324
Awad, F.S.	361
Ayoubi, S.	295
Azizi, O.	321
Azo, W.M.	80

B

Bach, H.	383
Bachmann, J.	52
Backman, J.	74
Baguena, E.M.	8
Bahr, C.	214, 266
Bakker, T.	87, 100
Baltissen, A.H.M.C.	25
Baltussen, W.	349
Balzer, H.-U.	236
Barabás, J.	143
Bareth, G.	33
Barioni, L.	373
Barkema, H.W.	109, 252
Barreiro, P.	8
Bartelme, N.	190
Barthalos, P.	284
Basahi, J.M.	361
Basden, T.J.	158
Basso, B.	57, 246, 254, 255, 261
Bauer, P.	341
Bauer, S.D.	26
Bauer, U.	237
Bauriegel, E.	23
Beasley, J.	282
Bechar, A.	86, 149
Becker, T.C.	176
Bee, P.	198
Beegle, D.	11
Beek, A.	95
Bélec, C.	151
Belmonte Calera, A.	302
Ben-Gal, A.	279
Ben-Shachar, O.	86
Bennett, R.M.	117
Berckmans, D.	214, 217, 242, 266, 323, 324
Berducat, M.	43
Berenstein, R.	86
Berg, Van Den, W.	56
Berruto, R.	178, 195
Berzins, A.	311
Betteridge, K.	183, 330
Beuche, H.	23
Beulens, A.J.M.	129, 174, 363
Bewley, J.M.	116
Biber, P.	97
Biemans, H.	365
Binnendijk, G.P.	239
Blaauw, S.K.	87
Blackmore, B.S.	245
Blackmore, S.	246, 254, 255

Blanke, M. 99, 234
Blaschka, A. 146
Bleumer, E.J.B. 211, 212, 225, 265, 320, 333
Bochtis, D.D. 29, 63
Boehlje, M.D. 116
Boffety, D. 355
Boksai, D.B. 314
Bonfil, D.J. 356
Boniecki, P. 394, 395
Bontsema, J. 100, 125
Boote, K. 89
Borgonovo, F. 242, 323
Börjesson, T. 16, 17
Bosch, D. 142
Bouloulis, C. 169
Bourgain, O. 114
Bouroubi, M.Y. 151
Boxberger, J. 389
Boyce, R.E. 232
Brandimarte, P. 195
Bravo, C. 7
Bright, K. 229
Briz, J. 139
Brunsch, R. 319
Bueno, J. 110
Bundgaard, E. 208
Bunte, F. 268
Burai, P. 399
Bure, V.M. 359
Burgos-Artizzu, X.P. 101
Burks, T. 30
Bürling, K. 24
Burose, F. 241
Burriel, C. 206
Busato, P. 178, 195
Büscher, W. 222, 317, 341

C
Caballero-Novella, J.J. 354
Caceres, V. 338
Calcante, A. 2, 46
Calus, M.P. 318
Camp, F. 21
Campen, J.B. 219
Campisi, R. 362
Canakci, M. 343
Canavari, M. 137, 140
Cangar, O. 214
Cannon, N.D. 80
Cantore, N. 140
Caramori, P.H. 152
Cardei, P. 377
Cariou, C. 43
Carter, M.L. 330
Casa, R. 4
Casadesus, J. 94

Castillejo-Gonzalez, I. 354
Castrignano, A. 57
Cepicky, J. 260
Chanet, J.P. 355
Chanet, M. 301
Chapman, D.F. 173
Charnomordic, B. 382
Charvat, K. 253, 260, 310, 396
Chatzinikos, A. 246
Chen, X. 33
Chizari, M. 365
Choi, D.H. 296
Chosa, T. 18
Ciraolo, G. 4
Clarke, M.L. 13
Clevers, J.G.P.W. 308
Cockram, M.S. 233
Coffion, R. 155
Cohen, S. 279
Cohen, Y. 10, 35, 35, 39, 147, 279
Cointault, F. 5, 77
Conte, F.S. 141
Corelli Grappadelli, L. 381
Corner, R.J. 70
Corre-Hellou, G. 79
Costa, A. 242, 323
Costall, D.A. 330
Costopoulou, C.O. 122, 201
Cristia, V. 183
Crow, S. 142
Crozier, C.R. 55
Csiba, M. 284
Csótó, M. 388
Cugnasca, C.E. 387
Cui, D. 14

D
D'Urso, G. 4
Da Silva, O. 373
Dag, A. 279
Dainis, N. 45
Dal Fabbro, I.M. 145
Dammer, K.H. 23
Danilova, T.N. 286
Dassanayake, D. 325
Dassanayake, K.B. 325
Davies, W.P. 80
Davis, J. 164
De Boer, I.J.M. 215
De Boer, J. 150
De Bruin, S. 189
De Buisonje, F.E. 218
De Felipe, I. 139
De Ketelaere, B. 214
De Mol, R.M. 216, 225, 263, 265
De Sousa, G. 355

Debain, C.	43
Decuypere, E.	214
Deghani, A.A.	295
Dehne, H.-W.	38
Dekker, S.E.M.	215
Del Solar, D.E.	184
Dela Rue, B.	229, 335
Dellinger, A.	11
Destain, M.F.	5
Devos, W.	262
Deákvári, J.	399
Dieleman, J.A.	275
Dietrich, P.	59
Díez-Galera, Y.	400
Dillon, C.R.	191, 350
Dinaburga, G.	311
Dminriev, A.	27
Dobers, R.	135
Doluschitz, R.	67, 175, 177, 179, 345, 346
Domsch, H.	51
Dooren, H.J.C.	219
Dorado, J.	42, 303
Dorna, M.	97
Douglas, P.	325
Dowdy, T.	102
Dreger, F.	248
Dryslova, T.	31, 34
Dubois, J.	77
Dunn, G.M.	325
Düpjan, S.	244
Dywer, C.	233

E

Eastwood, C.R.	173, 231
Eben Chaime, M.	235
Edan, Y.	86, 149
Efimov, A.E.	286
Efron, R.	39
Ehlert, D.	22
Eicher, S.D.	116
Eizenberg, H.	39
Elias Correa, F.	182
Endler, M.	192
Engler, B.	175
Engström, L.	16
Erdélyi, É.E.	68, 314
Eren Özcan, S.	217
Ericson, S.	75
Esau, T.	6
Esau, T.J.	45
Escolà, A.	20, 21
Evans, J.R.	78
Everaert, N.	214
Exadaktylos, V.	217

F

Faber, N.	200
Fallast, M.	224
Farajallah Hosseini, S.	154
Farajjolah Hosseini, S.J.	366
Farajollah Hosseini, J.	131
Faria, R.T.	152
Faure, P.	230
Feelders, A.	249
Fenyvesi, L.	399
Fernandes, L.M.	136
Fernandes, V.M.	108
Fernandez, C.	139
Fernandez-Quintanilla, C.	42, 303
Fick, J.	345
Fiorentino, C.	57
Fleming, E.	193
Forghani, A.	360
Förstner, W.	26
Fountas, S.	29, 169, 246, 247, 254, 255, 256, 261
Francoy, T.	110
Franke, J.	271
Frazzi, E.	3
Frentrup, M.	180
Frier, I.	92
Frisch, J.	119, 120, 121
Fritz, M.	96, 137, 138, 140, 181, 194
Frost, M.	179
Fukatsu, T.	93, 316
Fullen, J.T.	158, 351
Fülöp, G.Y.	143
Furuhata, M.	18

G

Galanis, M.	169
Gallmann, E.	221
Gandonou, J.	350
Gang, L.	294
Garcia-Torres, L.	354
García-Ferrer, A.	354
Gasteiner, J.	224
Gebbers, R.	163, 165
Gee, C.	76
Gehl, R.	168
Gelb, E.	267, 402
Gemtos, T.A.	29, 169
Gent, M.	88
Gerhards, R.	40, 41, 170
Gerlich, R.	28
Giebel, A.	23
Gilad, D.	235
Gimenez, L.M.	280
Giordano, L.	4
Glasbey, C.A.	275
Gnip, P.	253, 396

Gnyp, M.L.	33
Gobor, Z.	84
Godsk, T.	315
Goense, D.	150, 223
Gogliano Sobrinho, O.	387
Goh, H.	227
Golbach, F.	28
Goldstein, E.	147
Gomes, C.D.	152
Gómez-Candón, D.	354
Goo, S.	227
Gottschalk, R.	101
Goulevant, G.	79
Govers, M.H.A.M.	87
Gray, A.W.	116
Green, O.	226
Gremmes, H.	97
Grethe, H.	344
Griepentrog, H.W.	99
Griffin, T.	257
Groeneveld, R.	268
Groot Koerkamp, P.W.G.	125, 215
Grove, J.H.	69
Grzebellus, M.	259
Guarino, M.	242, 323, 324
Guggenberger, T.	146, 190, 224
Guilleume, S.	382
Gusakova, L.P.	291
Gushiken, I.Y.	128
Gutjahr, C.	170

H

Haapala, H.E.S.	353
Haas, R.	137, 140
Hadders, J.	47
Hadders, J.W.M.	47
Hagner, O.	17, 58
Haixia, L.	297
Hakimhashemi, M.	217
Hakojärvi, M.	264, 276
Halachmi, I.	210, 232, 235, 327
Halas, J.	273
Hammarberg, J.	135
Handcock, R.	231
Hank, T.B.	32
Hänninen, L.	264
Hansen, J.P.	159, 208
Hansen, N.F.	208
Happich, G.	72
Har Gil, D.	356
Harms, H.-H.	72
Harms, J.	237, 328
Harrison, M.T.	78
Hartmann, K.	364
Haskell, M.J.	228
Hasselmann, L.	321

Häusler, J.	224
Hautala, M.	276
Havard, P.	144
Havlicek, Z.	203, 209
Hedges, S.	153
Heijting, S.	189
Heinemann, O.	99
Heiniger, R.W.	55
Heisig, M.	22
Hejndorf, P.	159
Hennen, W.H.G.J.	106, 349
Hennig, S.D.	33
Herbst, R.	163, 165, 185
Herd, D.	221
Herdon, M.	202, 205, 367
Herppich, W.B.	23
Herzog, F.	281
Hetzroni, A.	92, 147
Hijazi, B.	77
Hillnhütter, C.	339
Hirafuji, M.	379
Hirakawa, A.R.	386
Hoefer, G.	52
Hoffmann, C.	177
Hoffmann, W.	270
Hofstede, G.J.	137, 138, 268
Hofstee, J.W.	36, 62, 87, 403
Hogeveen, H.	109, 116, 249, 250, 251, 252
Hogewerf, P.H.	225, 239, 263, 265, 333
Højsgaard, S.	234
Holl, C.	155, 206
Honarmand, M.	360
Hoogenboom, G.	89
Hoogmoed, W.B.	50
Horak, P.	310
Hørning, A.	208
Hou, K.M.	355
Houin, R.	183
Houwers, W.	223
Huang, Y.	270
Hunsche, M.	24
Huzsvai, L.	329

I

Igras, J.	304
Ijken, H.	62
Imperatriz-Fonseca, V.	110
Inan, H.I.	385
Intreß, J.	23
Ipema, A.H.	225, 239, 263, 265, 333
Ivanova, K.F.	288

J

Jacquot, A.	355
Jaggard, K.W.	13
Jago, J.	229, 251, 335

Jahanshae, P.	368	Kochegarov, S.F.	286
Jais, C.	332	Koenderink, N.J.J.P.	28
Jalalian, A.	295	Köhler, S.	236
Jamieson, R.	144	Komasilovs, V.	370
Jansen, F.	324	Kopmanis, J.	311
Jansen, R.	87	Korc, F.	26
Janssen, H.	95, 188, 225	Kostamo, J.	98
Jarolimek, J.	203, 209	Koutsostathis, A.	29
Jarysz, M.	111	Kovács, L.	399
Jasper, J.	1	Kozłowski, R.J.	391, 392
Jassar, S.	398	Kozyra, J.	305
Jensen, A.L.	135	Kozyreva, L.V.	286
Jensen, R.	229	Krause, A.	390
Jeuffroy, M.-H.	79	Kravchenko, A.	168
Jezek, J.	396	Kren, J.	31, 34, 278
Jia, L.	33	Kristensen, E.F.	196
Johnson, R.	277	Kristensen, K.	226
Jombach, S.	143	Kristensen, L.M.	135
Jones, H.G.	4	Kroulik, M.	73
Jonsson, A.	397	Kuhlmann, A.	221
Jónsson, R.I.	234	Kulkarni, S.	344
Jørgensen, R.N.	226	Kultus, K.	236
Journaux, L.	5	Kunisch, M.	119, 120, 121
Jukema, J.N.	56	Kviz, Z.	73
Jung, J.	306	Kwong, K.	227
Jungbluth, T.	221, 241		
Jurado-Exposito, M.	354	**L**	
		Lacerda, T.	373
K		Lacey, R.	270
Kadaja, J.	81	Lagacherie, P.	65
Kafka, S.	260	Lamaker, A.	83
Kaivosoja, J.	64, 348	Lambooij, E.	239
Kalmari, J.	82, 348	Lammers, R.J.H.	263
Kämmerling, B.	199	Lan, Y.	270
Kamphuis, C.	249, 251	Lane, G.P.F.	80
Kanash, E.V.	12	Lang, T.	72
Karetsos, S.	134	Langford, J.	325
Karydas, C.G.	187	Lanir, T.	149
Kassahun, A.	390	Lapins, D.	311
Kastrantas, K.	207	Lashgarara, F.	366
Kaufmann, O.	321	Lee, B.W.	296
Kay, S.	262	Lee, K.J.	296
Kempenaar, C.	83, 105	Lehmann, R.	96
Kenny, S.	231	Leithold, A.	190
Khademi, H.	295	Lenain, R.	43
Khosla, R.	164, 337	Lénárt, C.S.	399
Kilian, M.	237	Lengyel, P.	202
Kjær, K.B.	208	Lerink, P.	189
Kjærgaard, M.B.	315	Leroy, T.	242, 323
Kleinhenz, B.	90, 306	Levi, A.	39
Klompe, A.	189	Li, F.	33
Klopcic, M.	232	Li, L.	297
Klose, R.	97	Li, M.	14, 287, 289, 290
Kluge, A.	37	Li, X.	14
Kluza, T.	391	Lilienthal, H.	364
Knapen, M.J.R.	126, 401	Lima, H.	373

Lindstrøm, J.	123	Matocha, D.	129
Lindén, B.	16	Mayer, W.	347
Link, A.	1	Mayus, M.	281
Linkolehto, R.	82	Mazzetto, F.	2, 46
Linz, A.	97	Mačuhová, J.	332
Lisker, I.	27	Mcbratney, A.B.	112
Liu, G.	287, 293	Mcclement, I.	117
Llorens, J.-M.	114, 115	Mcfarlane, I.D.	117
Loch, T.	73	Mclaughlin, N.B.	54
Lokers, R.M.	258	Megna, A.	362
Lokhorst, C.	225	Meijer, A.D.	55
Loonstra, E.H.	49	Mein, G.	251
Lopes, C.	136	Meixner, O.	137, 140
López-Granados, F.	354	Melse, R.W.	220
Lorén, N.	16	Mena, A.	2
Lotz, L.A.P.	83	Menz, G.	271
Lowenberg-Deboer, J.	257	Meron, M.	10, 92, 279
Lowrance, R.	142	Meuleman, J.	36
Lubbe, S.	161	Mewes, T.	271
Lücke, W.	48	Meyer, C.	194
Lueck, E.	53, 307, 309	Miao, Y.	33
Lukas, V.	34, 278	Michie, C.	227
Lund, I.	104	Michielsen, J.M.G.P.	91, 105
Lupia, F.	302	Mickelåker, J.	85
		Miedema, H.M.	233
		Miedema, J.	200
M		Migdall, S.	383
Maack, C.	317	Milenov, P.	385
Maassen, J.	197	Milics, G.	284, 399
Macrae, A.I.	233	Mirdamadi, M.	131, 375
Madani, A.	144	Mirdamadi, S.M.	366
Madsen, T.	99	Miron, J.	327
Maertens, W.	266	Misa, P.	31, 34
Magrisso, Y.	92	Misopolinos, N.L.	187
Mahlein, A.-K.	38	Miteran, J.	5
Mahmood, H.S.	50	Moghadasi, R.	131
Maia, J.	136	Moiroux, L.	230
Maidana, N.	326	Mokhtari, H.	107
Malagoli, P.J.	79	Mokhtari Aski, H.	374
Malano, H.	325	Molin, J.P.	103, 280
Maltese, A.	4	Mollenhorst, H.	249, 250
Maltz, E.	210, 327	Monod, M.O.	230
Manfrini, L.	381	Moore, A.D.	78
Manouselis, N.	122, 201, 207	Morales, L.A.	313
Manteuffel, G.	243, 244	Mortazavi, M.	107, 369, 374
Manzini, V.	362	Moshia, M.	164
Maria Koukouli, M.K.	384	Moshou, D.	7
Marinelli, M.A.	70	Mosquera, J.	220
Marionneau, A.	301	Mueller, R.A.E.	193
Marte, W.E.	118	Mueller, W.	391
Martin, R.	76	Mulder, M.	365
Martini, D.	119, 120, 121	Müller, H.C.	223
Martínez-Casasnovas, J.A.	166, 186, 400	Müller, J.	100
Masip, J.	21	Munksgaard, L.	234
Masselin-Silvin, S.	206	Munoz, J.D.	168
Massruhá, S.M.F.S.	108, 373	Muñoz, R.E.	162
Mata, G.	231		

Murakami, E.	127, 128
Muraru, V.M.	377
Muraru Ionel, C.	377
Murphy, D.J.	380
Murphy, P.	142
Murray, B.B.	240
Musil, M.	396

N

Namdarian, I.	157
Nannen, C.	222
Nanos, G.	169
Nanseki, T.	93, 118, 316
Naor, A.	279
Naraghi, M.	360
Narciso, M.	373
Nash, E.J.	246, 247, 256, 261
Nasonov, D.V.	376
Nassetti, F.	4
Naudin, C.	79
Neményi, M.	284, 399
Neto, M.C.	136
Neudert, L.	31, 34, 278
Nguyen, C.	298
Nielsen, J.	99
Nieróbca, A.	305
Nieter, J.	51
Nieuwenhuizen, A.T.	36, 87, 403
Nikkilä, R.	256, 348
Niknamee, M.	154
Nino, P.	302
Ninomiya, S.	93
Noga, G.	24
Nordmeyer, H.	37, 312
Noroozi, O.	365
Norouzi, M.	360
Nørremark, M.	104
Norros, L.	133
Nouri, H.	167
Nouruzi, M.	295
Nováková, M.	273
Nowakowski, K.	111, 393, 394
Nüsch, A.-K.	59

O

Oerke, E.-C.	38, 339
Ogink, N.W.M.	218, 220
Ogundele, B.A.	334
Ohl, N.	383
Ojanne, A.	82, 348
Oksanen, T.	63, 74, 82, 98, 348
Olsen, H.J.	104
Omidi Najafabadi, M.	131
Omine, M.	18
Oosterkamp, E.	138
Oosting, S.J.	125

Oppelt, N.M.	32
Oppermann, P.	332
Orlowski, A.	160
Ortega, R.A.	162, 184
Ortiz, B.	9, 89, 282
Ortiz-Laurel, H.	213
Ørum, J.E.	254
Osann Jochum, M.A.	302
Osipov, Y.A.	12
Otten, G.	28
Otto, J.	179

P

Paindavoinc, M.	77
Paine, M.S.	173
Palacín, J.	20
Palavitsinis, N.	207
Palma, J.	281
Palombo, A.	4
Panneton, B.	151
Paoli, J.N.	112
Papp, Z.	399
Paraskevopoulos, A.	169
Paree, P.G.A.	172
Pascucci, S.	4
Pastell, M.	264
Patzold, S.	269
Pauli, S.	238
Paz-Ferreiro, J.	274, 283, 313
Pecile, A.M.	324
Pedersen, S.	246
Pedersen, S.M.	99, 254, 255, 261
Pedroso, M.	382
Pena-Yewtukhiw, E.M.	69, 158, 351
Percival, D.C.	6, 45
Pereira De Almeida, V.	274
Perrin, O.	115
Perry, C.	9, 89
Pesonen, L.	133, 246, 254, 255, 256, 261
Petrushin, A.F.	285, 300
Pettersson, C.G.	17
Peña-Barragán, J.M.	354
Piekarska-Boniecka, H.	394
Pirna, I.	377
Piron, A.	5
Piron, E.	76
Pisante, M.	57
Pizzigatti Corrêa, P.L.	182
Planas, S.	21
Pluk, A.	266
Plume, A.	311
Plumeyer, C.H.	336
Polak, P.	232
Põldaru, R.	204
Polder, G.	25, 83, 275
Pompe, J.C.A.M.	263, 320

Poulsen, N.K.	234	Roumet, P.	15
Povh, F.P.	103, 280	Ruckelshausen, A.	97
Preisinger, R.	238	Rucker, K.	282
Prosek, V.	73	Rueda Ayala, V.P.	40
Przybylak, A.	395	Ruehlmann, J.	53, 307, 309
Pudełko, R.	304, 305	Ruhrberg, Y.	149
		Rumbles, I.	240
Q		Rydberg, A.	58
Quanten, S.	324		
Queiroz, D.M.	44	**S**	
Quendler, E.	389	Sagris, V.	385
		Saha, C.K.	322
R		Salomoni, F.	46
Raatjes, P.	47	Sambra, A.	196
Rabatel, G.	15	Samsom, J.	83
Rafiee, H.	331	Sandt, N.	312
Rahe, F.	97	Santana, F.S.	127, 128
Rahnama, M.B.	368	Sanz, R.	20, 21
Rajabbeigi, M.	107, 374	Sapounas, A.A.	219
Rakut'ko, S.	272	Saraiva, A.M.	110, 127, 128, 182, 386
Rameau, P.	230	Sartori, L.	255
Ramon, H.	7	Sato, M.	118
Ramos, M.C.	166	Saue, T.	81
Ranaie, H.	369	Sauer, U.	59
Ravn, O.	99	Sawhney, B.	398
Rayburn, E.B.	158, 351	Schepers, H.T.A.M.	91
Razavi Najafabadi, J.	167	Schiefer, G.	96, 181, 194
Razgale, V.	372	Schirrmann, M.	51
Reich, R.	164	Schmidt, J.	11
Reiche, R.	181	Schmilovitch, Z.	210
Resch, R.	97	Schmitz, M.	119, 120, 121
Reusch, S.	1, 19	Schneider, K.	224
Reynaldo, E.F.	103	Schneider, M.	185
Ribeiro, A.	42, 101, 303	Schnug, E.	364
Richter, O.	312	Scholtz, P.	273
Ringdorfer, F.	146	Scholz, C.	269
Ritchie, G.	9	Schuiling, H.J.	239
Roberts, D.J.	232	Schulze Lammers, P.	84
Robinson, P.H.	357	Schumann, A.W.	6, 45
Rodemann, B.	23	Schutz, M.M.	116
Rodenburg, J.	240	Schwarz, J.	248
Roehrig, M.	192	Schön, P.C.	244
Rogacki, P.	111	Seger, M.	301
Rogge, C.B.E.	60, 176	Seginer, I.	88
Rong, L.	322	Sester, M.	79
Roosenschoon, O.	390	Sfaru, R.	377
Roots, J.	204	Shani, U.	39
Rosa Pereira, V.	145	Shapiro, A.	86
Rose, S.	319	Shaver, T.M.	337
Rosell, J.R.	20, 21, 400	Shearer, S.A.	44, 102
Rosenkranz, S.	224	Sherlock, R.	251
Ross, D.	228	Shneider, B.	235
Rossel, D.	213	Shockley, J.M.	191, 350
Rößler, B.	221	Shoshani, E.	327
Roth, A.	67	Sideridis, A.	122, 134, 201
Roth, M.	346	Sikora, R.A.	339

Silerova, E.	203, 209	Tanaka, K.	379
Silleos, N.G.	187	Taragola, N.	402
Silva, D.A.B.	152	Tartachnyk, I.	24
Silva, F.F.	152	Taylor, J.	382
Simek, P.	203, 209	Taylor, J.A.	65, 153, 381
Singh, S.	161	Teixeira Filho, J.	145
Slinchuk, S.G.	300	Ten Napel, J.	125
Smits, M.A.	318	Ten Voorde, H.	352
Smits, M.C.J.	219	Ternes, S.	124, 326
Snapp, S.	168	Teye, F.	82
Söderström, M.	16, 17, 378	Theuvsen, L.	180, 336
Sökefeld, M.	170	Thiel, M.	97
Solanelles, F.	21	Thiemann, F.	193
Solomon, R.	327	Thurner, S.	238
Sonck, B.	266	Thysen, I.	132, 135
Sørensen, C.	133, 246, 255	Timmerman, M.	148
Sørensen, C.G.	123, 196, 226, 254, 256, 261	Tisseyre, B.	65, 112, 382
Soroker, V.	35	Tiusanen, J.	98, 264
Sotiriou, S.	201	Top, J.	268
Souza, K.X.S.	108	Topcu, S.	281
Spada, V.	362	Trautz, D.	97
Spangenberg, U.	307, 309	Tremblay, N.	151
Speetjens, S.L.	87	Troccoli, A.	57
Sprinstin, M.	279	Tsatsarelis, K.A.	187
Spätjens, L.E.E.M.	62	Tsipris, J.	10
Sripada, R.	11	Tsror, L.	147
Stallinga, H.	91, 105	Tsukahara, R.Y.	152
Stange, R.L.	127, 128	Tzikopoulos, A.	122, 201
Staritsky, I.	188		
Steeneveld, W.	109, 252	**U**	
Steiner, U.	38	Ulén, B.	378
Steinmeier, U.	48	Umstatter, C.	228
Stenberg, M.	378	Unsenos, D.	223
Stern, H.I.	30	Uskov, A.O.	113
Steven, M.D.	13	Uskov, I.B.	286, 376
Stigter, J.D.	87	Uyterlinde, M.	95
Stombaugh, T.S.	44, 102, 191		
Stone, K.	341	**V**	
Strauss, O.	112	Valero, C.	8
Ströbel, U.	319	Vallès, J.M.	20
Strøm, J.S.	322	Vallés-Bigordà, D.	166
Sugahara, K.	93, 316	Van Asseldonk, M.	349
Sugiura, R.	18	Van Asselt, C.J.	87, 100
Suomi, P.	82, 133, 348	Van De Zande, J.C.	36, 91, 105, 403
Svensson, S.A.	85	Van De Zedde, R.	28
Svobodova, I.	31, 34	Van Der Gaag, L.C.	109
Svoray, T.	356	Van Der Heijden, G.W.A.M.	25, 83, 275
Swain, K.C.	6, 45	Van Der Hoeven, R.	130
Szenas, S.Z.	367	Van Der Lans, A.	91
Szilagyi, E.	340	Van Der Schans, D.	105
Szilagyi, R.	340	Van Der Schoor, R.	25
Székely, L.	388	Van Der Tol, P.P.J.	250
		Van Der Veen, A.A.	125
T		Van Der Wal, T.	95, 189, 225, 258, 262
Tailleur, C.	228	Van Der Zalm, A.	83
Tamás, J.	399	Van Der Zijpp, A.J.	125

Van Dooren, H.J.	83	Weiss, U.	97
Van Doorn, J.	25	Welp, G.	269
Van Duinkerken, G.	211, 212	Wendl, G.	84, 237, 238, 328
Van Egmond, F.M.	59	Wenkel, K.-O.	163, 165
Van Evert, F.K.	83	Werban, U.	59
Van Gurp, H.	172	Weres, J.	111, 391
Van Harn, J.	218	Werner, A.	255
Van Hattum, T.G.	218	Werner, A.B.	248
Van Henten, E.J.	36, 50, 87, 403	Westfall, D.G.	164, 337
Van Horne, P.	349	Whelan, B.M.	153
Van Huylenbroeck, G.	344	Wigham, M.	28
Van Keulen, H.	281	Wijnholds, K.H.	56
Van Lierde, D.	402	Winkel, A.	218
Van Oosterom, P.	385	Wolfert, J.	129, 172, 174, 363
Van Riel, J.W.	148	Wolfert, S.	268
Van Rossum, J.	95	Wozniakowski, M.	160
Van Straten, G.	87, 100	Wozniakowski, T.	160
Van Velde, P.	91, 105	Wright, G.	70
Van Willigenburg, L.G.	87	Wu, T.	227
Vanek, J.	203, 209	Wulfsohn, D.	99, 169
Vangeyte, J.	77, 266		
Vanino, S.	302	**X**	
Vanmeulebrouk, B.	188	Xiangjian, M.	294
Vatsanidou, A.	247	Xu, Y.	118
Veerkamp, R.F.	318		
Velikanov, L.P.	291	**Y**	
Vellidis, G.	9, 89, 142, 282	Yakushev, V.P.	359
Vercesi, A.	2	Yakushev, V.V.	358
Verdouw, C.N.	130, 352	Yang, F.	77
Verloop, C.M.	129, 130, 174, 352, 363	Yang, H.	326
Viator, H.	277	Yang, W.	14
Vieira, S.R.	283	Ye, Z.	322
Vigneau, N.	15	Yehuda, Z.	39
Vigneault, P.	151	Yermiyahu, U.	279
Vijn, M.	83	Yilmaz, D.	343
Vilichko, A.K.	291	Yohn, C.W.	158, 351
Villette, S.	76, 77	Yomralioglu, T.	385
Vincini, M.	3		
Visala, A.	74	**Z**	
Visoli, M.	124	Zacepins, A.	371
Vlk, M.	310	Zacharias, S.	66
Von Hörsten, D.	48	Zaharian, J.G.	113
Von Wulffen, U.	51	Zalidis, G.C.	187
Vonk, M.B.	189	Zaman, Q.U.	6, 45, 144
Vougioukas, S.	7, 71	Zandonadi, R.S.	44, 102
		Zauer, O.	51
W		Zeuner, T.	90
Wachs, J.P.	30	Zevenbergen, J.	385
Wagner, P.	61, 185	Zhang, F.	33
Waksman, G.	155, 206	Zhang, G.	322
Wallenhammar, A.-C.	397	Zhang, H.	270
Wang, M.	287	Zhang, J.	290
Wang, M.H.	298	Zhang, L.	13
Warren, M.F.	156	Zhang, M.	287, 290, 293, 298
Wehren, W.	223	Zhao, Y.	290
Weis, M.	41, 170	Zheludkov, A.G.	291

Zheng, L.   289
Zhou, J.J.   293
Zig, U.   147
Zurita, L.   171
Zähner, M.   241

Printed in the United States
by Baker & Taylor Publisher Services